AF606776

SAP® Signavio®

Johannes Strasser, Michael Sokollek, Manuel Sänger,
Maike Spierling, Marvin Schönwälder

SAP® Signavio®

Business Process Transformation

Editor Megan Fuerst
Acquisitions Editor Hareem Shafi
German Edition Editor Nicole Hohmann
Copyeditor Julie McNamee
Cover Design Graham Geary
Photo Credit Shutterstock: 370485713/© Martin Mecnarowski
Layout Design Vera Brauner
Production Kyrsten Coleman
Typesetting SatzPro, Germany
Printed and bound in the United States of America, on paper from sustainable sources

ISBN 978-1-4932-2562-0

1st edition 2024
1st German edition published 2023 by Rheinwerk Verlag, Bonn, Germany

Library of Congress Cataloging-in-Publication Control Number: 2024009862

Contents at a Glance

Contents

Part III How to Successfully Navigate the Transition to SAP S/4HANA with Business Process Transformation

Preface

In a constantly changing world, it's of great importance for companies to understand, improve, and continuously change their business processes to remain competitive. Only those who can adapt quickly to changing market conditions and customer requirements on the basis of efficient processes and effective decisions will be successful and remain competitive in the long term. Particularly since 2020, companies around the world have been facing major challenges—some of them highly unpredictable—that require existing business processes to be adapted. Supply chains disrupted by COVID-19 and geopolitical conflicts are just two examples. To actively meet these challenges, process management isn't just a nice-to-have but a need-to-have for every company.

Business process transformation enables companies to understand, adapt, and implement change in their business processes. *SAP Signavio Process Transformation Suite* is a platform that supports companies in exactly these aspects. This book provides a detailed overview of the SAP Signavio product portfolio and shows which benefits the use of the suite or individual solutions can provide for your company.

Another trigger that can cause companies to take a hard look at their business processes is the transition to SAP S/4HANA. If you're about to make the transition to SAP S/4HANA, you're probably asking yourself the following questions:

- How can you ensure that the transition to SAP S/4HANA adds value to your business?
- How can the transition to SAP S/4HANA be made as quickly as possible?
- How does your company ensure that the transition to SAP S/4HANA is feasible?

Business process transformation management supports you with these and other issues. It should therefore be a central component of your conversion to SAP S/4HANA. We describe in detail how to proceed during the transformation, what happens after the transformation, and how you can get started right away. Let's dive into the world of business process transformation management with SAP Signavio together!

Target Audience

This book is intended for executives, business process managers, IT experts, and anyone interested in business process optimization and transformation. It offers a practical guide for the effective use of SAP Signavio Process Transformation Suite and provides valuable insights and best practices for a successful business process transformation. Finally, it also addresses consultants who accompany companies in their digital transformation and implementation of SAP Signavio solutions.

Other target groups include the following:

- *Business executives* who want to better understand the added value of SAP S/4HANA, RISE with SAP, and intelligent technologies with SAP Signavio
- *Project managers* who want to support their projects (e.g., SAP S/4HANA migration, RISE with SAP, business transformation) with SAP Signavio
- *Line of business managers* who want to monitor performance and identify areas for improvement in their field
- *Process owners* who want to monitor performance and identify areas for improvement in their end-to-end process
- *Transformation drivers* looking for enterprise-wide automation benefits

How This Book Is Organized

This book is divided into three parts. Part I introduces you to SAP Signavio and explains typical motivations for using business process management and SAP Signavio Business Process Transformation Suite. Part II introduces you to the functions of the individual components and explains their application with an example in each case. Part III helps you successfully make the transition to SAP S/4HANA by putting the individual solutions in context and showing how they can be used in the context of an SAP S/4HANA transformation and beyond. We describe how companies can use SAP Signavio Business Process Transformation Suite to optimize their business processes while being prepared for future challenges.

The chapters of the book are built on each other, so it's advisable to first understand the *why* (see Part I, Chapter 1 and Chapter 2), then the *what* (see Part II, Chapter 3 through Chapter 9), and finally the *how* (see Part III, Chapter 10 through Chapter 14).

Chapter 1 aims to help you understand the urgency and opportunities of business process transformation. To this end, the framework conditions for business process transformation are discussed, including new technologies and business models, market developments, globalization, and complex product dependencies.

Business process transformation is a comprehensive framework for holistic business process transformation. In Chapter 2, you'll learn what this framework consists of and what core capabilities it offers.

Chapter 3 explains the first SAP Signavio solution in detail. This chapter provides you with a functional overview, the solution's benefits, and a practical application example for using SAP Signavio Process Insights.

The SAP Signavio Process Intelligence solution is introduced in more detail in Chapter 4. In this chapter, you'll get an overview of the functions, possibilities, and a practical application example of SAP Signavio Process Intelligence.

The SAP Signavio Process Manager solution is introduced in Chapter 5. This chapter provides you with a functional overview, the solution's benefits, and a practical application example for using SAP Signavio Process Manager.

Chapter 6 introduces you to the SAP Signavio Journey Modeler solution. In this chapter, you'll also get an overview of the functions and benefits of the solution. In a practical application example, you'll then see how you can use SAP Signavio Journey Modeler in practice.

In Chapter 7, we dedicate ourselves to the SAP Signavio Process Collaboration Hub solution. After an overview of the functions and benefits, we show you in a practical application example how the solution can improve collaboration.

The SAP Signavio Process Governance solution and its features are introduced in Chapter 8. Get to know the workflow functions and find out how to set up a release process in the practical application example.

Although SAP Build Process Automation belongs to the SAP Build product family, it's functionally integrated into the SAP Signavio portfolio, which is why we introduce you to the solution in Chapter 9. You'll learn about the functions and benefits of the solution. You'll then learn about possible application scenarios for the solution in a practical application example.

The transition to SAP S/4HANA is the ideal time for companies to get started with business process transformation management. In Chapter 10, we describe the basic principles of SAP S/4HANA and change scenarios when transitioning to SAP S/4HANA. We also describe the success factors and dimensions of business process transformation management as well as the opportunities and challenges that arise when companies decide to use business process transformation management.

Important aspects such as process organizations, process roles, process architectures, variant management, and the linking of process management with application lifecycle management (ALM) and enterprise architecture management (EAM) are described in Chapter 11.

The SAP Signavio methododology for end-to-end business process transformation along the analyze, enhance, process design, solution design, build, test, and enablement phases is described in Chapter 12. We also use customer examples to illustrate how the method supports customers in their transformation project.

After the transition to SAP S/4HANA is complete, it's necessary to align business process transformation management with the continuous process transformation phase. Chapter 13 is dedicated to this phase of continuous process transformation.

There are various offers available for getting started with business process transformation management with SAP. In Chapter 14, we describe options, solutions, and basic concepts for analyzing productive systems and identifying potential for improvement.

Information Boxes

In highlighted information boxes, you'll find content in this book that is worth knowing and helpful but falls outside of the core content. So that you can immediately classify the information in the boxes, we've marked the boxes with symbols.

[»]

Note

In boxes marked with this symbol, you'll find information on *further topics* or important content that you should remember.

[!]

Attention

This symbol draws your attention to *special features* that you should be aware of. It also *warns* you of common mistakes or problems that may occur.

[+]

Tip

This symbol marks *tips* and *hints* from professional practice that provide practical recommendations to make your work easier.

Acknowledgements

A big thank you goes to our colleagues from SAP Signavio and Westernacher Consulting, who actively supported and motivated us during the creation of this book. We would also like to thank the editorial office of Rheinwerk Verlag and Rheinwerk Publishing for their active support and publication of this work.

We hope you enjoy reading it.

Johannes Strasser, **Michael Sokollek**, **Manuel Sänger**, **Maike Spierling**, and **Marvin Schönwälder**

PART I

Introduction to Business Process Transformation

Chapter 1
Why Business Process Transformation?

The aim of this chapter is to convey to you the urgency of and opportunities with business process transformation. To this end, we address the framework conditions of business transformation, which include new technologies and business models, market developments, globalization, and complex product dependencies.

Our business world today is completely different than it was 10 or even 20 years ago. It has become more fast-paced, more digital, more globalized, and more unpredictable and volatile than ever before. To be able to hold their own in the long term, successful companies of tomorrow must have special competencies to remain agile in an environment of increasing complexity and changing market conditions. Tomorrow's companies will be intelligent, customer-centric, and networked. Only in this way can the current challenges be effectively seen as an opportunity to sustainably differentiate themselves from the competition and to continuously secure market shares.

In this context, we first examine the current challenges facing companies in Section 1.1. In Section 1.2, we'll then discuss the competencies of today's successful companies. Finally, in Section 1.3, we look in detail at the demands of the intelligent enterprise of tomorrow.

1.1 Current Challenges for Companies

Today's business environment is changing faster than ever before. Since 2020 alone, all companies around the world have had to adapt to new and sometimes unpredictable conditions: from the COVID-19 pandemic and related lockdowns, to rising inflation, volatile markets, changing consumer behavior, and geopolitical conflicts that keep the world on edge. At the same time, new technologies and trends have asserted themselves—most notably Web 3.0 and the opportunities it presents for companies across a wide range of industries.

Our globalized world has become so insanely fast-paced that it even has its own term: *VUCA*. VUCA stands for volatility, uncertainty, complexity, and ambiguity. Figure 1.1 illustrates this concept.

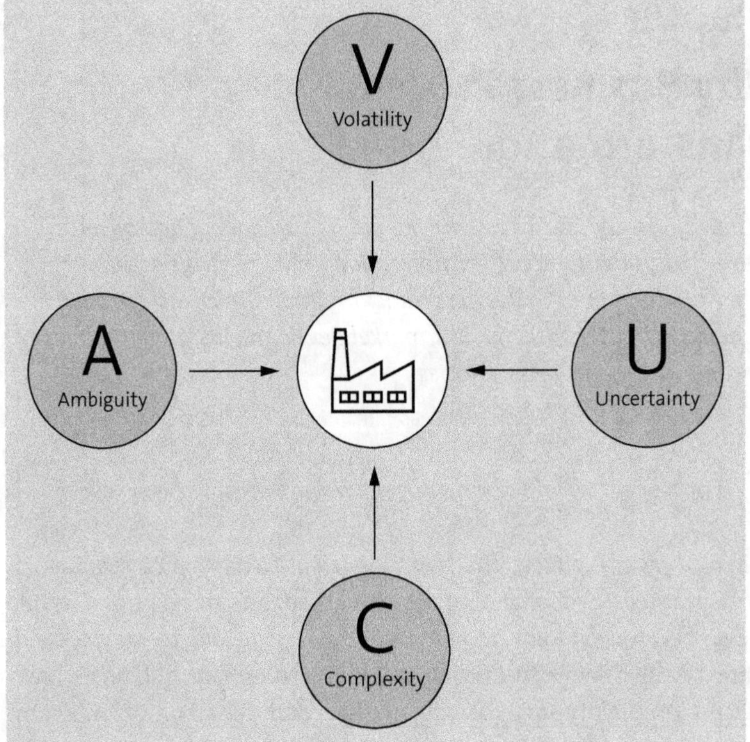

Figure 1.1 The Influences of the VUCA World on Companies

VUCA is the umbrella term for the various challenges that companies face today. Put simply, VUCA is a byproduct of the fast-paced and digitized business world we find ourselves in. Meanwhile, we're seeing increased complexity in terms of product diversity, business processes, and stakeholder relationships. The term originated in the 1990s when the US Army used it to describe the major changes within the military after the end of the Cold War. Today, however, the terminology is used primarily in a business context. Volatility, uncertainty, complexity, and ambiguity are being emphasized due to major environmental changes such as ongoing digitalization or globalization, which in turn puts pressure on changing the way we work. All these aspects are shaping the world we live in today and posing huge challenges. Established companies or even entire industries are collapsing; at the same time, new companies are springing up. Restructuring and optimization are becoming the new standard.

As if all of this weren't enough, new technologies and business models are also taking hold at a rapid pace in almost every conceivable area of business. Before we get into the technologies, let's first look at and better understand the current challenges facing today's businesses.

1.1.1 Volatility

Volatility comprises fluctuations and often unpredictable changes in general conditions, which often occur rapidly and on a large scale. Accordingly, increasing volatility results in accelerated and strong changes in the framework conditions and the environment. This also includes the emergence of new trends. These provide for new, unexpected competitors as well as for substituting products and ever shorter product lifecycles. Uncertainty, in turn, describes the uncertainty associated with the impeded unpredictability of events, as many variables are unreliable or insufficiently known. Increasing uncertainty leads to a future that is more difficult to predict, as well as to a greater probability of the occurrence of surprising events. These include a reduced validity of past-oriented analyses, more difficult decision-making and planning, and a reduced understanding by employees of what is happening around them. Customer needs and demands can also change overnight. In addition to volatility and uncertainty, today we're also exposed to a variety of complex conditions. This makes it difficult to assess the impact of one changing factor on other factors. Increasing complexity includes many unknown elements that are interdependent as well as interacting with each other. This is associated with a large amount of unmanageable information as well as a higher number of factors influencing the market, the company, or the individual employees. The increasing ambiguity of information ultimately results in contradictory information being interpreted differently. On one hand, this is caused by the flood of data and the enormous number of different information channels, and on the other hand, by the different evaluations of information based on individual values and interests. This ambiguity of information consequently leads to more difficult orientation for all parties involved, from which misunderstandings and even conflicts can arise. When information is ambiguous, this leads to different interpretations, which in turn leads to a higher risk of wrong decisions.

You may be wondering what all this means for your company. The answer to this question is complex: In the past, it was still possible to make long-term decisions with assessable risks based on years of experience. In today's VUCA world, risks as well as decision factors can only be estimated in the very short term, if this is possible at all. Companies that want to enforce their usual decision-making principles in the VUCA world often make decisions incorrectly, too late, or not at all. Whether it's societal change or global power shifts, you can't avoid accepting the unpredictability of our times to some degree. That said, every business needs solid strategies and effective tools to compete in an ever-changing world. Focus on your core competencies, and address the fundamental questions that matter to you, such as the following:

- How do you weigh your information?
- How do you stand out from your competitors?
- How can you promote long-term loyalty among your customers, suppliers, and employees, taking into account the factors just mentioned?

A clearly defined, customer-oriented vision is helpful to be aware of realistic goals and to provide yourself with both structure and orientation. Similar to disruption, it's also a good idea to approach the unpredictability of our VUCA world in a correspondingly positive mood and to see the associated challenges as an opportunity for your company to constantly evolve. Addressing the individual characteristics and devising suitable strategies for them will also dampen the "everything used to be better" feeling. In any case, it's important to position your company well for the future. One way to achieve this is through agility and enterprise-wide transparency. Both of these approaches, as well as others, are discussed in Section 1.2 in more detail. To understand these even better later, we'll first dive a little deeper into our current challenges and devote ourselves to the topic of uncertainty.

1.1.2 Uncertainty

Another challenge is the *uncertainty* and the associated permanent *unpredictability* of today. The volatile situations already discussed, including pandemics, lockdowns, shifting societal values, geopolitical conflicts, and volatile market developments, are just a few examples. This unpredictability encompasses four features of our modern world, all of which require new approaches to thinking and acting. The already introduced acronym VUCA is frequently used in business management to describe precisely this independence.

The concept of *disruption* also contributes to the unpredictability of our time. Disruption means that existing structures and processes are broken up and, if necessary, completely redefined. In the process, the established solution is replaced by a simpler, cheaper, or faster innovation. The term "disruption" has gained popularity mainly through startups. In fact, disruption was the business word of the year in 2015 and has been virtually omnipresent in our society ever since. Whether it's electromobility or online payment, we're all familiar with examples that have turned traditional business models upside down and reinvented them.

What is the difference between these and typical innovations, and why does disruption represent a challenge for companies today? To understand this, we must first distinguish between incremental and radical innovations. By definition, innovations are inventions, ideas, or transfers from other industries that are intended to renew or improve individual products and companies or even an entire industry.

Incremental innovations are innovations that involve incremental change. Changes are accordingly identified and/or implemented gradually. They are usually the result of applying a continuous improvement process, for example, when a car is developed with lower CO_2 emissions compared to the previous model.

In contrast, *radical innovations* aim at the complete transformation of the existing status quo, including customer and supplier relationships. An example of this is a new market player in the automotive industry that focuses exclusively on electric-powered vehicles. This radical innovation is now a disruptive innovation. Note that an innovation only becomes disruptive when the majority of users or customers also accept it and prefer it to the conventional variant. In any case, innovation always starts with the customer and his needs. Radical innovations are therefore not per se worse or better than incremental ones. Because radical innovations usually don't happen within companies but are brought to them from the outside, the consequences of these innovations are therefore often associated with major restructuring—whether at the level of individual employees, within the company, or even in entire markets. Effective disruptions can lead to surprising, rapid, and also harsh upheavals in both existing and new markets. In both cases—whether in an existing market, such as the automotive industry, or in a newly developed market, such as the metaverse—disruptions pose significant challenges to traditional, innovation-averse companies. Well-known victims of disruptive technologies include Nokia and Kodak. Both companies were once market leaders in their respective segments but have since been overrun by the success of smartphones and digital photography.

You may think that none of this affects your business, and you may be right at the moment, but there is no industry that won't be affected by disruption in the future. As a result of ongoing digitization, more disruptions are taking place today and at a much faster pace than a decade ago. In addition, familiar job profiles are already disappearing, and completely new ways of thinking are establishing themselves in global markets. As hard and surprising as the upheavals of disruptive forces often appear, the emergence of disruption is at the same time an extremely gradual process. Although companies want to be innovative, disruptive ideas are often initially neither perceived nor taken seriously. Apple AirPods, for example, were initially ridiculed because of their similar appearance to electric toothbrush attachments, but they have now established themselves on the market. Manufacturers of cord-based headphones are accordingly left behind and consequently at a considerable competitive disadvantage.

Another major risk is that actual changes only become noticeable gradually. Accordingly, these are only discovered belatedly by (still) established companies. Sales initially remained constant until then, only to collapse from one day to the next. So, if you don't move with the times, sooner or later, the market will move on without you.

Not every company can develop a culture of innovation that is disruptive. Nor would this culture be beneficial at all for some companies, especially if radical innovations neither advance the business model nor create real added value for customers or employees. At the same time, companies and organizations should use the digital transformation for their own benefit and see innovations—in whatever form—as an opportunity. If you successfully use digitization for yourself, you can strengthen your own core competencies as a result and take a step forward.

To do this, it's important to have a good feel for developments and trends. If you're always up to date, you can adapt to innovations more easily. For example, it's worth taking a regular look at research institutions for innovative developments, such as the

Fraunhofer Society in Germany (Fraunhofer-Gesellschaft), which highlight future-relevant key technologies and the utilization of their results in business and industry. There is no universal template for the emergence of innovative ideas, products, or concepts that can be applied to all situations, industries, and market conditions. However, continuous evaluation of your strategy and business processes can contribute to sustainable success and help counter disruptions. If disruption is occurring in your environment, business restructuring may be an appropriate path. Take the time to redefine your business model in a completely new and revolutionary way. Proceed as if you had to found your company again according to the current challenges. Use the following questions as a guide:

- What customer problems do you solve, how, and in what ways?
- How do you promote communication between your employees?
- How do you avoid silo mentality?
- What is your organizational structure?
- How do you make effective decisions quickly?
- How can your business processes best support your strategic goals?
- How do you adapt your business processes to any regulatory changes?
- Which processes can you automate?
- How do you ensure comprehensive customer excellence?

In this way, you're essentially cannibalizing your own business area by both defending it from external influences and acting strongly on the market from within. Even if you can't or don't want to change your business model from the ground up, you can experiment in this context and identify your priorities. By doing so, you build a mechanism that protects you from disruption in the future. In the course of this, you can see disruption as an opportunity to develop further and become even better.

1.1.3 Complexity

As described in Section 1.1.1, the "C" in today's VUCA world brings with it considerable *complexity*. Complex issues in the wake of economically uncertain times have particular significance. Changing regulations, geopolitical uncertainties, highly volatile capital markets, and the availability of key skills are just the beginning. From mergers and acquisitions, the decentralization of tasks and decision-making powers to IT infrastructure, and increasing business requirements (e.g., based on new regulatory requirements or to ensure competitiveness), the causes of business complexity are multifaceted and affect business processes, data, and systems, as well as employees. For example, new environmental standards affect your manufacturing operations, and, as a result, you need to adapt your production processes as well as your products to remain competitive in the future. Due to the significantly increased scope and diversity of products, processes, and stakeholder relationships, combined with the constant pressure for growth, business complexity has increased dramatically worldwide over the past decades.

Growth is generally a sign of economic success. Driving profitability and market presence is essential for most companies to survive in the long term. For companies to grow, they have to deal with diverse markets, an increased number of product categories, compliance requirements, diverse customer channels, and segments. Now, increasing growth can also lead to more complexity, especially in a globally interconnected economic environment. At the same time, the resulting complexity can in turn have a negative impact on the company's success and damage core competencies, such as productivity, innovative strength, decision-making power, or customer service. This complexity can lead not only to a burden for companies but also to a significant overload.

As organizations grow, their IT systems and business processes can reach a point where employees and business partners face various hurdles on their way to achieving their goals. In this context, silos can develop as a result of increasing business complexity. Once silos form, individual employees focus exclusively and sometimes narrowly on their own activities. The big picture and company-wide collaboration to promote shared business success are neglected. This approach not only harms the corporate climate but also inhibits innovation and the company's long-term competitiveness. In short, companies can no longer afford a silo structure today.

The fact that complexity and silos in the company counteract innovations has a problematic effect on the company's success in the long term. In fact, it's the innovations that are most important for the company to constantly develop new products and generate revenue streams. Complexity is associated with the risk that company-wide collaboration and the exchange of information between teams and departments will be drastically reduced or even completely eliminated. This can negatively impact other functions, such as product development, supply chain, human resources, or customer service, and it results in delayed *time-to-market* (i.e., it takes longer to get products to market). In addition, opportunities may not be exploited, and the competitive advantages that were the only ones for the company soon evaporate. Nevertheless, business complexity isn't exclusively a technical problem. For example, ineffective IT governance can cause a significant number of problems. If many different IT systems are used in the company while IT governance is weak, this can have serious consequences. IT-related project initiatives can even fail completely under these circumstances. It's therefore essential that such initiatives aren't just handled by the IT department but with the entire business, taking into account the business objectives.

Too much complexity, for example, with regard to your IT systems or business processes, can harm your company and therefore must not be ignored. It can impair your growth, your productivity, and your effects to secure competitive advantages. How do you deal with increasing complexity in your company, and what factors underlie it? To answer these questions, it's useful to differentiate between external and internal complexity. *External complexity* is caused by the outside, for example, by environmental changes, market developments, regulations, new technologies, and so on. You have only limited influence on these factors; you can only adapt as best you can. *Internal complexity*, on the other hand, is directly internal to your organization and is driven by factors such as a siloed culture or the nature and number of your business processes. It can be directly influenced by you and reduced by decisions on structures, products, and business processes. For the latter point, you can benefit, for example, from digital tools and tools that are easy to use and help you generate added value as well as increase your value creation.

In conjunction with this, it also makes sense to simplify your business processes and systems. You'll find inspiration for this in Part III of this book. It's up to you whether you allow complexity to undermine your company's profitability and effectiveness by simply accepting the individual factors, or whether you actively address complexity on a company-wide basis. How you achieve a lasting competitive advantage depends on the extent to which you can reduce, avoid, or eliminate unnecessary complexity.

1.1.4 Ambiguity

Ambiguity is a state of uncertainty in which a decision-maker lacks a clear understanding of the probability of many possible outcomes. In this context, information can be interpreted and weighted differently. This ambiguity often occurs in economic decision-making situations, as decision makers rely on subjective assessments of environmental conditions and their underlying causes.

We also undoubtedly live in an era characterized by information overload and constant innovation. Technological progress is leading to profound change in many business-relevant areas, such as communications technology, capital markets, and transportation. At the same time, this progress also enables a high degree of global collaboration at the business level—in real time. This international collaboration, in turn, has resulted in numerous changes in a wide range of areas, including individual economic and corporate structures, the possibility of global trade, and financial markets. The expansion of the international division of labor has increased significantly and has led to restructuring in almost all sectors of the economy. The increased international cooperation between companies has also led to an unmistakable increase in the transportation of goods around the globe. Likewise, it's possible to transfer billions of dollars around the globe in a matter of seconds.

Globalization and the associated fast pace of life are encouraging the emergence of global players with economic power comparable to that of entire countries. These companies optimize their value chains to perfection, thereby lowering their production costs. This use of economies of scale in conjunction with the exploitation of locational advantages for individual areas of the value chain provides optimal opportunities to build up a dominant market position. In this context, smaller companies can't withstand the price pressure in the long term and are squeezed out of the market.

The increasing interlinking of companies and value chains across national borders is unstoppable. Competition now no longer consists only of competitors in the immediate vicinity but also includes competitors from all over the world. At the same time, there have hardly been so many profound changes in such a short time as today in the history of mankind. We're in the midst of a globalized age of innovation, driven by digital transformation that constantly pushes us to be faster and better. This fast pace is creating a certain pressure on our business world. Shorter product lifecycles, increasing import competition, and strong global competition make it more and more important for domestic companies to act quickly and innovate—at ever shorter intervals. They are being forced to constantly offer new products and services, and, as a result, attract new customers and retain existing ones. Companies are now striving to continuously keep up with this technological progress. New technologies are constantly finding their way into our daily work, communication channels are increasingly being shifted to virtual spaces, and inefficient, repetitive tasks are being replaced by automated processes.

To counteract the challenges of ambiguity and to view it as an opportunity in the best possible way, it makes sense to consider ambiguity in your corporate strategy and vision. In doing so, ask yourself the following questions, for example:

- What does globalization mean for your company?
- How do you deal with today's fast pace?
- How do you counteract surprising changes?
- Do you have real-time insights into your data, systems, and business processes?
- Are your distributed offices all at the same level of knowledge regarding business data and business processes?
- Do you simulate alternative business scenarios to be prepared for possible environmental fluctuations?
- How do you identify the rapidly changing needs and requirements of your customers?

Reflecting on such issues will help you maintain your international trade, strengthen the global distribution of your products, and even open up new sales markets. On the basis of these expansion opportunities, you can in turn achieve corresponding growth. Saturated markets in your home country and the goal of increasing sales have a positive influence on your expansion plans. You can generate added value through profitable partnerships and enter into meaningful cooperations, which can again have a beneficial effect on your own economic performance. It's also advantageous that you can increase the number of your potential customers thanks to global reach. Export orientation combined with innovative strength are ultimately fundamental criteria for your business success in the course of globalization. By cultivating existing as well as new foreign markets, you can further expand your success. In this context, solutions from SAP's business process transformation portfolio will help you with your plans for international orientation—you'll find out exactly how in the course of this book. The current challenges facing companies, above all VUCA, result in process-related challenges for companies, in particular:

- Ensuring real-time insight into and understanding of enterprise-wide as well as locally defined business processes and best practices
- Identifying and minimizing local and global process inefficiencies, such as workflow delays or long lead times
- Dealing with error-prone, repetitive, and manually executed business processes
- Promoting transparency of the maturity level of a documented business process and the associated risks
- Creating a sustainable increase in customer, employee, and supplier satisfaction and loyalty
- Avoiding the misalignment of decisive transformation activities of the company
- Performing cross-divisional knowledge exchange with regard to process initiatives
- Ensuring continuous process compliance
- Adhering to internal and external governance requirements and regulatory compliance

Meeting all of these challenges remains exciting in the business world, and in the course of this book, we'll present you with a proposed solution for each of the individual challenges.

1.2 Competencies Required by Companies Today

The challenges in the wake of today's VUCA world are real and pervasive. Successfully adapting to these realities will make the difference between market winners and losers. Established companies must nevertheless function and build their competitive advantage to the best of their ability—no matter how unpredictable market conditions may be. At the same time, ambitious visions for the future are being set, for example, in terms of sustainability, efficiency, digitization, growth, or attractiveness as an employer. For companies that want to be successful in the long term, it's therefore essential to rethink not only their strategies and structures but also their business processes and adapt them to current conditions. This requires a certain degree of agility and company-wide transparency and traceability of all business activities. Finally, the focus is also on people—even though we keep talking about processes, data, and systems. Leading companies of tomorrow will bring empathy to their customers, employees, and suppliers to increase the satisfaction and the associated loyalty of all stakeholders involved.

Figure 1.2 summarizes five success-critical competencies that are worth cultivating for a contemporary company: agility, transparency, collaboration, transformation, and customer excellence. Let's now take a closer look at these.

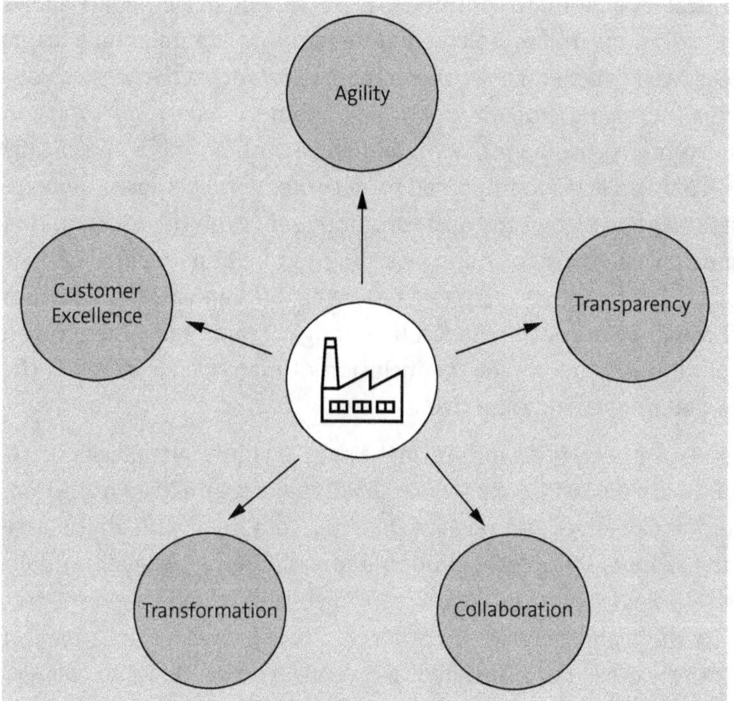

Figure 1.2 Success-Critical Competencies of Contemporary Companies

1.2.1 Agility

By *agility*, we mean a fast, active, and flexible adaptation of strategies, corporate structures, and business processes to the current challenges from Section 1.1. More and more companies are using the concept of agility to thus also meet the VUCA challenges. This is because one of the greatest challenges of our VUCA world is the short planning horizons coupled with a high degree of unpredictability. In the meantime, there are a number of agile work forms such as *Scrum*, *Kanban*, and *design thinking* that increase flexibility and reduce any risks.

This modern way of working is being driven forward because teams are much better able to deal with complex framework conditions thanks to agile methods, structures, and mindsets. The introduction of agile methods is particularly suitable when the environment of a team or even the entire company becomes complex. This is usually the case when both the problem and the appropriate solution are unclear, and/or the framework conditions are constantly changing. Agile forms of work help to organize companies more efficiently and promote rapid reactions to possible market changes.

Being able to quickly adapt one's own actions and react quickly to changing circumstances definitely helps in the VUCA world.

Agile methods can also be used to manage the complexity of our VUCA world more effectively. This is critical if you want to continue to be successful. In fact, agility allows you to achieve better solutions in a complex environment. For example, agile teams allow you to stay close to the market. You're therefore in constant exchange with your target group and thus increasing empathy for your customers. As a result, you can identify potential customer problems early on and address them more specifically with your solutions. This also leads to improved team competence because you regularly gather new customer knowledge and also constantly reflect on the collaboration of your team. This means that you're constantly learning and are best prepared for new challenges and changes. As a result, you can react faster and differentiate yourself from your competitors through greater competence. If your agile teams collaborate on an equal footing, this additionally allows for the inclusion of many perspectives in the development of new and innovative solutions.

At the same time, agile ways of working minimize the risks of failure. A team of experts from diverse disciplines can deal with a greater variety of topics than a team from a single area. The different perspectives and ways of thinking, combined with an iterative approach, increase the chances of success of your initiatives. Likewise, engaging different stakeholders also eliminates harmful silos. When you're able to generate responses to your challenges in short and iterative intervals, it also becomes easier to better assess the framework as well as the associated risks. Thus, you can react to changes even more effectively. In this context, it's useful to integrate flexibility and agility, for example, to better adapt your business processes as needed.

Note that complex situations are usually best solved in collaboration and by taking into account a wide range of views and ideas. Chapter 7 explores this topic in more detail using an application example of the business process transformation solution SAP Signavio Process Collaboration Hub.

So, ask yourself where your company stands in terms of agility and for what reason you actually want to become agile. What is the purpose? Your aspiration is most likely to keep up with your competition and to adapt to the speed and influences of a never-still, fluid, and digital world while maintaining an attractive, trust-based business environment. How do you successfully move into the future? Agility can't be implemented overnight, of course. Agility must fit or be made to fit your company's individual needs. It's a transformative process that takes time. Critical to its success is the willingness of everyone involved to be open to new things and to break down slowing structures and outdated hierarchies. However, you don't have to compulsively and aggressively turn your entire company upside down. Just by having accurate, real-time insights into success-relevant information, you have an agile basis for making active decisions quickly. Building on this, it's also helpful to rethink any encrusted business processes and inefficient workflow management. Subsequently, business areas emerge in which agility is particularly worthwhile. For all this, there is reliable technological support. *SAP Signavio Process Transformation Suite*, which will be introduced in Chapter 2, will give you more detailed information about agility.

Agility ultimately stands for vitality and energy. It's about implementing changes that are decisive for success in an anticipatory and proactive, effective and efficient, and proactive and reactive manner. Agility is based on trust and communication, among other things. Decisions are attributed to each other. The best possible agreement is reached, rather than issuing orders. The basis for this is holistic, company-wide transparency.

1.2.2 Transparency

Transparency creates understanding for all those involved and increases trust in the company's decisions. To enable employees to react quickly and identify new solutions, it's necessary to make all relevant information available to them. After all, companies can get into difficulties through no fault of their own, for example, if a supplier fails to meet their delivery deadlines and production comes to a standstill as a result. This connection is often not obvious to employees. Accordingly, it's important to inform all parties involved. In practice, however, many companies lag behind when it comes to providing employees with clear insights into strategy, business goals, and processes. Employees in these companies don't feel valued, which in turn hurts motivation, loyalty, and satisfaction. How are the challenges of our VUCA world to be met if employees aren't fully engaged and informed about critical risks? Open, active, and regular communication within the company is therefore indispensable for sustainable success. If transparency is taken into account and firmly anchored in the corporate culture, this will subsequently increase responsiveness, efficiency, and effectiveness, especially in dynamic markets. In addition to employees, customers also rate such companies as more trustworthy because transparent companies are perceived more positively by a wide range of stakeholders.

In times of crisis, the need for transparency and open communication is even greater. After all, crises not only bring economic uncertainties but also changes for employees. For example, due to the COVID-19 pandemic, many teams experienced and continue to experience significantly less personal contact. To ensure that all employees know where they stand and how the company is doing in general, companies should therefore always ensure that no information is lost—even if work is done from home or distributed. Of course, it's impossible and also completely unreasonable to pass on unfiltered company information. In every company, there is information that is classified as internal, critical to competition, or even a company secret. Accordingly, there can be no such thing as unlimited transparency—nor is this what is meant in this context. Rather, it makes sense to develop strategies for the voluntary and continuous publication of relevant information. Transparency here doesn't mean sharing every single detail with everyone, but rather providing the context to make good decisions.

Transparent business processes are also a core element in being able to grow at a healthy rate in the future. But how do you obtain full process transparency? The following questions are relevant here:

- Which process details are particularly relevant for your employees?
- Which business units could benefit from increased transparency?
- Which operational processes run efficiently? Which processes do not?
- Which activities can be improved or automated?
- How do you keep your employees informed about process changes?

Until now, such questions were laboriously clarified by means of assumptions, workshops, and lengthy interviews, which mostly drew an incomplete and above all subjective picture of the actual process events. Next-generation perspectives are offered by SAP Signavio Process Transformation Suite. It provides you with data-driven real-time insights into your business processes, identifies weak points and improvement potentials, visualizes and simulates various variants with regard to your process changes, and controls the traceability as well as the governance of all workflows. Thus, you can derive targeted transparent measures for process optimization and automation because fact-based insights promote transparent decision-making.

SAP Signavio Process Transformation Suite also enables you to actively support the collaboration of your teams across departments. This way, you not only achieve transparency about relevant process flows and changes, but you can also easily maintain them company-wide. This leads to more clarity, effectiveness, and focus on what matters. While SAP Signavio Process Transformation Suite ensures holistic transparency regarding your process initiatives, SAP Signavio Process Collaboration Hub is especially suited to spreading those initiatives across all areas of your company. To learn more about the exact features of this hub, see Chapter 7.

First, let's consider in detail why collaboration is an important competency of successful companies in the first place.

1.2.3 Collaboration

Collaboration is another answer to the challenges of our VUCA world because needs and markets are constantly and rapidly changing, technologies are more accessible and usable, and—at the same time—at least three very different generations are actively shaping our society and companies. This is an excerpt of factors that companies must take into account to identify their customers' wishes early on and respond to them swiftly and accurately. Contemporary work without collaboration and teamwork is no longer imaginable. Of course, companies have been collaborating since time immemorial, but often only within the respective department and with serially linked activity steps. After all, it's characteristic for most companies to formalize, regulate, and regiment tasks with the goal of providing orientation and relief for all actors. However, such rigid definitions and regulations tend to be counterproductive in our VUCA world, especially when surprises are to be expected.

Collaboration is understood to mean purposeful, voluntary, and—above all—proactive. It arises from the intention of the team members to make a valuable contribution to the individual team or to the entire company. True collaboration can't be ordered or agreed upon, and it knows no departmental boundaries. It only recognizes the inner drive of every person involved to contribute in the best possible way with their strengths.

Accordingly, new, agile ways of thinking are needed that promote collaboration away from limiting departmental thinking. These thought models help to influence complexity in a targeted manner to continue to achieve added value. Today's tasks are for the most part so complex that they can only be completed in a functioning collaboration. Collaborative and open cooperation is therefore one of the success factors for companies in our digital age. In most cases, individual team members are located in different parts of the world—in other branches or in the home office—or they work at different times of the day, as is the case with globally distributed teams. Then, solid collaboration can only succeed with technical support. To improve or simplify this, suitable collaboration tools can be used. Since the COVID-19 lockdowns in 2020/2021, we appreciate them even more. They make it possible to work from anywhere. Digital collaboration tools can generally help to eliminate data silos, share relevant information company-wide, and make processes more transparent. In this way, they help to find orientation in larger groups. For team members, the current status is always visible.

When we think of such tools, applications for document processing and storage, virtual meetings, project management, or planning come to mind first and foremost. At this point, it's therefore appropriate to also refer to a collaboration tool for process management. As already mentioned in Section 1.2.2, SAP Signavio Process Collaboration Hub offers an interesting solution. With this solution, teams can realize and continuously expand extensive process initiatives together worldwide. This includes the sharing and further development of business processes as well as their joint optimization—across distributed teams, of course. Business process standards are accessible, understood, and implemented uniformly throughout the company, regardless of location or language. This means that even new employees can quickly understand and record activities, including all process-related content such as diagrams, comments, risks, or responsibilities, which in turn reduces the time and effort required for familiarization.

In a world where everything has to move faster, collaboration must also be made possible. Lengthy coordination and email chains have had their day. They are replaced by intuitive workflows and commenting and editing functions, always with the aim of actively involving all participants in the collaboration and promoting it on an ongoing basis. From a process perspective, this brings you one step closer to operational excellence. To learn more about how to achieve and maximize operational excellence with SAP's business process transformation portfolio, see Chapter 4.

How do you promote collaboration in your company and in your individual teams? Are there any blockages that might inhibit proactive collaboration? If so, which ones and how can you reduce or avoid them? In-depth reflection in this regard is recommended so that you become even more aware of your strengths and potential for improvement with regard to collaboration.

1.2.4 Transformation

Companies that can always successfully redefine themselves in the face of constant change have a significant advantage over their competitors. The starting point for change is developments in the market environment. Changes alone require a high degree of transparency based on constant communication and collaboration with all stakeholders. *Transformation* goes one step further and takes into account perception and intention as well as direct dialog with employees, customers, and suppliers.

Changes are past-oriented: they are concerned with making something from the past better. Transformation is future oriented: it's concerned with actively switching the future and promoting the company's ability to innovate and its uniqueness. In addition, as old paradigms lose relevance and new approaches gain traction, companies also renew themselves with the aim of becoming more customer-centric, employee-friendly, participative, sustainable, and digital. In this section, transformation describes the company's intention to redesign existing business processes, data, and systems and then adapt them to new technologies to meet the current challenges from Section 1.1 in a sustainable manner. However, the application of new technologies alone doesn't add value to the company. Technology in itself doesn't deliver direct value, but with the right application, it supports and enables us to create success-critical added value. The most important thing in transformation therefore isn't the technology, but the people who handle it and use it continuously, taking the corporate vision into account.

Transformation represents a continuous process and is therefore in constant motion. Even when one has "transformed," it's important to continuously and proactively maintain the status quo. During transformation, it's also advantageous to reflect on the inherent dynamics in the course of the process and to gain an awareness and sensitization with regard to the complexities. In most cases, companies have to redefine a considerable part of their environmental relationships in the course of their transformation process. Above all, a holistic business transformation influences the course of the entire organization. It's therefore a good idea to first develop a clear vision and direction for the future for each company and each individual team. This serves as an orientation compass, and a commonly understood goal helps to avoid getting lost in our turbulent VUCA world.

Transformation has to do with change and revolution. Consequently, it's essential for all stakeholders to have a strong change competence, that is, the ability to react and manage change as it comes. In addition to the implementation of changes in the daily work routine, the principle of openness and the desire to actively shape changes in the company are decisive. It's also important to maintain your authenticity during your transformation process in order to maintain all-around relationships with your employees, customers, and suppliers even in times of upheaval. All of these characteristics, along with the competencies covered, agility, transparency and collaboration, are core elements for effectively managing transformation. You'll find suitable starting points for successful business process transformation management in Chapter 11.

An active examination of the current future viability of your company, together with the establishment of holistic transformation processes and the willingness to also analyze the problem areas in detail, represents a further basis with regard to your transformation. The following questions could be relevant here:

- What does your company stand for?
- What values do you cultivate?
- What makes you great?
- What purpose do you serve?
- What barriers can you eliminate, and how?
- Where can you create more freedom in the company and reduce any fears to favor your ability to innovate?

Although you may aim to become more digital, innovative, sustainable, and so on through transformation, the real motivation is probably to remain competitive or become even more competitive in the future. The top priority of the complete transformation initiative is therefore to be able to offer your customers what they are asking for, even under changing framework and environmental conditions and market developments. As a result, it's essential to take your customers and their wishes and needs into account throughout your transformation project. In the following section, you'll learn why customer excellence is another valuable competence for modern companies and what elements it consists of.

1.2.5 Customer Excellence

To keep up with our VUCA world, it's no longer enough to operate in a purely product-oriented and profit-driven way. Rather, it's about identifying the needs, challenges, and problems of customers and, based on this, developing products and solutions. This approach is more than in keeping with the times. After all, successful companies must always adapt to the wishes of their target group to remain competitive in the long term. In the past, the product came first, and only then came the clientele. Today, it's the other way around because the greatest products are useless if they don't offer a solution for the clientele. This customer orientation is understood as process orientation, which helps to put the customer at the center of business decisions. Due to the constantly changing competition as well as customer behavior, customer focus is thus becoming an increasingly significant approach to continuously drive the company's growth. Customer focus per se aims not only to perceive the expectations and wishes of the clientele, but also to take them seriously. The decisions, actions, and behavior of the company are geared to the customer and their requirements and needs. Customer wishes and expectations are systematically recorded, documented, analyzed, and evaluated. As a result, products, services, and/or individual process flows can be individually adapted to the respective customer.

If you're able to actively respond to your customers' needs, you can increase your overall customer satisfaction in the long term because it enables you to support your customers even better in their purchasing decisions and in achieving their goals. As a result, you build even more loyal customer relationships. By resolving customer issues, proactively implementing communication with your clientele, and quickly responding to feedback, you build further trust. The ultimate goal of a customer-centric approach is to retain the customers and thus not lose them to the competition.

Of course, how intensively you cater to your clientele depends on your mission statement. In any case, it serves to anchor the optimal internal and external customer focus in your corporate practice and not only to meet customer expectations but also to exceed them if possible. This is because customer focus promotes an increase in customer value as well as your economic success and also helps you to outpace your competitors. Taking customer perspectives into account also helps you deliver products that your customers love. In this context, it's a good idea to involve your customers in your day-to-day processes—from recording experiences to communicating feedback. Your customers will become part of the process and feel even more valued as a result. The resulting *customer excellence* is the necessary foundation for an outstandingly successful company—regardless of the industry. Those who are able to serve the wishes as well as the needs of their clientele in the best possible way and at the same time influence them with a view to future trends will very likely prevail significantly in their segment. For more detailed information on customer excellence and how you can achieve it with SAP Signavio, see Chapter 6.

To analyze the individual steps from the first contact of the customer with the respective product up to the moment of the purchase decision, *customer journeys* can be used. These describe all the phases that a customer goes through before making a purchase decision. The customer journey links all the customer's points of contact with the company, be it the brand, the product, or the service. Accordingly, it takes into account a wide variety of factors to promote effective customer loyalty. For more details on the customer journey, see Chapter 2, Section 2.2.4.

Even though we're dealing with the topic of customer excellence in this section, it's first and foremost about the human factor. Therefore, it's also essential for modern companies to pay attention to the experiences of their employees and suppliers.

Increasing employee dissatisfaction due to the constant double burden of family and work in a home office, for example, also has an impact on process management. In this sense, it's important to holistically link operational business processing with experience management in process management. In the future, it will no longer be just about efficiency and costs but also about the possibility of including the experiences of customers, employees, and suppliers. Relevant functions, benefits, and application examples for this can be found in Chapter 6 when discussing SAP Signavio Journey Modeler.

1.3 The Intelligent Company of Tomorrow

Traditional approaches and conservative approaches don't adequately support fast-moving change and lack meaningful business process insights, resulting in decisions based on outdated information, which ultimately leads to dissatisfied customers and consequently reduced competitiveness.

Competitive advantages in the VUCA world are achieved primarily by companies acting proactively and anticipatorily, focusing on positive customer experiences (CXs). The intelligence of companies no longer lies solely with the employees, but in the technology and networking of all people, business processes, and systems throughout the entire value chain. Smart technologies such as artificial intelligence (AI), blockchain, cloud computing, Internet of Things (IoT), and robotic process automation (RPA) are helping organizations pursue this mission to drive sustainable real results in the era of big data and Industry 4.0. SAP has therefore created *RISE with SAP*, which is a bundle of products, services, and tools to help companies on their journey to becoming intelligent enterprises. The main idea of RISE with SAP is to deliver business transformation as a service, covering different areas, including infrastructure, technology, applications, and business processes.

The business process transformation solution portfolio focuses on business processes and, as a result, represents an essential component when it comes to harmonizing business processes, data, and systems with each other, and putting people at the center. Business process transformation from SAP lays the foundation for the intelligent enterprise of tomorrow. The solutions in the portfolio provide next-generation end-to-end capabilities for strategic process transformation as well as reinventing CXs.

As shown in Figure 1.3, the concept of an intelligent enterprise in itself generally comprises three essential components:

- Holistic digitization
- Customer centricity
- Networking

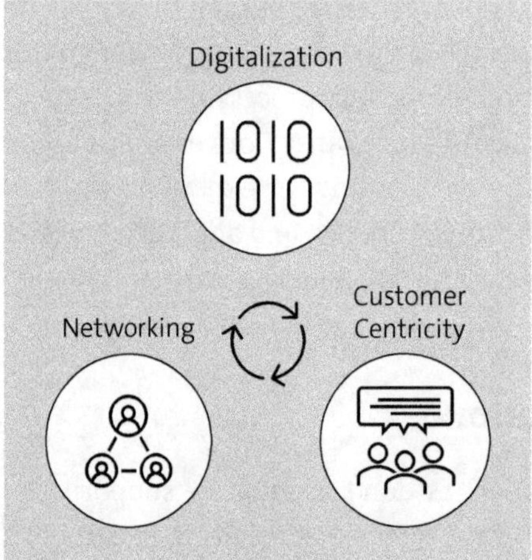

Figure 1.3 Core Components of an Intelligent Enterprise

Next, we'll highlight each component of an intelligent enterprise to deepen our basic understanding.

1.3.1 Holistic Digitization

Digitization is the process of completely transforming information from analog or manual form into digital form—for example, converting handwritten records into computerized records. Digitization also involves the application of digital technologies and capabilities to perform activities in new and more efficient ways that lead to better results. It's the foundation for digital transformation, and digital transformation enables companies to rethink how they use their technologies, people, and processes to drive their business forward.

Digital transformation in itself takes a customer-centric, digital approach to all aspects of an organization, from its business models to CXs to processes and operations. The approach embraces AI, automation, cloud, and other digital technologies to leverage data and drive intelligent workflows, faster and smarter decision-making, and real-time responses to market disruptions. Today, digital transformation means moving from manual and analog processes to digitized processes in all aspects of business—including supply chain, enterprise resource planning (ERP), operations, customer service, and more. Essentially, digital transformation helps businesses and other organizations achieve better results by connecting people, places, and things.

Every business has its own reason for digital transformation. Some may be spurred by growth opportunities or increased competitive pressure, while for others, it's changing regulatory standards. Whatever the reason a company takes this path, the results can include purposefully designed products, services, and experiences that meet or exceed customer expectations. Increased efficiencies and other business benefits can drive profitability and foster innovation.

The story of digital transformation begins with the first computers that converted handwritten notes into computerized information that could be processed, analyzed, and shared. With the advent of networks and the internet, these capabilities evolved, and data sets became extremely large. Big data required more robust digital data management and analysis processes supported by data centers, data warehouses, and now data lakes. The evolution of the cloud was in part a response to ever-increasing data volumes and the need for ever-expanding capabilities to manage, analyze, and process that data. IoT, AI, and machine learning have since provided even more sophisticated technology options for companies looking to transform their operations to drive better outcomes. Virtual and augmented reality and blockchain-based technologies are part of the mix of innovative capabilities contributing to widespread digital transformation.

Customer expectations have always been the key driver of digital transformation. It began when an onslaught of new technologies made new types of information and capabilities accessible in new ways, such as smartphones, social media, IoT, and cloud computing. Disruptors such as Amazon and Tesla also snatched market share from their competitors by adopting these technologies to reinvent business models (e-commerce, electronic delivery), optimize processes (supply chain management, new feature development), and work to continuously improve the CX (contextual customer reviews, personalized recommendations). Competitors adapted to offer even more options and convenience (or they struggled and perhaps even disappeared). Today, and especially after the COVID-19 pandemic, customers expect more than ever to do all their business digitally, wherever and whenever, on any device, with all the supporting information and content they need close at hand.

While many companies undertake digital transformation in response to a single competitive threat or market change, it's not about finding a one-off solution. The goal is to create a technical and operational foundation to evolve and best respond to unpredictable and ever-changing customer expectations, market conditions, and local or global events. Only when all steps in the value chain are digitally traceable and holistically networked has the technological basis for an intelligent company been created. This is a constant process that must always be carried out and expanded flexibly with regard to future developments. As a result, intelligent business applications aren't isolated solutions as we know them from traditional technologies such as ERP or customer relationship management (CRM). They are integrated into a platform that provides central business process functions as well as innovative technologies. This enables smart companies to access data and information, create new business opportunities, and make better decisions at all levels of the organization to ultimately target customer expectations.

1.3.2 Customer Centricity

As a result of digitization, the number of touchpoints increases with every new device. Customers can contact a company from anywhere at any time via a wide variety of channels. Customers are therefore more present than ever for companies. This is because customers have freedom of movement and choice like never before. On a global level, offers are obtained, purchases are made, and items purchased are critically evaluated. As a result, customers are placing higher demands not only for the product itself but also for the entire customer journey and thus for the overall shopping experience. This applies to both consumers and business customers. Another challenge here is the ironclad price war in global competition, which even small and medium-sized enterprises (SMEs) can't escape. So, if you want to be successful, you have to consistently and continuously focus on the needs of customers while ensuring that you not only sell them a product or service but also generate lasting emotions in the process.

Customer centricity in this context describes an organization's ability to understand the situations, perceptions, and expectations of its customers. Customer centricity requires that customers be at the center of all decisions regarding the delivery of products, services, and experiences to ensure customer satisfaction, loyalty, and advocacy. For too many companies, however, this is mostly hype—a rebranding of traditional marketing, sales and customer service that doesn't result in fundamental change, and, consequently, little benefit for customers. In this context, it's also important to differentiate between customer centricity and customer orientation. The latter usually pursues a sales and revenue objective. Customer centricity, on the other hand, has the goal of creating added value for customers and retaining customers in the long term. For individual customers, this means an individual, custom-fit CX and ultimately satisfaction, loyalty, and brand loyalty. As a result, true customer centricity requires transforming all business functions that affect customers, breaking down the silos between these functions, and building a culture that rewards behaviors focused on customer success.

All areas of the business—from research and development to product management to marketing and sales—must thus expand their traditional priorities and measures of success to include metrics focused on customer value and success. That's because too often, employees who don't have direct customer contact feel disconnected from and unaccountable for customer outcomes. This leads to decisions and prioritization based on the needs of suppliers rather than customers. It behooves companies to help their employees understand how their work contributes not only to the delivery of specific products and solutions but also to the customer's overall business strategy.

Tomorrow's successful companies strive for competitive differentiation through every aspect of the customer journey and CX. They distinguish themselves in two ways. First, unlike many competitors, such leaders treat solution implementation, service delivery, and customer service not as cost centers but as value drivers. Rather than focusing primarily on the internal efficiencies of these elements, they continually look for ways to improve the CX at every touchpoint. They agree to accept internal costs if it reduces complexity for customers, shortens the time to actual sales, and increases customer retention and cross-selling opportunities. Customer service and support teams in customer-centric organizations look for ways to deliver unexpected value—rather than simply solving the immediate problem. Such an approach requires a close partnership between sales leaders and operations executives, who are responsible for many of the systems and resources that ultimately shape the CX.

Second, successful businesses of tomorrow also balance investments in automation with a continued focus on human interactions with customers. While customer journeys are increasingly automated, a large proportion of consumers worldwide still want to interact with people. For what reason? While customers appreciate the ability to quickly select from a catalog of online offerings or configure a solution themselves, for example, many prefer to contact a knowledgeable, consultative sales team for complex or unique requirements. Self-service tools are proliferating as they become more powerful, but human judgment, ingenuity, and empathy when things go wrong can't be replaced. Best-in-class companies understand the interdependence between systems and people, and they optimize the CX by making complementary improvements to both and ensuring their seamless integration.

As with many business disciplines, customer centricity is therefore about more than just systems and tools. Effective implementation ultimately depends on people. That's why it's imperative to foster the satisfaction, motivation, creativity, inspiration, and commitment of individual employees. After all, at the end of the day, they are the ones who create sustainable added value for customers. In addition to any cultural change, the networking of people, processes, data, and systems within the company is also fundamental to this.

1.3.3 Networking

More than ever, it's critical to harmonize people, business processes, data, and systems for a company to operate successfully. If adequate processes aren't in place, employees can be ineffective, and technology can fail. At the same time, inaccurate data that isn't available in real time can lead to flawed decisions. Technology alone won't eliminate existing problems if people and business processes are lacking and out of sync. In this regard, business processes here create the framework for people to drive their making and decision-making. The technical systems and tools are subsequently used to execute these very processes more efficiently. So, for efficiency and effectiveness, there needs to be a good balance so that the business processes and systems with the right data can really support people in the end.

In the past, a company's intelligence lay exclusively with its employees. Today, it can be found in the combination of people, business processes, data, and systems. The technological aspect is taking on more and more tasks because innovative technologies now learn faster and more efficiently than we humans do. Such smart technologies are capable of learning through the use of AI. They are able to deal with certain rules, or they can be trained in such a way that they can react depending on the situation. This may seem daunting, but it makes the work much easier. Tasks that are time-consuming for humans can be accomplished within a very short time using technological means. They make it possible to master the enormous amounts of data from big data because they create intelligence without capacity limitations. Technological elements also function 24/7, unlike humans.

Despite all of this, it's humans who control these smart technologies. Thus, despite its capabilities, technology alone doesn't solve problems. Even when automation is operated, someone must carefully monitor the effort and intervene if there are any blockages. Technology doesn't eliminate existing problems without the intervention of people and processes. In this sense, technology can make people's jobs easier without taking them away. In addition to automation functions, employees can, for example, work even more productively, committedly, and efficiently through suitable communication and collaboration tools. Ultimately, the focus is on people. After all, they possess valuable empathy, creativity, and decision-making skills that are critical to success.

Far too often, however, companies invest mainly in technology and try to retrofit processes and people. In this context, many business transformation strategies focus predominantly on technology and processes, while almost completely ignoring the people involved. However, this involves backward logic, and isn't a good strategy if you want to successfully implement and drive a transformation. The real value of any technology is to get buy-in and support from all stakeholders to fully realize the potential of the technology and business processes. It requires *networking* of processes, systems, and data—with people at the center (see Figure 1.4).

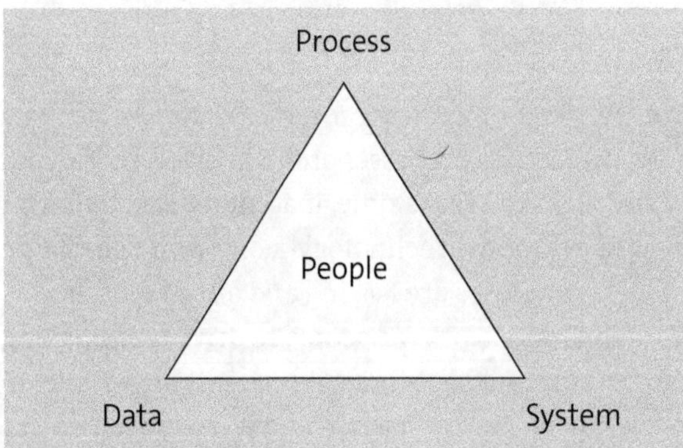

Figure 1.4 Focus on People

True business transformation is based on people executing and continuously innovating business processes while benefiting from smart technologies and systems to intelligently process business data. This networking enables, among other things, the right, fact-based business decisions to be made, full process transparency to be ensured, business and IT to be fully aligned, and business transformations to be completed as expected and—above all—securely.

1.4 Summary

In this chapter, we've presented the framework conditions that companies have to deal with today and that can be countered with the use of business process transformation. These include new technologies and business models, general developments such as market trends, globalization or pandemics, and the significantly increased scope of products, processes, and stakeholder relationships. Process activities must therefore be traceable from A to Z, and decision-making processes must be traceable as well. Open communication and company-wide collaboration are necessary for the further development of business processes. Experience on the customer side must also be taken into account.

2

Chapter 2
What Is Business Process Transformation?

Business process transformation is a comprehensive framework for future-proofing your business processes from an end-to-end perspective. In this chapter, you'll learn about the parts of this framework and its core capabilities.

Section 2.1 provides a general overview of the *business process transformation* portfolio from SAP. In Section 2.2, we present the individual product solutions of SAP Signavio Process Transformation Suite.

2.1 Business Process Transformation at a Glance

Business process transformation aims to harmonize your business processes, data, and systems while putting people at the center. It provides you with a wide range of tools to cope with the challenges of our world of volatility, uncertainty, complexity, and ambiguity (VUCA) and thus remain competitive in the long term.

Business process transformation is a hot topic for modern organizations. In this context, SAP Signavio's comprehensive product solutions help you confidently navigate your business through your day-to-day operations, processes, and decisions in a continuously changing market environment.

[«]

The Acquisition of Signavio by SAP

SAP acquired Signavio, a leader in process management, in early 2021. As a result, SAP can now more easily help companies understand, improve, transform, and manage their overall business processes more quickly and at scale.

First, let's take a brief look at the target groups of business process transformation:

- *Business executives* who want to better understand the value proposition of SAP S/4HANA, RISE with SAP, and intelligent technologies in the course of business process transformation
- *Project managers* who want to support their projects (SAP S/4HANA migration, RISE with SAP, and business transformation) with business process transformation

- *Line-of-business managers* who want to monitor performance and identify areas for improvement in their area
- *Process owners* who want to monitor performance and identify areas for improvement in their end-to-end process
- *Process managers* who are looking for company-wide automation benefits

In this section, we'll now show you exactly what business process transformation from SAP does and how this solution package came about in the first place.

2.1.1 What Business Process Transformation Does

Business process transformation solutions enable strategic end-to-end transformations of your business processes and provide the foundation for tomorrow's intelligent enterprise. They allow you to shape your digital transformation at your individual pace and on your own terms, regardless of your starting situation. Successful business transformations require a solid understanding of customer issues, in-depth process analysis, industry benchmarking, redesign of relevant business processes, and smooth collaboration among all stakeholders—just to name a few factors. Business process transformation from SAP is a holistic solution portfolio that combines these components in a unified, cloud-based tool suite. In addition, business process transformation combines granular process analysis with the tools that are essential for successfully revising and adapting business processes or developing completely new and innovative processes. In addition to business process design and benchmarking, SAP's business process transformation portfolio also covers, among other things, critical gap analysis, improvement management, and process change management.

[»]

The Importance of Business Process Transformation

By integrating Signavio's process suite, SAP extended its in-house business process transformation solution. Business process transformation today represents the latest process platform within the SAP portfolio and offers comprehensive as well as holistic product solutions for the transformation of your entire business processes.

All these functions support you in your transformation process. If this is done continuously, new competitive advantages will be created as a result. The benchmarks for time-to-insight and time-to-adapt are decisive factors in this respect and help you on your way to becoming the intelligent company of tomorrow. *Time-to-insight* measures how quickly you can turn your data into actionable insights. *Time-to-adapt* subsequently measures how quickly you can practically implement or adapt your insights. Figure 2.1 illustrates the basic idea behind business process transformation.

In short, data from various enterprise resource planning (ERP) systems is mapped with regard to your business processes. At the same time, the experiences of your customers, suppliers, and employees are also included. Finally, you can use the business process transformation portfolio to implement sustainable process improvements and innovations. Details on this can be found in Section 2.2.

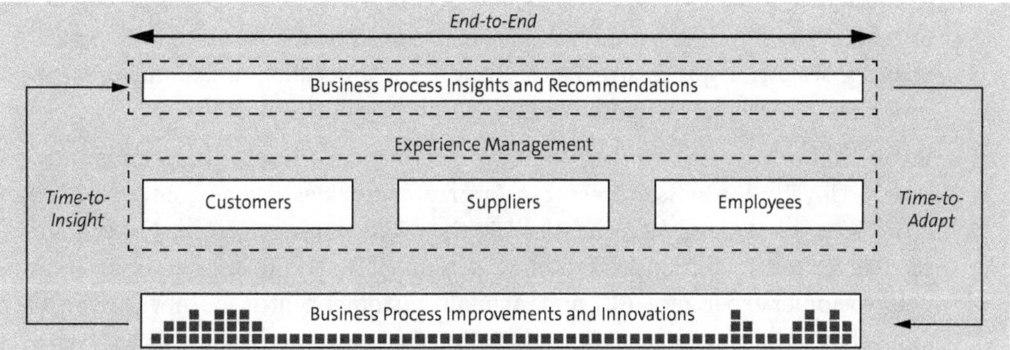

Figure 2.1 SAP's Business Process Transformation Vision (Source: SAP)

The complete suite of business process transformation in itself includes the following business process transformation solutions:

- SAP Signavio Process Insights
- SAP Signavio Process Intelligence
- SAP Signavio Process Manager
- SAP Signavio Journey Modeler
- SAP Signavio Process Collaboration Hub
- SAP Signavio Process Governance
- SAP Build Process Automation

With *SAP Signavio Process Insights*, you can ensure immediate value. The tool gives you an overview of your company's process performance, you can dive deep into business processes with next-generation analytics, and you can benefit from improvement recommendations tailored to your needs (Section 2.2.1). For example, you can view key performance indicators (KPIs) with above-average outstanding receivables against industry benchmarks and aim for even better receivables management based on tailored recommendations such as setting up automatic reconciliation of incoming payments and invoices, credit management, or dunning procedures.

SAP Signavio Process Intelligence enables you to visualize and scan your as-is processes, including possible bottlenecks from detailed process analysis for business transformations (including SAP S/4HANA migrations), to the examination and interaction with business data, to the use of process mining capabilities for fact-based changes to your business processes to fully uncover their true potential (Section 2.2.2). *Process mining* is a technique to identify, analyze, and optimize business processes. Based on existing process data, process mining can be used to automatically show a dynamic visualization of real processes, their performance, and their compliance.

When it comes to modeling and completely redesigning your business processes, you can turn to *SAP Signavio Process Manager* (Section 2.2.3). This business process transformation solution helps you react quickly to unpredictable business and regulatory changes, scale your process repository across the enterprise, and simulate business processes in such a way to define and test alternative business scenarios.

If you want to increase your customer excellence and the resulting customer satisfaction and loyalty, business process transformation provides a helpful tool with *SAP Signavio Journey Modeler* (Section 2.2.4). With it, you can develop customer journey models for touchpoint optimization, align your business processes specifically with your customer experience (CX), and maintain customer-centric journeys for feedback and potential improvement actions regarding your customers' experiences.

SAP Signavio Process Collaboration Hub enables you to collaborate across your company in a customized way by creating a central source of information for all your teams, breaking down any business silos and fostering an improved understanding about KPIs, tasks, and projects (Section 2.2.5).

To ensure and also track the success of your process initiatives, you can use SAP Signavio Process Governance (Section 2.2.6). This tool helps you achieve complete visibility and control over your organization's workflows, prototype workflows for rapid implementation and scaling without coding, and reduced business process variation and possible rework.

Ultimately, as the name implies, with SAP Build Process Automation, you can automate manual, repetitive, and/or time-consuming activities of your business processes thanks to robotic process automation (RPA) and artificial intelligence (AI). SAP Build Process Automation thereby promotes the productivity of your employees as well as the process efficiency and agility of your company (Section 2.2.7).

[»]

Produce Excellence

Excellence is an essential keyword when it comes to business process transformation. The diverse product solutions help you on your way to process excellence, operational excellence, and customer excellence.

You'll certainly remember the process-related challenges from Section 2.1. SAP's business process transformation portfolio delivers all the capabilities to permanently strengthen the performance of your business processes. The individual challenges and solutions within SAP Signavio Process Transformation Suite that you can use to address them are shown in Table 2.1.

Challenge	Business Process Transformation Solutions
Ensure real-time insight into and understanding of enterprise-wide as well as locally defined business processes and best practices.	SAP Signavio Process Insights and SAP Signavio Process Intelligence
Identify and minimize local and global process inefficiencies, such as delays or long lead times.	SAP Signavio Process Insights and SAP Signavio Process Intelligence
Deal with error-prone, repetitive, and manually executed business processes.	SAP Build Process Automation
Promote transparency about the maturity of a documented business process and the associated risks.	SAP Signavio Process Manager and SAP Signavio Process Collaboration Hub
Create a sustainable increase in customer, employee, and supplier satisfaction and loyalty.	SAP Signavio Journey Modeler
Avoid misalignment of decisive transformation activities of the company.	Complete Business Process Transformation suite
Perform cross-divisional knowledge exchange with regard to process initiatives.	SAP Signavio Process Collaboration Hub
Ensure continuous process compliance.	SAP Signavio Process Governance
Adhere to internal and external governance requirements and regulatory compliance.	SAP Signavio Process Governance

Table 2.1 Functions and Solutions of the Business Process Transformation Suite

Table 2.1 includes an excerpt of potential challenges that you can overcome using business process transformation solutions. However, before we drift too much into the details of the individual solutions, we'll first look at their development.

Accelerate Transformation Projects with SAP Signavio Process Explorer

The ability of any business to adapt quickly to new market conditions and customer behaviors is a key pillar of long-term success. Accordingly, companies are rethinking the way they work, sell, deliver, and interact with employees, partners, and customers. To facilitate business process change, SAP has announced the general availability of *SAP Signavio Process Explorer*. The solution organizes and centralizes collective knowledge from thousands of transformation projects delivered by SAP and its partner ecosystem. As a result, customers are able to operate faster and with greater confidence by referencing more than 7,000 process models, capability maps for more than 20 business areas, value accelerators for more than 13 industries, and a wide variety of process metrics and product recommendations.

This solution is designed to shorten the time-to-value of any business process change by providing a single gateway for exploring and accessing value accelerators and other resources. Time-to-value per se characterizes the lead time that elapses between an intention and its initial creation of value. Available resources include business capability and solution maps, process models, metrics and industry best practices, and SAP product best practices and recommendations. For example, a utility can quickly access an industry-specific process model for the end-to-end order-to-cash process and identify where process changes will have the greatest impact.

To access the content of SAP Signavio Process Explorer, you only need a user registered with SAP.

2.1.2 Evolution of Business Process Transformation Solutions

When Signavio was founded in 2009, the industry was still all about business process management systems (BPMS). Comprehensive technology solutions were developed that would solve all problems at once: from capture to modeling and automation to business process monitoring. At first glance, of course, this sounded appealing. After some time, however, numerous software providers discovered that there was a world of difference between analyzing a problem and solving it. Responsibilities were also fundamentally different: professionals didn't want to leave process improvements solely to IT managers, and these in turn modeled processes that didn't reflect the professionals' day-to-day business. Today, many BPMS providers have disappeared from the market, and new competitors have discovered the industry for themselves. Lean solutions with little code entered into competition with comprehensive development environments, while open-source providers conquered the market.

The Name Signavio

What is behind the name Signavio? Signavio was derived from the Italian *Segnavia*, which means signpost. Today, Signavio is a reliable signpost for numerous companies all over the world and navigates them through the constant change of the markets.

As one of the first serious solutions for process modeling, *Signavio Process Manager* conquered the market a few years ago. At that time, the industry was still discussing the benefits of the Business Process Model and Notation (BPMN) standard. In 2010, research firm Gartner published the groundbreaking Magic Quadrant for Business Process Automation (BPA) solutions and featured 14 leading software vendors. As of 2019, only 8 software vendors from this list have remained in the market. Many companies eventually turned away from it or didn't make the list of most popular vendors. In the beginning, process modeling tools were aimed only at experts. As a technology for a select few, they could only be used after extensive training. This had consequences for the market: as demand fell, so did the sales of the software providers. The prices for these solutions were correspondingly high. So, it wasn't surprising how much enthusiasm Signavio triggered. After all, in addition to pure process experts, there are many other professionals within the companies who are interested in further developing the operational processes in their daily work. They benefited from an intuitive solution that could be used without any training effort to model and improve processes within a short time. Of course, the cloud as a trending IT topic also plays an important role: people could suddenly access an application without involving their IT department—revolutionary! In addition, to being able to capture, document, compare, and simulate the customer's experience and journey, Signavio launched the intuitive tool *Signavio Journey Modeler.*

Signavio was the first software vendor in the industry to develop a collaborative platform that gives companies an end-to-end view of their process landscape. This also contributed to the company's success, as previously static websites were still the norm: as a result, users of these older solutions didn't always get access to up-to-date information and could not participate in improvements without costly licenses. It was at this time that the *Signavio Collaboration Hub* was developed.

Of course, the past few years have not only been about modeling and further developing business processes. A lot has also happened in the area of automation. Today, workflow management acts independently and integrates numerous different types of products: many solutions for content management, service management, and company-wide collaboration now also have workflow functions. After all, software manufacturers and their customers have known for some time that fast and user-friendly solutions can achieve better results than extensive enterprise solutions. With comprehensive workflow technology that revolves around everyday work tasks and is embedded in a modeling environment, Signavio is unique: the software manufacturer included *Signavio Process Governance* in its product portfolio a few years after its founding.

Even at a time when processes weren't in vogue, Signavio enjoyed triple-digit growth rates. Finally, more than three years ago, exciting changes were announced that turned the world of process management upside down. One important change was the growing interest in the method of process mining. While some BPMS solutions already had features reminiscent of this method, many saw process mining primarily as an academic approach that had no clear value to companies, rather than as an innovative way to make business processes more transparent and to compare actual processes with optimized target processes. Some software vendors specialized in process mining tools, and start-ups such as Celonis also entered the market in this segment. Inspired by a growing interest in process mining, the Signavio team also launched an analysis tool: *Signavio Process Intelligence*. Due to its cloud technology and collaborative approach, this tool is not only more cost-effective than other solutions, but it also has another important advantage: it's not only targeted at employees with specialized knowledge but also at business users from different departments of the organizations. As a result, the methodology of process mining has also evolved.

Another important change was announced by the newly created *Signavio Business Transformation Suite*: it brought together process modeling, automation, and evaluation and then linked them with collaborative functions in a single solution. For the first time, process mining technology was combined with a comprehensive process modeling solution. The result was an end-to-end view of everyday business processes in the enterprise that could be analyzed, modeled, and improved (both manually and automatically) without further ado. Organizations were able to build a bridge to the internal IT department and integrate innovative methods of automation into everyday operations. In this way, it should be possible to understand and measure processes along different IT systems and to keep them in view through monitoring.

For some time now, a new trend has also been stirring the industry: robotic process automation (RPA). RPA is an approach to process automation in which repetitive, manual, time-consuming, or error-prone activities are learned and automated by software robots (known as *bots*). This technology competes with all the software vendors that enter the market with a solution for automation. However, like any technology hype, RPA comes with certain risks. Frequently, the following questions arise in relation to RPA:

- How can an RPA initiative be scaled and deployed across the enterprise?
- How can the performance of software robots be controlled?
- Can solutions from different RPA vendors be combined with each other?
- How do I protect my company from faulty software robots?

At present, it looks as if software robots can't be stopped on their way into companies. But the challenge is to use RPA today in such a way that it won't be regretted tomorrow. Today, process modeling is equally interesting for business users and RPA providers: The methodology offers valuable support in everyday business and for the implementation of RPA solutions. This represents a turnaround—after all, process modeling in the BPMS microcosm was considered a sore point of comprehensive automation projects. Moreover, the link with process mining is interesting for RPA vendors because it offers important advantages in the implementation and execution of RPA projects, such as the identification of tailored use cases for RPA initiated by process mining. This can massively promote the potential of RPA throughout the enterprise. In retrospect, it can be observed that modeling, monitoring, and cross-system evaluation of processes belong together. But what was underestimated in the past was the important role that collaborative technologies should play in enabling the evolution of business processes. What wasn't foreseen was the development that process modeling was to overtake numerous solutions for automation.

[«]

Process Management as a Dynamic Initiative

Broaden your view of your process landscape: stop thinking of your process management as a project and start thinking of it as a dynamic initiative. Instead of simple documentation, SAP Signavio provides you with valuable insights that enable you to plan and profitably manage your day-to-day business even under changing conditions.

In 2021, SAP SE acquired Signavio, the leading company for business process transformation, and expanded the Signavio Process Transformation Suite with the in-house solutions SAP Process Insights, now called SAP Signavio Process Insights, and SAP Build Process Automation (formerly SAP Process Automation). Since then, business process transformation, then known as *business process intelligence*, has been, among other things, an important part of SAP's new offering *RISE with SAP*, which is designed to help companies achieve holistic digital transformation. RISE with SAP makes companies resilient, agile, and intelligent, and it supports moving core ERP processes to the cloud. Consequently, the integration of Signavio's cloud-native process suite with its in-house business process transformation solution allows SAP to offer a holistic suite of flexible process transformation solutions that enable customers to adapt their business processes end to end. This includes business process analysis, design and improvement, and process change management. The suite also enables customers to monitor the long-term success of these process changes.

[«]

The Positioning of SAP Signavio

Process modeling, governance, or analysis alone are no longer sufficient to achieve effective process transformation results in today's world. Rather, it's about combining different tools, content, and methodologies to generate sustainable added value for business process transformation. SAP Signavio positions itself in the area of business process transformation and encompasses this combination of methodology, product solutions, and process content.

Through the acquisition, the business process transformation suite was born. SAP Signavio now performs standardized out-of-the-box process KPIs, comprehensive benchmarking data, process mining, user behavior mining, and CX analysis to provide customers with a 360-degree view of every business process. At the same time, customers are given tools to fully understand and transform processes.

2.2 SAP Signavio Process Transformation Suite

As mentioned in the introduction, business process transformation from SAP enables companies to understand, improve, and transform their business processes from an end-to-end perspective. Because the solutions are web-based, they enable rapid, scalable transformation of all business processes, across all areas of the enterprise.

Accordingly, SAP Signavio Process Transformation Suite includes the following core capabilities:

- **Process analysis and process mining**
  End-to-end process analysis for business transformations and initiatives regarding operational excellence.
- **Process and journey modeling**
  Management, modeling, and simulation of business processes and diverse journeys.
- **Process governance and automation**
  Ensuring organizational and regulatory compliance of all documented processes.
- **Process collaboration**
  Interactive and cross-divisional collaboration in real time.

To become even more aware of the power of SAP Signavio Process Transformation Suite, let's take a look at Figure 2.2.

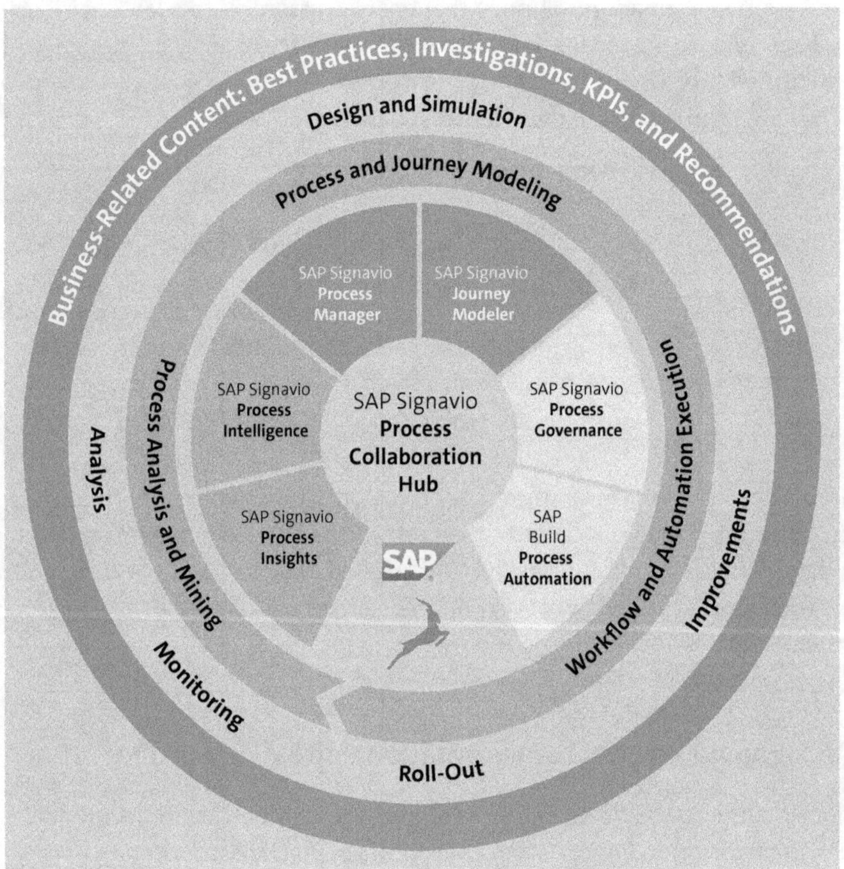

Figure 2.2 SAP Signavio Solutions at a Glance (Source: SAP)

Figure 2.2 visualizes how the individual business process transformation solutions interact—always taking into account SAP's best practices and industry-specific suggestions for improvement. SAP Signavio Process Insights and SAP Signavio Process Intelligence tools are primarily used for process analysis and process mining activities. SAP Signavio Process Manager and SAP Signavio Journey Modeler, on the other hand, focus on process and journey modeling. SAP Signavio Process Governance and SAP Build Process Automation, in turn, focus on the management of workflows and automations. The key to success—as shown in Figure 2.2—is and remains the human aspect, which is cultivated holistically with SAP Signavio Process Collaboration Hub. You can also see in Figure 2.2 all SAP Activate project phases, which are consistently supported based on the respective business process transformation solutions. We'll address this topic in more detail in Chapter 12. First, however, we would like to introduce you to the business process transformation solutions in more detail in this section.

2.2.1 SAP Signavio Process Insights

With SAP Signavio Process Insights, you quickly gain an overview of your company's performance. You benefit from targeted improvement recommendations and can secure direct business value by systematically examining your business processes end to end with advanced analytics. The solution enables you to monitor, evaluate, and improve your business processes holistically and continuously. It analyzes the performance of your business processes and supports you in evaluating them quickly and easily on the basis of data- and industry-based facts, evaluating transformation opportunities, and implementing process adjustments directly.

[«]

Your Benefit from SAP Signavio Process Insights

SAP Signavio Process Insights not only supports visibility and understanding across all business processes but also provides insights into where optimization is needed and how to achieve desired business outcomes. This accuracy of insights helps your teams discover, analyze, improve, and optimize processes.

You can link SAP Signavio Process Insights almost effortlessly with your SAP ERP or SAP S/4HANA system and uncover all process weaknesses and optimization opportunities in real time within a very short time. The tool generates tailored improvement recommendations and shows you automation potential—always taking industry-specific best practices into account.

SAP Process Insights includes 40+ business process flows for seven different business areas as well as six end-to-end processes, corresponding to more than 400+ individual metrics, 130+ performance indicators, and 15 of the most important business goals and value drivers. More than 100 corrective recommendations, such as master data corrections or configuration changes, are already integrated in the solution. In addition, SAP

Signavio Process Insights delivers more than 300 tailored innovation recommendations for SAP S/4HANA as well as SAP Fiori apps and SAP offerings, including RPA, machine learning, and situation management.

With SAP Signavio Process Insights, you can support the shift away from process transformation based on intuition or gut feelings to deep and data-driven process improvement. You can use it to create dynamic, real-time reports that assess the performance and efficiency of existing business processes. As a result, these reports can be considered for quick performance comparisons and KPI benchmarking.

Through integration with SAP Intelligent Robotic Process Automation (SAP Intelligent RPA) and automatic bot creation, the tool enables you to identify bottlenecks at an early stage and take into account improvement suggestions recommended by the system, including corrective actions. In this context, you can then further fine-tune your processes through automated monitoring functions. This results in both quick time and cost savings as well as long-term benefits in terms of efficiency and process optimization.

Advanced and, most importantly, intuitive visualization of process data provides a live picture of how your business is actually operating. Combined with real-time monitoring of business process performance, you can increase your operational visibility to break through any silos and meet your customers' needs and concerns even more expeditiously.

SAP Signavio Process Insights provides rapid visibility into sources of value within the SAP ERP application or SAP S/4HANA system you're using, helping to maximize your revenue across all areas of your business. You can also drive continuous improvement initiatives with next-generation performance views and process drilldowns, resulting in an even clearer understanding of your business processes.

You can leverage innovative process data extraction and continuous data updates to gain real-time insights into your process performance, making your overall operations even more effective. With built-in root cause analysis, you can narrow your process focus, pinpoint the source of any problems, and analyze them in depth. By incorporating data-based process improvement recommendations, you can get started immediately with your optimizations and focus on the essential corrective actions. Additionally, SAP Process Insights' advanced, highly user-friendly interface enables you to continuously foster complete collaboration between all of your decision-makers, subject matter experts, and IT professionals.

Note

For more in-depth information about the features and how to use SAP Signavio Process Insights, see Chapter 3.

2.2.2 SAP Signavio Process Intelligence

From performing detailed process analyses and interacting with business data to using process mining for fact-based process optimization, SAP Signavio Process Intelligence offers you a reliable product solution to uncover, for example, bottlenecks or hidden benefits of your business processes in a targeted manner.

SAP Signavio Process Intelligence is a key component of SAP's business process transformation portfolio. The tool includes next-generation process mining capabilities and supports smarter business decisions by providing powerful, fact-based insights into conceivable risks and sustainable improvement opportunities. This enables you to identify and maximize the business value hidden in your processes, data, and systems over the long term.

[«]

Benefits of SAP Signavio Process Intelligence

By automatically applying process mining algorithms, SAP Signavio Process Intelligence provides visibility and understanding into your organization's actual business processes. You gain insights into how and why your end-to-end business processes are currently running in a certain way and can subsequently align your operations even better with your processes.

SAP Signavio Process Intelligence allows you to perform detailed business process analysis for continuous improvement and create smart diagnostics about process performance. These diagnostics are then used to test potential bottlenecks and process simulations and to evaluate process change alternatives considering best practice scenarios. Detailed analytics can also be used to derive business process optimizations in terms of RPA, hyperautomation, AI, and machine learning, as well as precise customer behavior for improved customer journeys. If you choose the RISE with SAP solution, SAP Signavio Process Intelligence additionally takes your company's SAP S/4HANA transformation to the next level.

SAP Signavio Process Intelligence also helps you gain a richer understanding of your data to make fact-based process changes. With this tool, you get a clearer end-to-end view of your business processes—from event logs to individual user actions. From a technical perspective, you can manage your data, generate new process models, integrate content from third-party applications, and define metrics for in-depth process analysis to better understand the current state, activities, and workflows. Coupled with this, you can benefit from out-of-the-box capabilities to reduce time to actual process analysis using extract, transform, and load (ETL) connectors and transformation templates for the most common business processes, including order to cash and procure to pay. ETL is a process of unifying data from various data sources into a target database. First, the relevant data is extracted from various sources (extract), then transformed into the format of the target database (transform), and finally loaded into it (load).

Interacting and analyzing the most essential information is extremely efficient, as SAP Signavio Process Intelligence immediately uncovers inefficiencies and bottlenecks in business processes and thus ensures smooth operations. In the course of this, individual process activities can be examined granularly to compare them with predefined KPIs, perform impact analyses, and identify variants for streamlining individual process flows. With SAP Signavio Process Intelligence, you can dynamically map and intuitively connect your process control flows to ensure compliance with corporate regulations and process deviations. By performing root cause analysis, the tool enables you to investigate any delays, skipped activities, and event sequences while dramatically reducing cycle times and potential rework.

With SAP Signavio Process Intelligence, you can also achieve added value related to widgets for accelerated performance analysis and results sharing. These widgets deliver an engaging user experience while providing detailed extractions for instant insights into performance-related KPIs, cycle time analysis, and general indicators about the current state of your business. In addition to the widgets, process performance and compliance metrics can be created and combined with process mining analytics to further accelerate your business results.

SAP Signavio Process Intelligence also allows you to focus on information that is highly relevant to you—bundled on a single platform. This gives you an even more accurate data-driven foundation for your decision-making. By extracting and connecting data with advanced integration capabilities, it's also possible to link to multiple source systems to identify relevant widgets for instant information. In this way, you can transform your data into an environment where business processes and IT processes cooperate harmoniously to promote the long-term efficiency and effectiveness of your business.

Note

For more in-depth information about the features and how to use SAP Signavio Process Intelligence, see Chapter 4.

2.2.3 SAP Signavio Process Manager

SAP Signavio Process Manager helps you react immediately to unpredictable business and regulatory changes. With this tool, you can also scale your company's entire process repository in a targeted manner and, for example, use process simulations to define and constantly reinvent the most diverse business scenarios. This is how agility works.

SAP Signavio Process Manager comes into play if you want to comprehensively model, analyze, simulate, and continuously optimize your business processes. This intuitive business process solution not only allows you to visualize the flows of your business activities using advanced process modeling capabilities but also to fully monitor them. From finance and human resources, to purchasing and manufacturing, to logistics and sales, your employees will benefit from the efficiency and effectiveness that comes from a single cohesive business process landscape that also crosses IT system boundaries. SAP Signavio Process Manager makes all business processes within your entire organization much more easily accessible. This promotes enterprise-wide process transparency and ultimately enables process modeling for every employee in your organization.

However, SAP Signavio Process Manager is much more than a process modeling tool. This business process transformation solution also includes extensive functionality to fully organize your process repository. For example, it enables you to link your business processes in a way that improves the customer journey and thus enhances the CX, and an improved CX is just one of the benefits of this powerful tool. In fact, SAP Signavio Process Manager offers a multitude of other benefits and potential for your company. For example, the tool allows you to gain detailed insights into all of your business areas, increase productivity, and reduce costs; better connect processes to the overall strategy; implement a strategic and comprehensive business plan; and quickly respond to regulatory changes. All of these useful skills provide you with a great opportunity. They will make your company even more sustainable, competitive, and resilient in our fast-paced, globalized business world.

In addition, SAP Signavio Process Manager includes an effective potential to quickly improve your operational processes by simplifying model creation with the *Quick-Model* feature. This allows you to easily create business processes in tabular form, add new activities, edit them, and then review them separately before publishing the process model. Of course, we'll go into detail about this function later in this book (see Chapter 5). In any case, with SAP Signavio Process Manager, you keep full control over your business activities and holistically improve process output with a wide variety of editing functions, no matter how many thousands of activities your process landscape comprises.

[«]

Benefits of SAP Signavio Process Manager

Process models are already useful individually and clarify how your organization works together. However, when these models are connected with other factors, they only unfold their true potential. As part of the SAP Signavio solution portfolio, your process models can be viewed from different angles with SAP Signavio Process Manager in the context of the customer journey, risk and decision management, resource planning, IT implementation, and many other aspects. SAP Signavio Process Manager is the foundation to bring your processes to life.

Using SAP Signavio Process Manager, all employees in your team can share their expertise and always get the support they need. The resulting shared input as well as feedback raises the standard of working practices in your company. What does this look like in detail? It's simple: relevant employees can add comments to task messages and respond in real time to colleagues from other departments and company locations. They can also solicit feedback and have conversations that lead to faster and better results. This way, you can stay at the forefront of your business environment by eliminating silos and encouraging everyone to perform at their best.

[+]

Connecting Process Models with SAP Signavio Process Intelligence

With SAP Signavio Process Intelligence, you can build an operational process mining cockpit that connects your data to process management and transforms static process models into dynamic and responsive dashboards. This way, you're informed, guided, and warned at an early stage, so you can easily decide which processes and activities you want to monitor or better understand within them. Place indicators, choose thresholds, and let SAP Signavio Process Intelligence connect your process models to the relevant data.

To help you prepare for alternative business scenarios, SAP Signavio Process Manager provides you with a simulation of your business processes. This creates a what-if environment to quantify and predict the potential return on investment of process changes. Finally, you can address the tasks that are most critical to you by enabling process simulation through setting and monitoring case scenarios. In this context, you can also identify hidden opportunities as well as uncover the best process alternatives for complex projects—always with the one goal of making your business smarter in the long run.

Note

For more in-depth information about the features and how to use SAP Signavio Process Manager, see Chapter 5.

2.2.4 SAP Signavio Journey Modeler

With SAP Signavio Journey Modeler, you have the ability to achieve high customer satisfaction through customer journey models, aligning business processes with CXs and sharing customer journeys for feedback and experience improvements. *Customer journey* describes the individual cycles of a customer, including all touchpoints with the company, from the first contact to the final purchase decision. Therefore, customer journey models provide, among other things, valuable insights to better understand customers' buying behavior and experiences.

SAP Signavio Journey Modeler is another helpful process visualization tool that is also part of the holistic suite of business process transformation. This solution helps you develop a centralized, real-time view of the CX by linking journeys to business processes, related IT applications, integration points, and data flows. With SAP Signavio Journey Modeler, you can operationalize your CXs by incorporating them into your business processes. This allows you to then adapt your organizational systems, metrics, and roles in such a way to positively drive your customers' experiences.

SAP Signavio Journey Modeler allows you to implement three essential steps:

1. Model the CX quickly and easily.
2. Manage models effectively within SAP Signavio Business Transformation Suite.
3. Link your customers' experiences to your business processes.

The journey models themselves are experiential and provide a view of your business from the perspective of your customers, employees, suppliers, and/or other critical stakeholders in your organization. For now, we focus on the customer scenario to narrow the scope of this first introduction, even though the basic idea of the journey models remains the same.

Because journey models differ from typical business processes that are mapped with BPMN, Table 2.2 illustrates the difference between the notation of business processes compared to journey models.

	Business Process	Customer Experience
Perspective	Inside out	Outside in
Modeling Type	Flowchart	Customer journey
Recommended Tool	SAP Signavio Process Manager	SAP Signavio Journey Modeler
Unique Selling Proposition	With SAP Signavio Journey Modeler, you link operational excellence from the inside out with the CX from the outside in, creating a harmonious match between experiences, processes, and data to delight your customers at scale.	

Table 2.2 Notation of Business Processes Compared to Journey Models

Your company's business processes are most appropriately represented using a process diagram with SAP Signavio Process Manager, while your customers' experiences are best modeled with SAP Signavio Journey Modeler. Related to this, all business processes already modeled in SAP Signavio Process Manager can be connected to SAP Signavio Journey Modeler. This includes the mapping of process models from SAP Signavio Process Manager with the journey models from SAP Signavio Journey Modeler on a specific task or step-based level. Subsequently, you also can open and share linked processes directly in SAP Signavio Process Collaboration Hub.

Customer journeys can be captured in tabular form, describing the individual phases of the CX in detail. SAP Signavio Journey Modeler helps you list all touchpoints, visualize the sentiment for each phase, link process maps in an experience-oriented way, and add data visualizations as widgets. Such widgets can be integrated directly from SAP Signavio Process Intelligence.

By integrating operational and external data with customer journeys, you can identify opportunities for improvement and ensure that you can permanently increase customer satisfaction and loyalty as a result. This *customer excellence* is the combination of the inside-out and outside-in perspective: Which of the business processes have touchpoints with the customers (inside-out), and—even more crucial—how do the customers perceive these touchpoints in the first place (outside-in)? As a result, such high-level insights lead to a wide variety of benefits and opportunities for your company.

Now, the tool allows you to translate your CX when interacting with specific business processes directly into quantifiable and manageable information. This allows you to quickly adapt to any changes in your customer expectations and subsequently delight your customers in the long run. This also results in smarter decision-making regarding which of your customers should be best served, when, and how. With SAP Signavio Journey Modeler, you can align your entire organization around critical customer outcomes using journey analytics and sentiment analysis. In addition, you'll better understand the interdependencies between customer sentiment, significant moments of truth, and underlying process flows. You can see what this might look like in Figure 2.3.

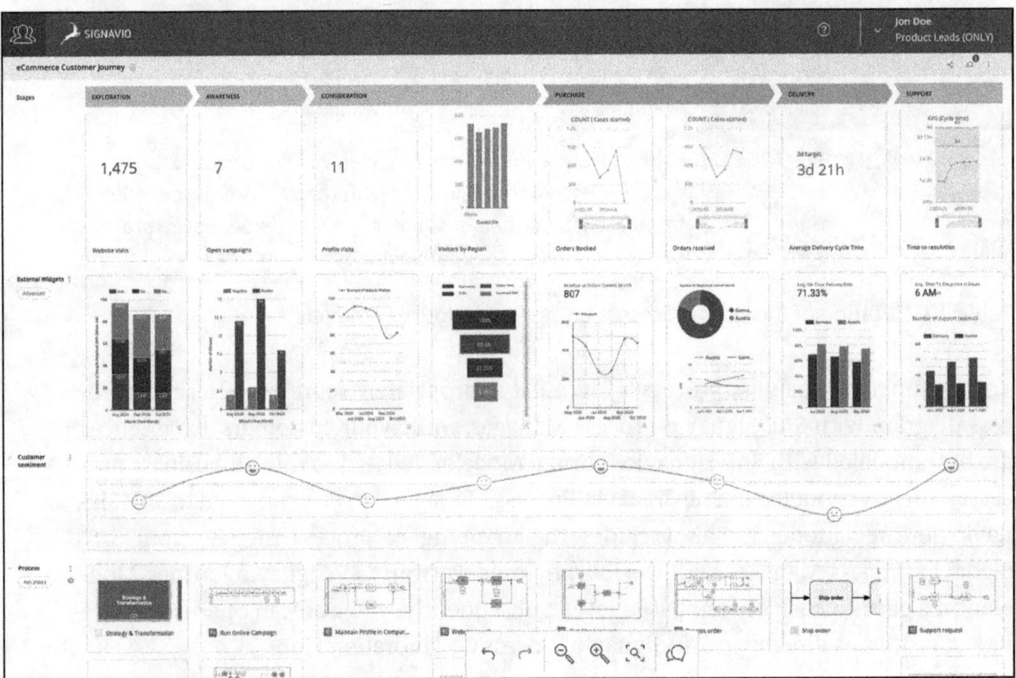

Figure 2.3 Example of an eCommerce Customer Journey

Benefits of SAP Signavio Journey Modeler

SAP Signavio Journey Modeler gives you the tools to do three things:

- Quickly and easily design engaging journey models.
- Manage these journey models in SAP Signavio Business Transformation Suite.
- Connect your journey models with your business processes to create new business values.

Further exciting potential lies behind the already existing business process landscape. If the process mining capabilities of SAP Signavio Process Intelligence are used, you can identify critical customer interaction points in even more detail. Customer data and process mining analyses help you understand the causes of customer frustration or satisfaction and then adapt business processes accordingly or even completely redesign them to effectively and, above all, sustainably, increase your customer satisfaction, retention, and loyalty.

Ultimately, you can also break down silos between your CX and process teams and focus on common goals and definitions of journey modeling. This allows you to both reduce operational complexity and manage variations in business process flows to ensure a holistic and consistent CX. Corresponding insights can be collected and shared across your organization via SAP Signavio Process Collaboration Hub to further enhance your customers' experiences.

Note

For more in-depth information about the features and how to use SAP Signavio Journey Modeler, see Chapter 6.

2.2.5 SAP Signavio Process Collaboration Hub

SAP Signavio Process Collaboration Hub allows you to use a single source of information for all of your teams, break down business silos, and create an improved understanding of your KPIs, tasks, and projects. The result is full process transparency.

In short, SAP Signavio Process Collaboration Hub changes the way your employees collaborate across your company. It provides a holistic collaboration platform to effectively leverage your employees' knowledge and continuously optimize and innovate business processes. This means that SAP Signavio Process Collaboration Hub acts as a single source of truth for your business processes, providing you with real-time insights and enabling employees to have a transparent and company-wide focal point where they can unite their activities and expertise related to business processes.

SAP Signavio Process Collaboration Hub helps you stay up to date on changes to your process models in real time and displays all process content in a clear and intuitive way. The tool helps you understand, track, and manage business process–related project activities. Moreover, this business process transformation solution promotes the organization of business process content across the entire SAP Signavio Business Transformation Suite. As a result, the SAP Signavio Process Collaboration Hub drives process initiatives throughout your organization by ensuring that all stakeholders are on the same level of knowledge, streamlining disparate views and providing understandable updates to all stakeholders.

First and foremost, the Hub maintains a central knowledge platform about the what, why, and how regarding your business processes, ensuring a common understanding across the enterprise. Everything you need for solid process communication and collaboration is in one place—the Hub. You can access data, process models, and analytics from SAP Signavio Process Intelligence, SAP Signavio Process Manager, and SAP Signavio Process Governance directly from the Hub. The streamlined search function across the entire SAP Signavio Business Transformation Suite ensures that you can find all process information relevant to your roles and see which topics and activities your employees are currently involved with.

[»]

Benefits of SAP Signavio Process Collaboration Hub

SAP Signavio Process Collaboration Hub is the heart of collaborative process management. The central knowledge base secures the know-how of all participants and ensures organization-wide collaboration and smooth communication—even across locations. This way, every employee in your organization is up to date and well informed not only about what but also about why and how.

The Hub's intuitive user interface (UI) makes interacting easy for users of all knowledge levels (see Figure 2.4). Because your employees are notified individually about relevant updates and conversations, you can close the gap between accountability and action. By accelerating the exchange of information among your employees, you now have the opportunity to generate more ideas, optimize processes faster, and significantly lower the inhibition threshold for process changes.

With the Hub's user-friendly presentation mode, your employees can see the big picture of your holistic process landscape. The tool can be used to create simple step-by-step walk-throughs of even the most complex business processes and provide detailed presentations in overview mode. SAP Signavio Process Collaboration Hub additionally promotes effortless navigation of your entire process repository as well as the presentation of truly significant interactions. Subsequently, it also supports end-to-end traceability and accurate final decision-making. This is supported by a seamless switch between a high-level overview and a granular detailed information view. The Hub's simple version control also helps you keep track of the status of your ongoing activities.

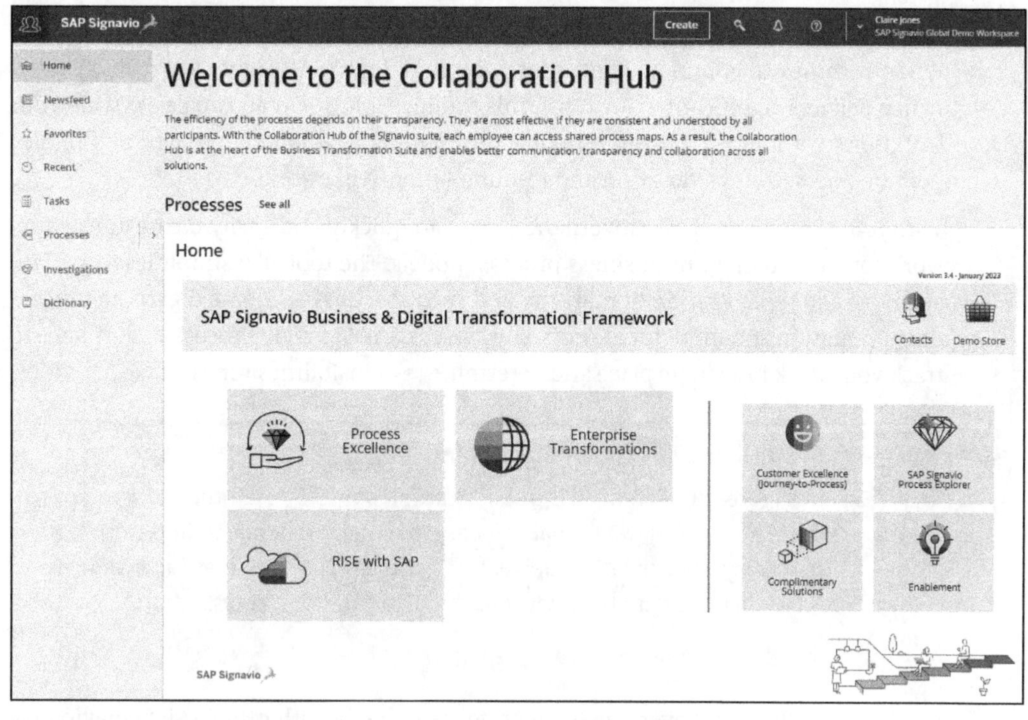

Figure 2.4 Initial Screen of SAP Signavio Process Collaboration Hub

When we briefly consider the challenges of successful business transformation, we think of organizational silos with associated problems such as conflicting priorities and agendas, diluted goals, and employee resistance to decisions based on incomplete or unclear data. SAP Signavio Process Collaboration Hub now offers a solution for each of these problems, helping to improve efficiency as well as effectiveness across your organization. As a result, you can set new milestones in operational excellence and positively impact the way your teams work with clearly defined goals and structured responsibilities.

Finally, the Hub can also be personalized. You can assign employees to specific groups. These group-based entry points mean that time no longer has to be wasted searching for information relevant to the actual work. You can control exactly what the hub looks like for specific user groups and what form it takes. You can also maintain your own newsfeed by subscribing to process models, conversations, or workspaces that are relevant to you. This way, tasks can be better structured, and the satisfaction and productivity of your employees can be improved.

> **Note**
>
> For more in-depth information about the features and how to use SAP Signavio Process Collaboration Hub, see Chapter 7.

2.2.6 SAP Signavio Process Governance

When it comes to complete control of your workflows, SAP Signavio Process Governance delivers significant value. With this product solution, you can easily scale your workflows without programming skills, prototype for rapid deployment, and reduce both business process variation and resource-intensive rework.

With SAP Signavio Process Governance, you can quickly and easily create automated workflows based on your business process models. The tool also simplifies task management and the initiation of maturity and regulatory assessments regarding process documentation and implementation. You can use SAP Signavio Process Governance to track your work in a single place and determine responsibilities for assigned activities.

[»]

Benefits of SAP Signavio Process Governance

SAP Signavio Process Governance is a solid starting point for even the most complex tasks or activities. The intuitive UI makes it easy to turn process models into workflows, so initially you only need the essential features of the process, such as tasks and decisions. You can add more details later (e.g., custom notifications, access control, or automatic reminders).

SAP Signavio Process Governance integrates seamlessly with other SAP Signavio Process Transformation Suite solutions to help you better manage and track the process lifecycle. For example, you have the option to export workflow data from SAP Signavio Process Governance directly into the SAP Signavio Process Intelligence tool. This data transfer supports internal analysis of business-related data using powerful process-mining capabilities that provide real-time insight into how your workflows and business processes can be optimized, among other things.

SAP Signavio Process Governance subsequently allows you to automate repetitive process governance workflows so your employees can focus on essential, value-added tasks. By creating forms, checklists, and decision points; setting up email notifications for deadlines; and collaborating on shared task lists, you can both identify unnecessary work and bypass it. In this way, you can significantly cut down on email traffic and unnecessary meetings. As a result, you can get the things that really matter to your business done faster and more efficiently. This saves you time and money, and you can be sure that your business processes are standardized and monitored.

SAP Signavio Process Governance also helps you keep pace with digital transformation and the challenges of today's VUCA world. The tool provides a perfect starting point for the most complex tasks and activities in your organization. Thanks to the intuitive UI, you can effortlessly transform process models into workflows, so you only need the most important features of the process, such as individual tasks and decisions, to get started. Your employees can add more information and details at a later stage, such as custom notifications, access controls, or automatic reminders. This establishes 360-degree process governance with expeditious approvals and maturity assessments, while eliminating spreadsheets and long email chains to track tasks directly. Because this solution is completely web-based, workflows can scale as your business grows.

SAP Signavio Process Governance is ready to use, automates and standardizes manual decisions using the integrated approval solution. This not only reduces the risk of possible wrong decisions but also allows business processes based on decisions made to be executed more efficiently. The automated approval process promotes a prompt response to process changes. Accordingly, it ensures that all business decisions are recorded to comply with the relevant internal and external regulations and guidelines.

SAP Signavio Process Governance additionally supports improved communication and coordination of various tasks. The tool gives a clear indication of how far a particular business process has progressed and what the next step will be. This results in fewer delays due to neglected or forgotten tasks, more accountability in decision-making, and more flexibility to consistently implement process improvements. Ultimately, onboarding new team members is also simplified, as they can see at a glance exactly what needs to be done and where each team member can find the resources they need. Teamwork is made easy with SAP Signavio Process Governance.

[«]

Note

For more in-depth information on the features and use of SAP Signavio Process Governance, see Chapter 8.

2.2.7 SAP Build Process Automation

SAP Build Process Automation is the next-generation no-code tool to drive process automation, support workflow enhancements, and increase process efficiency by automating repetitive work. In our fast-paced VUCA world, the need for process automation is constantly increasing. SAP Build Process Automation combines the capabilities of SAP Workflow Management and SAP Intelligent RPA in an intuitive, AI-powered, no-code solution.

SAP Build Process Automation simplifies process automation with visual drag-and-drop tools and prebuilt, industry-specific content. Using these tools, you can effortlessly create workflows and automate tasks and decisions, collaborating with in-house development teams as needed to meet all your automation needs. To kick-start projects, you can choose from a rapidly growing library of more than 340 prebuilt process flows, forms, business rules, dashboards, and bot automations for specific applications and industries. For example, SAP S/4HANA content is available for more than 100 automation scenarios in finance, manufacturing, sales, services and procurement, and supply chain. SAP SuccessFactors and SAP Ariba software development kits (SDKs) provide predefined activities to easily automate key tasks in these applications, complemented by SDKs for popular office solutions for desktop automation. All of this content is available directly in SAP Build Process Automation.

[»]

Benefits of SAP Build Process Automation

SAP Build Process Automation's RPA capabilities help you automate repetitive manual tasks by mimicking user interactions with each system. You can automate data transfers between legacy and web-based systems that lack sufficient integration capabilities. These features help you accelerate task processing, flexibly scale your systems to meet changing needs, and reduce error rates.

SAP Build Process Automation enables you to automate faster by giving you easy access to workflow management, task automation, and decision automation capabilities in a single tool. Built-in AI capabilities also allow you to make your processes smarter by leveraging machine learning for decision support, intelligent document processing, and more. The tool offers native integration with SAP applications as well as connectivity to non-SAP applications to enable automation of complex workflows holistically, even if they span multiple applications and business units.

With SAP Build Process Automation, you can securely manage your processes and automations in the cloud of your choice while leveraging your existing infrastructure. The solution gives you the ability to build process flows, automation, and decision models while protecting business operations with centralized governance, testing, and monitoring capabilities. Of course, to ensure ongoing compliance and automate processes in a scalable and reliable way, SAP Build Process Automation is designed to meet stringent service-level agreements (SLAs), compliance, and data protection regulations.

In a nutshell, SAP Build Process Automation helps you create automations and workflows using process automation capabilities without coding effort. The solution provides you with a foundation to quickly adapt, improve, and innovate your business processes. You can subsequently increase process efficiency as the tool automates recurring activities, as well as increase your business agility in responding to changing economic and industry conditions.

[»]

Note

For more in-depth information on the functions and use of SAP Build Process Automation, see Chapter 9.

2.3 Summary

This chapter serves as an introduction to the topic of business process transformation. You've learned what business process transformation is and what it can do. In addition, the individual business process transformation solutions from SAP were briefly introduced.

SAP's business process transformation offers a wide range of solutions to support frictionless and seamless business transformation from business process design to benchmarking, gap analysis, process improvement, and process change management.

PART II

SAP's Business Process Transformation Portfolio

Chapter 3
SAP Signavio Process Insights

With SAP Signavio Process Insights, you can generate dynamic process analyses that evaluate how fast and efficient your existing business processes are. This allows you to subsequently improve your processes in a data-driven manner, instead of relying solely on your intuition or gut feeling.

As introduced in Part I, SAP Signavio Process Transformation Suite represents the next-generation process platform in the SAP portfolio, providing a comprehensive and holistic offering for end-to-end business process transformation.

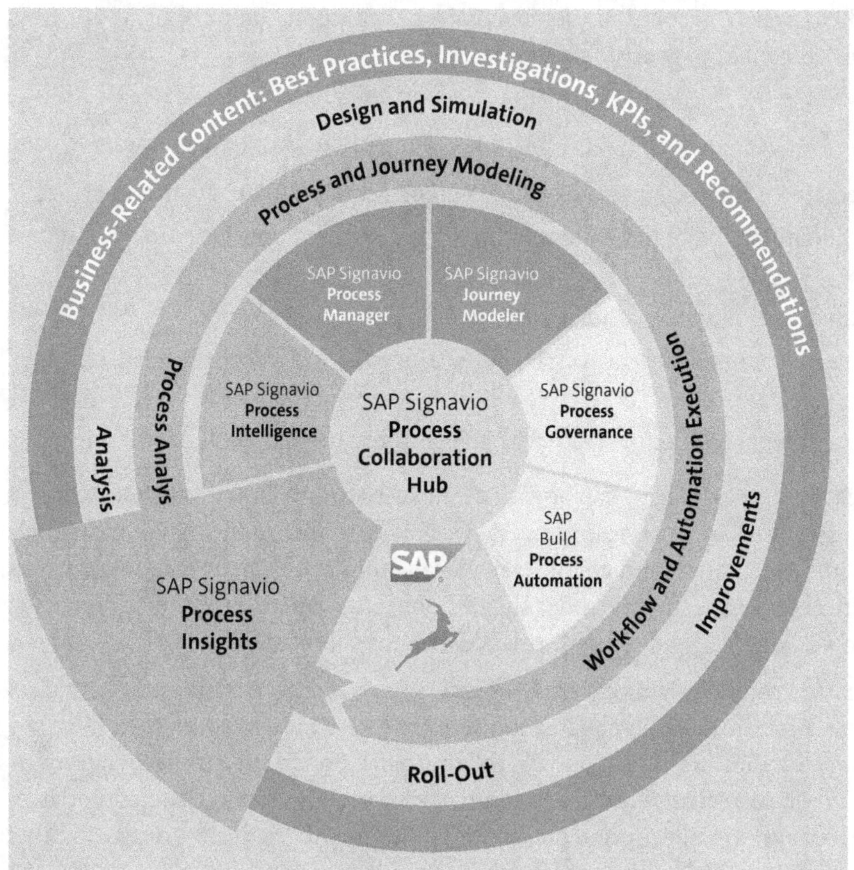

Figure 3.1 SAP Signavio Process Insights in Focus

Today's executives need to move faster; therefore, they need the means to better understand critical business processes and align all available resources across the enterprise. The business information required is complex, and analysis requires both time and expertise. This is where SAP Signavio Process Insights helps you achieve measurable results and meet your business goals faster and more effectively than ever before (see Figure 3.1).

In Section 3.1, we'll first look at the various features of SAP Signavio Process Insights. Section 3.2 shows the different benefits of the solution. Subsequently, we'll illustrate the use of SAP Signavio Process Insights in practice by means of an example (Section 3.3).

3.1 Functions

SAP Signavio Process Insights is an innovative process analysis solution that enables you to quickly identify areas of improvement and automation within your business processes and then automatically provide recommended actions to implement the required changes. The solution thereby enables continuous monitoring, evaluation, and improvement of your business processes.

The core functions of the solution include the following:

- Quick connection to SAP applications with fast data extraction functions
- Immediate insight into the core business processes of your SAP application
- Quick identification of focus areas for rapid and targeted optimization of your processes
- Drilldown capability to possible causes based on your key metrics
- Initiation of optimization of your business processes with recommended measures and technologies
- Comparison of your performance with the performance of your industry peers

SAP Signavio Process Insights can be easily connected to SAP ERP or SAP S/4HANA systems and immediately reveals process weaknesses and optimization potentials because implementation and go-live take place within a very short time. Meanwhile, the solution generates improvement suggestions to automate or optimize processes, for example, and thus increase performance within a short time.

SAP Signavio Process Insights covers 40+ process flows for seven business units and six end-to-end processes, which equates to more than 400 individual metrics, 130+ key performance indicators (KPIs), and 15+ of the most important business objectives. More than 100 corrective recommendations, such as configuration changes or master data corrections, are already integrated. In addition, SAP Signavio Process Insights delivers more than 300 tailored innovation recommendations for SAP S/4HANA and

SAP Fiori apps, as well as SAP offerings for situation management, robotic process automation (RPA), and machine learning.

Table 3.1 provides an overview of the numerous features and tools of SAP Signavio Process Insights.

Functions	Description
Focused entry points	This function allows you to do the following: ▪ Choose from predefined lists of end-to-end processes, modular processes, business units, or value drivers as entry points for process insights and recommendations. ▪ Navigate to predefined categories of interesting business content.
Process flow	This feature gives you an overview of available process flows and a visualization of process performance for typical business processes, based on data collected from your connected ERP system. Here's what you can do: ▪ Based on the implemented process performance indicators (PPI), select from a predefined set of process flows. ▪ Get an overview of the main phases of a process, specifying the business lines and business objects associated with each phase. ▪ View the progress of business object instances at each stage of a process phase. ▪ Obtain metrics, corresponding to a base set of business object instances in a given time period, indicating how many instances reached various stages and how long this took on average. ▪ Get additional information and metrics that provide more context or highlight identified blockers. ▪ Have lead times calculated for the transition between phases of the process. ▪ Use metrics for frequent blockers or other information displayed as boxes within a process flow. ▪ Select a link to go to more information about each process flow.
Standard PPIs	This feature allows you to obtain process performance metrics (based on data collected by your connected ERP system) that allow you to do the following: ▪ Get an overview of standard PPIs by category and by information about how many business object instances or entities an indicator is relevant for and what type of business object it is. ▪ Select a specific PPI to view more details, including a link to more information and a detailed list of affected business object instances.

Table 3.1 Functions of SAP Signavio Process Insights

Functions	Description
Standard PPIs (Cont.)	■ Access a filter option for a single PPI that allows you to drill down into details by selecting filters and filter values based on predefined characteristics for that specific PPI. ■ Access industry benchmarking information for standard PPIs for which external benchmarking information is available so you can compare your company's performance to industry peers.
Correction recommendations	This function allows you to do the following: ■ View a list of recommended corrections that outline steps your organization can take in your ERP system to resolve identified process performance issues. ■ Select a specific correction recommendation to get more information. This information includes a detailed list of business object instances for which the recommendation is relevant and access to a more detailed description of the recommended actions your company can take in your ERP system.
Innovation recommendations	This function allows you to do the following: ■ View a list of innovation recommendations that your organization can use to plan for long-term strategic improvements with SAP solutions and applications. You can choose from different categories of recommendations. You can also see which end-to-end processes or business areas a recommendation applies to and filter recommendations by end-to-end processes or business areas. ■ Select a specific innovation recommendation to view more information from other information sources outside of the solution.
Recommendations linked to standard PPIs	Correction recommendations for process metrics allow you to do the following: ■ View a list of recommended corrections relevant to a specific PPI. ■ Select a specific correction recommendation to get more detailed information on how to respond to the recommendation. Innovation recommendations for process metrics allow you to do the following: ■ View the available innovation recommendations for a specific PPI. ■ Select a specific innovation recommendation to get more detailed information about it from other information sources outside the solution.

Table 3.1 Functions of SAP Signavio Process Insights (Cont.)

Functions	Description
Monetary values	Monetary values for some PPIs allow you to do the following: ■ Authorized users are shown the cumulative total value of documents or items in their preferred currency. This information is available in the details displayed for a PPI when this monetary information is available. Monetary values in process flows allow you to do the following: ■ Authorized users can switch between the number of business object instances and their cumulative total value in their preferred currency. Only when this monetary information is available can values be represented in the process flows. Monetary values in filters allow you to do the following: ■ Users can toggle between the number of business object instances and their cumulative total value in their preferred currency in the filter values displayed when this monetary information is available.

Table 3.1 Functions of SAP Signavio Process Insights (Cont.)

3.2 Benefits

SAP Signavio Process Insights is a cloud solution that runs on *SAP Business Technology Platform* (SAP BTP) and delivers data-driven insights into business processes and their usage based on data from multiple SAP systems such as SAP ERP or SAP S/4HANA. The solution helps you achieve process excellence by identifying processes that need improvement, enabling users to dig deeper to understand root causes, and providing recommendations for improvement. This enables you to navigate through your business process transformation journey—from insight to implementation—to maximize your process quality.

By specifying your industry, you enable the solution to provide industry-specific innovation recommendations and content to help you make tailored decisions to improve your company's business processes. Industry-specific innovation recommendations are currently available for the industries listed in Table 3.2. If you don't specify an industry, you'll be presented with standard general recommendations.

Innovation Category	Innovation Type
SAP S/4HANA features	SAP S/4HANA
SAP Build Process Automation	■ Automations ■ Workflow management

Table 3.2 Industry-Specific Innovation Recommendations from SAP Signavio Process Insights

Innovation Category	Innovation Type
Smart technologies	▪ Machine learning ▪ Situation handling
User experience	SAP Fiori apps
Other SAP solutions	Industry cloud solutions

Table 3.2 Industry-Specific Innovation Recommendations from SAP Signavio Process Insights (Cont.)

[+]

Focus on Industry Popularity

If you specify your industry, your users can also see the industry popularity of some innovation recommendations. Industry popularity helps your organization understand how well an improvement recommendation is being adopted based on the number of industry peers using it.

As a result, with SAP Signavio Process Insights, you make the shift from transformation based on intuition or gut feelings to data-driven process improvement. You create dynamic, real-time reports that assess the speed and efficiency of existing business processes. These reports can be used for rapid performance comparisons and KPI benchmarking with competitors in your industry. We'll show you how this might look in practice in Section 3.3 using an application example for lead-to-cash process transformation.

Figure 3.2 shows the system landscape of SAP Signavio Process Insights and the SAP systems, applications, and components involved. SAP Signavio Process Insights uses the framework provided by *SAP Cloud ALM* to connect the managed SAP system to the cloud client. This means that the cloud connector isn't used to connect your managed SAP system to your SAP Signavio Process Insights application.

[»]

Trigger Data Transmission Yourself

The SAP Cloud ALM framework requires you to set up an HTTP connection to an external service. Using this framework and the HTTP connection, you can trigger the data transfer from your managed SAP system.

Basically, the ST-PI and ST-A/PI plug-ins are installed in your SAP ERP or SAP S/4HANA system. The performance data of your business processes are loaded into the *data sink* in the backend of SAP Signavio Process Insights. Finally, the data is transferred to SAP Signavio Process Insights via an *OData application programming interface (OData API)*.

SAP Signavio Process Insights offers you numerous possible functions based on your process data. The core potential activities are the following:

- Create full transparency through quick insights.
- Gain a clear picture of your process performance.
- Identify process problems quickly.
- Implement process improvements quickly.
- Ensure comprehensive process optimization.
- Accelerate process innovations.
- Improve decision-making.
- Promote collaboration between IT and business units.

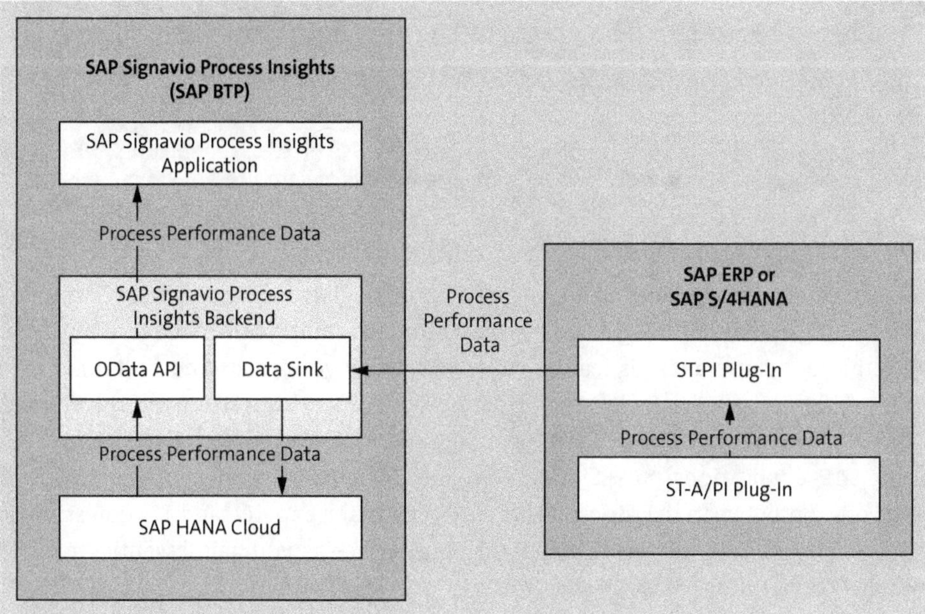

Figure 3.2 System Landscape of SAP Signavio Process Insights (Source: SAP)

[+]

Consider Benchmark Values from the Industry

For example, if you identify a KPI with a higher-than-average days sales outstanding (DSO), you can view industry benchmarks and work toward better receivables management based on recommendations such as setting up automatic reconciliation of incoming payments and invoices, credit management, or dunning procedures.

Thanks to integration with SAP Intelligent Robotic Process Automation (SAP Intelligent RPA) and automated bot creation, the tool helps you identify bottlenecks and make recommended improvements with corrective actions. You can then further fine-tune your operations through automated monitoring capabilities. This supports rapid time and cost savings and provides sustainable benefits in terms of efficiency and process optimization.

Automated and intuitive visualization of process data provides a real-time picture of how your business is really performing. Combined with live monitoring of process performance, you can increase your operational visibility to break through silos and meet customer demands faster than ever.

SAP Signavio Process Insights instantly creates visibility into sources of value within the SAP ERP application or SAP S/4HANA system, maximizing revenue across the business. You can also drive continuous improvement initiatives with enhanced performance views and process drilldowns, resulting in a clearer understanding of your business processes.

[»]

SAP Signavio Process Insights as the Basis for Specific Process Mining Initiatives

With SAP Signavio Process Insights, you can define use cases based on your live process data, which you can then analyze with the process mining capabilities of SAP Signavio Process Intelligence. With SAP Signavio Process Intelligence, you can consequently analyze your process data in even more detail as well as tailored to your requirements across systems.

You can then leverage automated, advanced process data extraction and continuous updates to gain instant insights and a clearer understanding of your overall business. With the root cause analysis feature, you can narrow your focus and find out where your problems really lie. By leveraging data-driven process improvement recommendations, you can start improving immediately and focus on corrective actions and relevant SAP applications. SAP Signavio Process Insights also enables you to foster collaboration between decision-makers, subject matter experts, and IT professionals through an intuitive, easy-to-use application interface as you gain essential insights into diverse business areas at a glance.

SAP Signavio Process Insights helps you quickly understand, innovate, and transform your business. Within hours, you can load, validate, and analyze your process data; identify improvement opportunities; and suggest remediation actions. In no time, you'll also quickly connect to your SAP application with automated advanced process data extraction and benefit from valuable process insights the same day. With daily updates, you gain a clearer understanding of the business processes in your SAP application, including finance, manufacturing, quality, maintenance, service, receivables, or capital expenditures. These rapid insights give you full visibility into what's happening in your processes.

Figure 3.3 shows a detailed list of transaction data for the **Sales billing documents created** PPI, which is part of the lead-to-cash process. We can see that we're below the median of our industry with an **Automation Rate** of **24%**. We could now apply SAP Signavio Process Insights recommended fixes and innovation recommendations to improve performance.

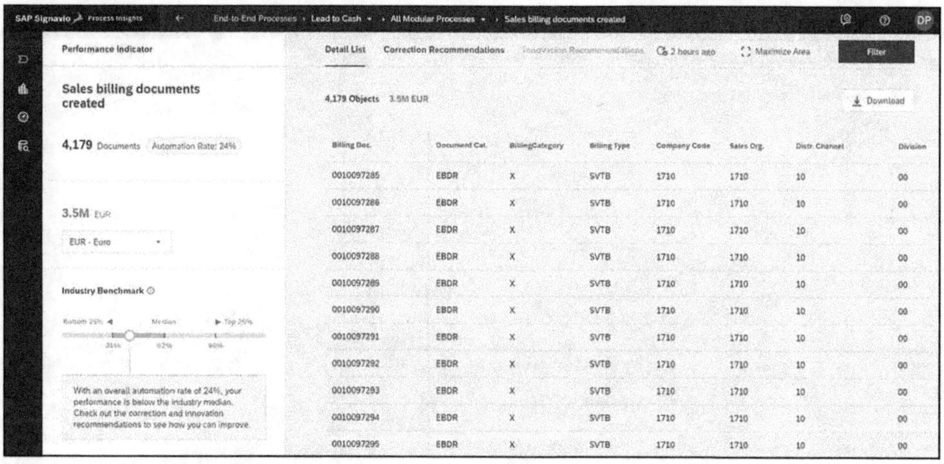

Figure 3.3 SAP Signavio Process Insights: Benchmarks

With ready-to-use *process flows* and preconfigured *performance metrics*, you shorten your time-to-insight to easily monitor the performance of process steps across business units and end-to-end processes. You'll benefit from built-in metrics that address more than 1,000 typical issues and inefficiencies, helping you quickly identify and prioritize what to improve first. Thus, on the one hand, you're now able to identify problems regarding your business processes in a timely manner. On the other hand, SAP Signavio Process Insights provides you with numerous tools to protect the integrity of your business processes and to react quickly to changing business results by investigating the impact of process inefficiencies to uncover data-driven improvements. Figure 3.4 shows an example of a process flow in SAP Signavio Process Insights. In the **Process Flow Performance** area, you can see the documents created in this process. In the **Most Frequent Blockers and Other Information** area below that, you see *blockers*, which inform you about process inefficiencies and should be investigated further. In our example, SAP Signavio Process Insights found several blockers in the financial process **Sales billing document creation to FI-AR clearing** (e.g., **Sales billing documents manually created** or **Open FI-AR items already dunned**).

SAP Signavio Process Insights helps you focus your transformation initiative on what really matters by making it easy to quickly locate and isolate any process deficiencies. In this context, you can find out where your process weaknesses really are in a simplified way, navigating to each individual document, such as an invoice, to find possible causes of low PPIs. Slice-and-dice data allows you to compare your business performance across the enterprise, while measuring your operational business performance and utilization against your industry peers. By using SAP Signavio Process Insights, you turn insights into immediate results. You focus on mission-critical processes that are at the heart of your business, as well as the areas that require your attention and deployment the most. Figure 3.5 shows you how to drill down to individual document types within a given company code to find possible causes of low PPIs.

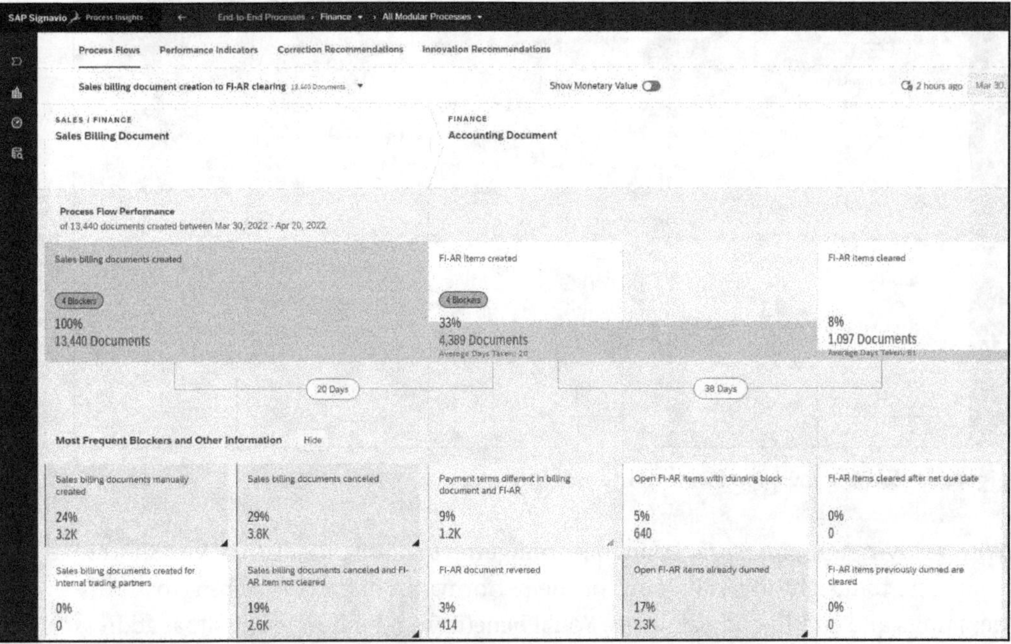

Figure 3.4 SAP Signavio Process Insights: Process Flows

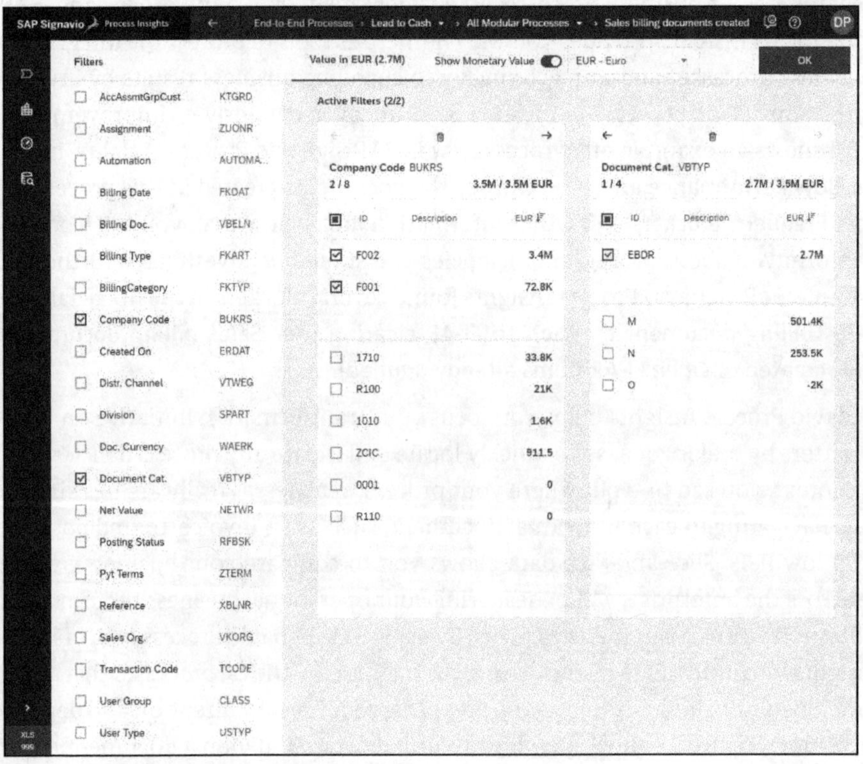

Figure 3.5 SAP Signavio Process Insights: Drilldown

With SAP Signavio Process Insights, you can quickly implement all business process improvements. You can follow the entire process optimization flow—from uncovering refinement potentials to implementing tailored improvement recommendations. The tool allows you to identify next-practice business processes and results. You can use the process improvement recommendations to address the root cause of poor process performance, for example. Correction recommendations such as integrated configuration changes or master data corrections can subsequently be used to make quick adjustments in your SAP system with clearly outlined action plans. In addition, SAP Signavio Process Insights helps you get the most value out of your SAP software investments with suggested SAP applications for best-practice long-term upgrades. Figure 3.6 shows the **Innovation Recommendations** tab with various innovation recommendations for the lead-to-cash process using intelligent technologies such as machine learning or SAP Intelligent RPA. In addition, you can see which business units are involved in the **Lines of Business** column.

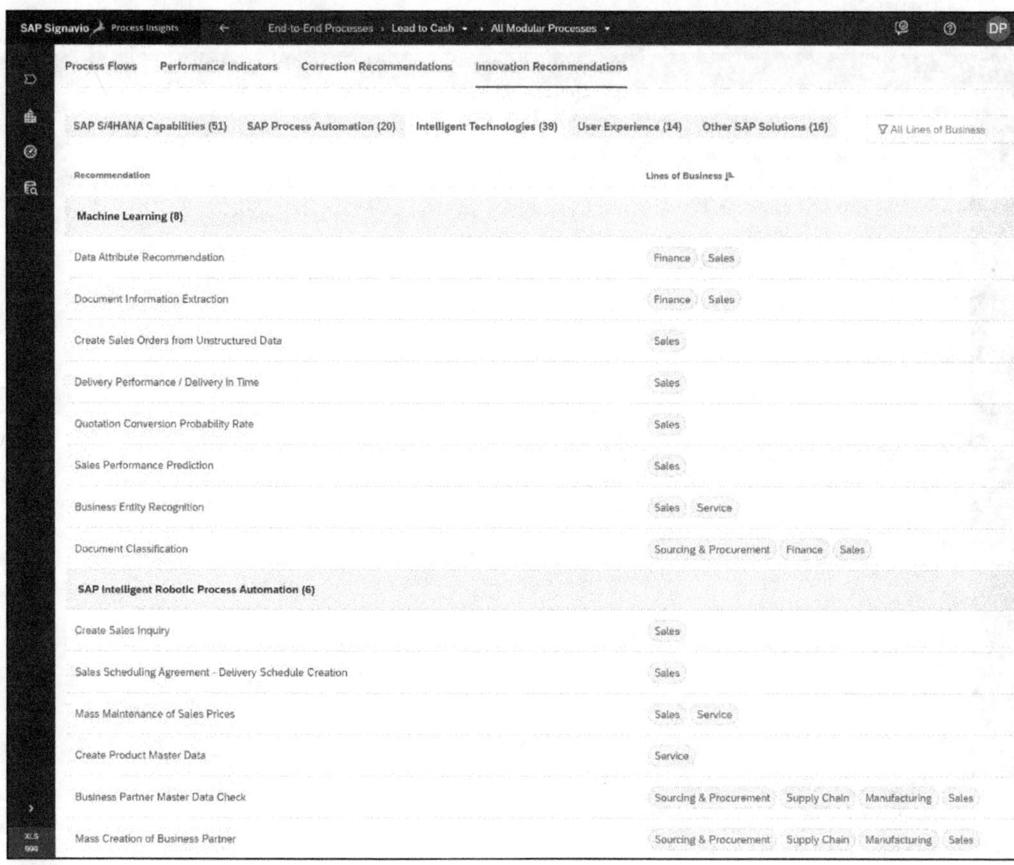

Figure 3.6 SAP Signavio Process Insights: Innovation Recommendations

Data-driven insights help you identify process improvement and automation opportunities faster and at scale, while fostering collaboration between IT and business units.

In this context, SAP Signavio Process Insights also breaks down any barriers between decision-makers, businesspeople, and IT experts through its intuitive, user-friendly application. You can prioritize where to focus your attention to see the biggest gains. In addition, you can easily assess the health and profitability of your business processes, investigate potential problems, and get built-in recommendations to start improving processes right away. Furthermore, SAP Signavio Process Insights helps you build the case for your SAP S/4HANA transformation, get stakeholder buy-in more easily, and foster collaboration between business and IT throughout your transformation process.

Figure 3.7 shows the **Correction Recommendation** page with correction recommendations for the lead-to-cash process. SAP Signavio Process Insights identifies not only potential root causes and which value drivers are affected but also the impact level and effort required to implement the correction. In this example, the solution has identified that standard automation potentials aren't fully exploited. With the correction recommendation to automate outbound deliveries, logistics costs could be reduced in the future.

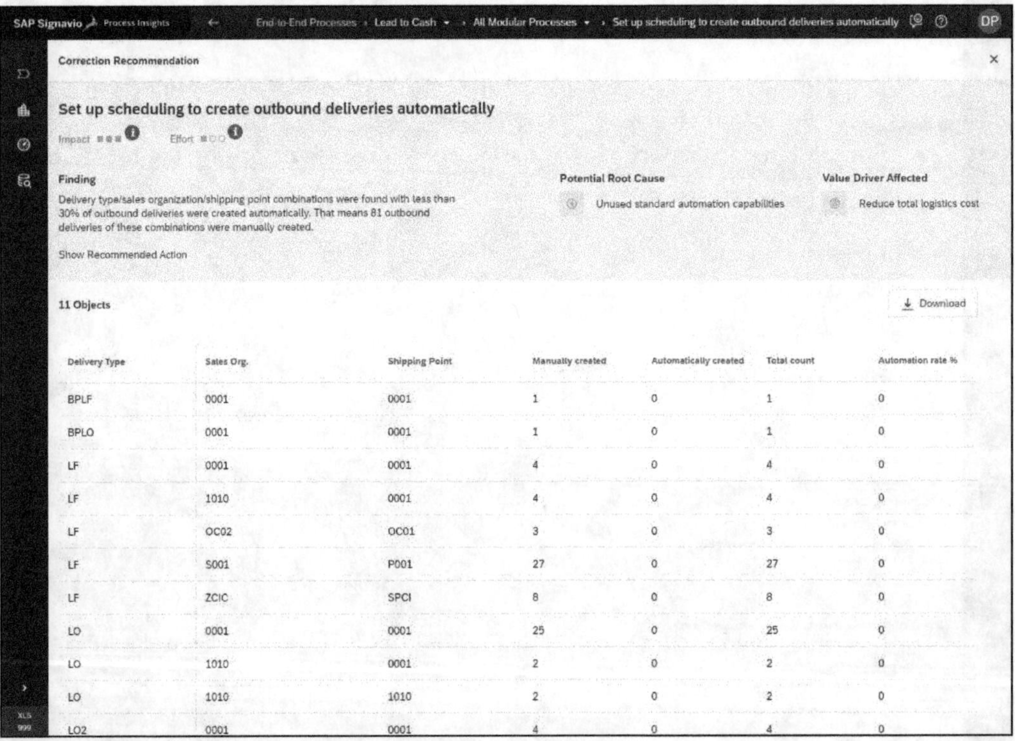

Delivery Type	Sales Org.	Shipping Point	Manually created	Automatically created	Total count	Automation rate %
BPLF	0001	0001	1	0	1	0
BPLO	0001	0001	1	0	1	0
LF	0001	0001	4	0	4	0
LF	1010	0001	4	0	4	0
LF	OC02	OC01	3	0	3	0
LF	S001	P001	27	0	27	0
LF	ZCIC	SPCI	8	0	8	0
LO	0001	0001	25	0	25	0
LO	1010	0001	2	0	2	0
LO	1010	1010	2	0	2	0
LO2	0001	0001	4	0	4	0

Figure 3.7 SAP Signavio Process Insights: Correction Recommendation

3.3 Application Example: Transformation of the Lead-to-Cash Process

To help you better understand the use of SAP Signavio Process Insights, we'll present it to you using a sample application. In this application example, we'll deal with the end-to-end process called *lead to cash*. You're the owner of the lead-to-cash process for a listed car manufacturer, and you want to uncover corresponding process inefficiencies. You also want to use SAP Signavio Process Insights to improve the DSO metric. This metric provides information on how long the period is between issuing the invoice (invoice date) and receipt of payment on the bank account. The DSO key figure is used not only to assess receivables management but also in liquidity management.

As shown in Figure 3.8, the SAP Signavio Process Insights start screen displays all end-to-end processes you can focus on. Alternatively, you can select the view per business unit in the left tab under **Lines of Business**. In our scenario, however, we'll first select the **End-to-End Processes** view to select the **Lead to Cash** business process there. We want to examine this process in more detail, improve it, and systematically transform it.

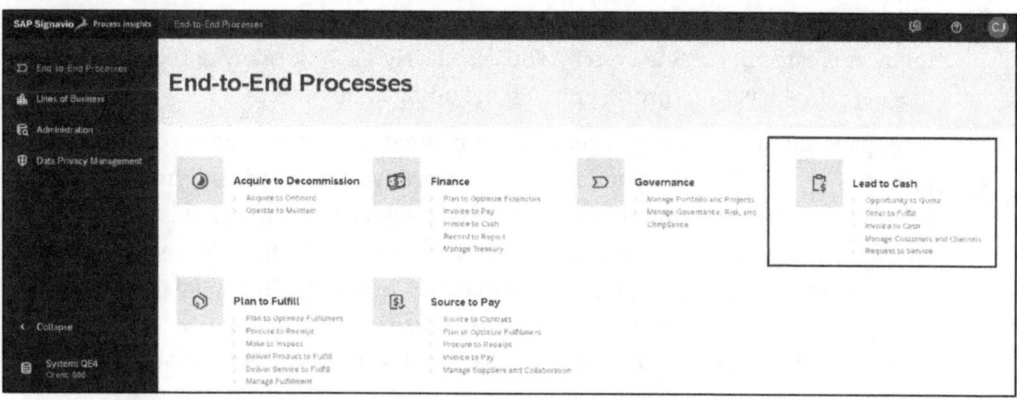

Figure 3.8 Analysis View for Each End-to-End Process

At the top of the screen, you have several options for analyzing your lead-to-cash process (see Figure 3.9). You can either dive deeper into the process flow, look at a wide variety of PPIs, or draw on tailored improvement recommendations as well as innovation potential.

When analyzing the process flow performance, you can see in Figure 3.9 that you have a considerable loss between the first and the second process step. This means that only one-third of the sales invoices created are created within accounting (**FI-AR**). Not only are a few documents transferred, but we also see that this takes far too long, averaging **20 Days**. To investigate this further, you can open the details from the **Most Frequent Blockers and Other Information** section by clicking on the **Show** button.

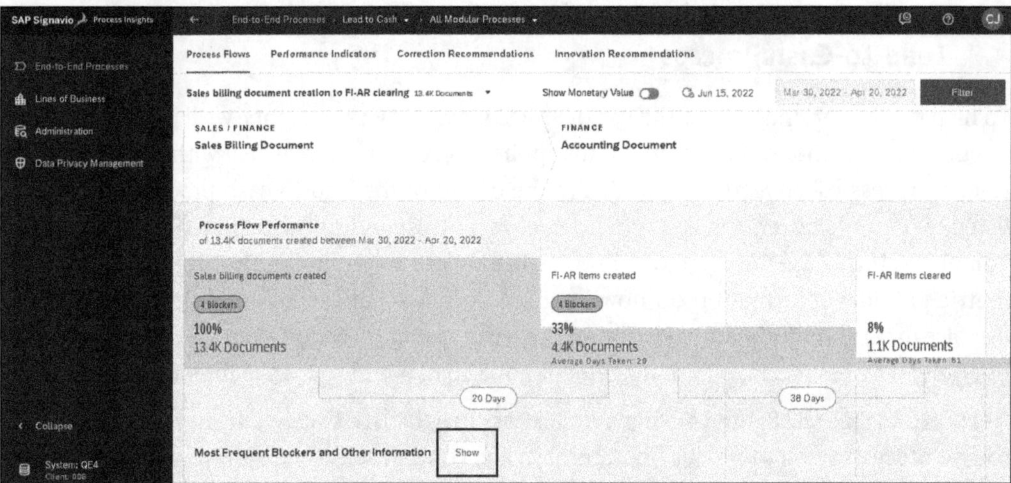

Figure 3.9 Analysis of Process Performance

By default, a process flow applicable to the selected business process is displayed. This helps you visualize the process performance using data from your SAP system. Within the process flow, you can immediately see the global process performance for all invoices created in the SAP system. You can clearly see how many of these documents make it into the various process phases, including cycle time.

By expanding the section, you get detailed contextual information on each individual process step. For example, you can see the respective, typical process inefficiencies. As shown in Figure 3.10, SAP Signavio Process Insights has highlighted several critical blockers in the first two process steps in red. For example, the first blocker indicates that **24%** of all sales settlement documents are created manually. This could potentially lead to high inefficiencies in the process. Let's analyze in more detail exactly which documents are affected. To do so, click on the **Sales billing documents manually created** tile.

[»]

Product Supplements

A full-size color version of Figure 3.10 is available for download at *https://sap-press.com/5855* under the **Product supplements** section.

The detail list opens in Figure 3.11 to show you a comprehensive overview of all manually created sales invoices, which is also indicated by the corresponding **Automation** column. If you scroll to the right, you'll get more information about these documents. We should further investigate why these documents have to be created manually and why the automation features of our SAP system aren't used. You have two options to do this: you can take a closer look at this insight together with your responsible key user and process experts, or you can make use of SAP's context-specific recommendations for improvement.

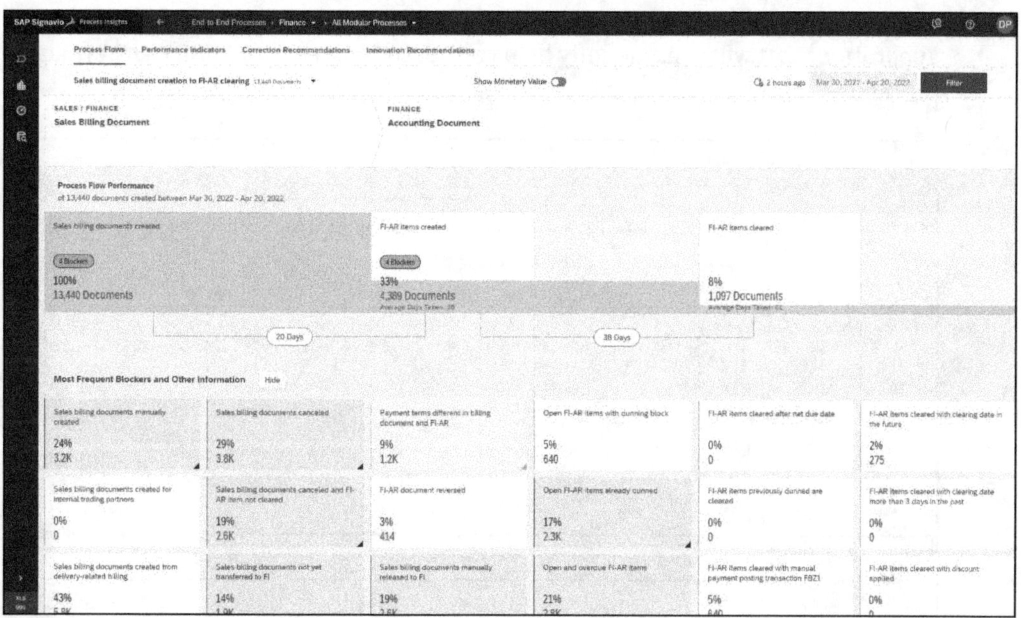

Figure 3.10 Analysis of Relevant Process Blocks

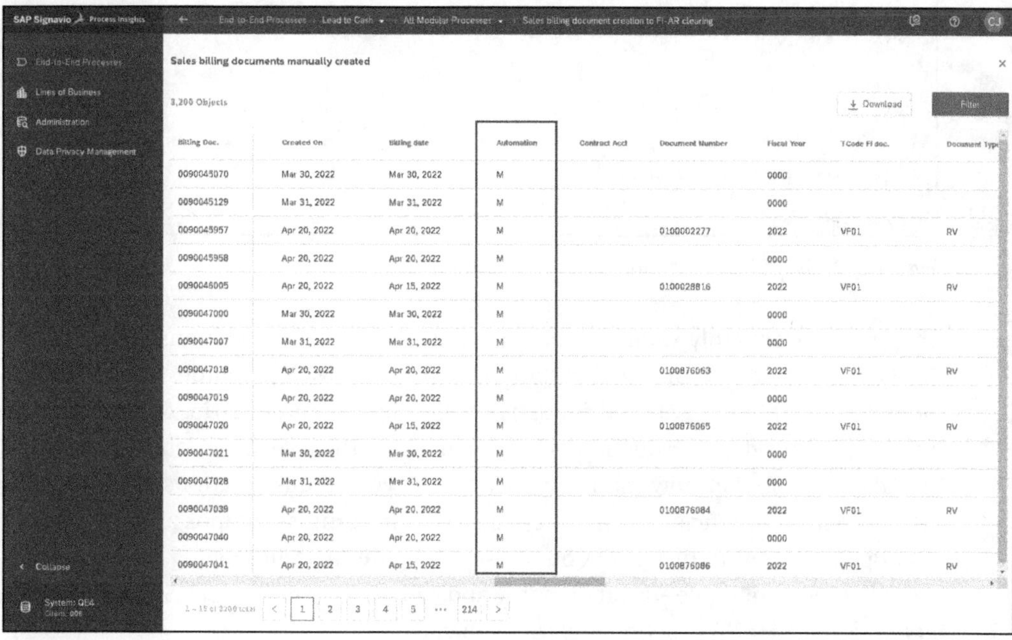

Figure 3.11 Analysis of the Degree of Automation

Now you can check if there are SAP solutions that can help you reduce manual effort and the inefficiencies that come with it. To do this, click on the **Innovation Recommendations** tab shown in Figure 3.12. These innovation recommendations will help your organization plan for long-term and strategic improvements with SAP solutions and applications that will consequently help you get started with your ERP transformation process.

Process Flows | Performance Indicators | Correction Recommendations | Innovation Recommendations

SAP S/4HANA Capabilities (48) | SAP Process Automation (21) | Intelligent Technologies (37) | User Experience (14) | Other SAP Solutions (10)

Recommendation	Industry Popularity	Lines of Business
SAP S/4HANA Capabilities (48)		
Accounts Receivable		Finance
Advanced Variant Configuration		R&D/Engineering
Available-to-Promise		Supply Chain
Periodic Billing Processes		Service
Sales Billing		Sales
Sales Contract Management		Sales
Sales Master Data Management		Sales
Sales Order Management and Processing		Sales
Service Billing		Service
Variant Configuration		R&D/Engineering
Advanced Available-to-Promise		Supply Chain
Advanced Incentive and Commission Management		Sales
Collections Management		Finance

Figure 3.12 List of Relevant Innovation Recommendations

In our case, we mainly want to look for innovation recommendations regarding the improvement of our sales invoice metric. SAP S/4HANA capabilities are ranked by popularity within the selected industry; according to our scenario, the automotive industry is considered. Industry popularity indicates the level of adoption of a recommendation based on how many competitors in your industry are using it. Because we've identified inefficiencies in the sales accounting process, we look at the most popular options in our industry. As you can see in Figure 3.12, **Sales Billing**—as we can tell from a three-point scale—is rated highly and sounds very promising. So, now we look at how SAP S/4HANA functionality can help you.

Looking at the detail page, notice that the **Sales Billing** feature may be relevant to you due to its high industry popularity, so click on the **Sales Billing** line. This component helps manage the entire sales order lifecycle. You get faster billing with less administrative work, allowing you to reduce manual errors and thus achieve higher customer satisfaction. Another great effect is the reduction of your DSO due to the integrated automated billing and invoicing functions. This is the solution you can bring up in your next strategic meeting to explore in detail. The window shown in Figure 3.13 provides you with the background knowledge you need. Based on the innovation recommendations, you can build or continuously adapt your global business plan to implement long-term strategic improvements with SAP solutions and applications.

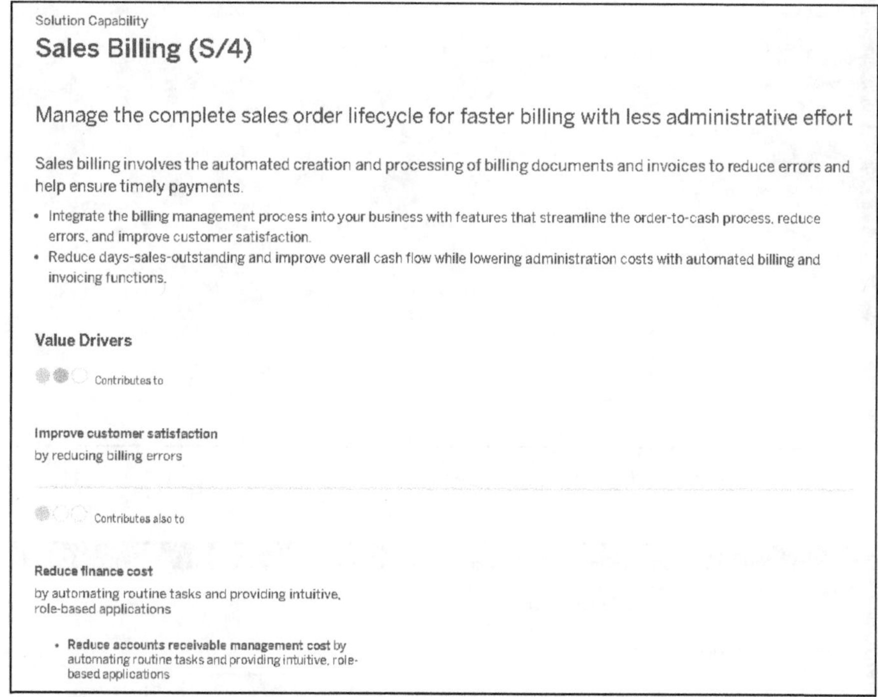

Figure 3.13 Analysis of the Proposed SAP Function

Now you can better understand where your inefficiencies are in the lead-to-cash process and how you can improve them. You may still wonder if there are more inefficiencies in your lead-to-cash process. To accelerate this process analysis, you can make use of the corrective recommendations automatically generated by SAP Signavio Process Insights. To do this, in Figure 3.14, we click on the **Correction Recommendations** tab. Correction recommendations are derived from data received from your SAP system, based on PPIs, and help you quickly resolve identified issues with immediate suggestions for your business. They are based on many years of experience within the SAP ecosystem.

In Figure 3.15, you can analyze the correction recommendations for your lead-to-cash process. You can see that there are two specific correction recommendations in this regard. The system shows you which value drivers are supported if you apply the suggested recommendation. It also gives you an indication regarding the benefits and the implementation effort. Let's now dive deeper into the meaning of these recommendations.

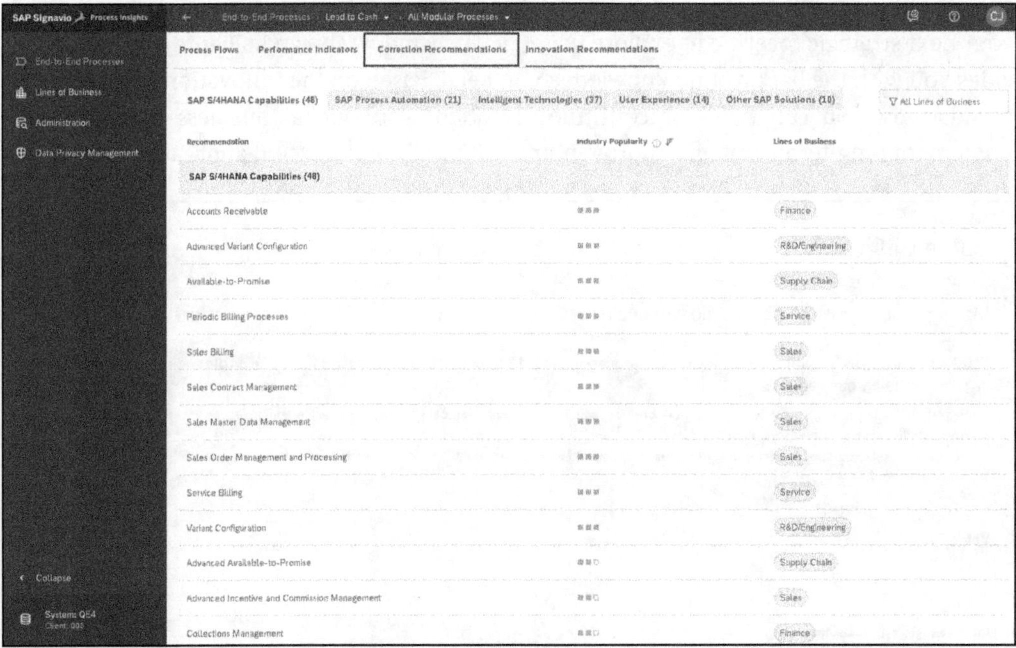

Figure 3.14 Display Correction Recommendations

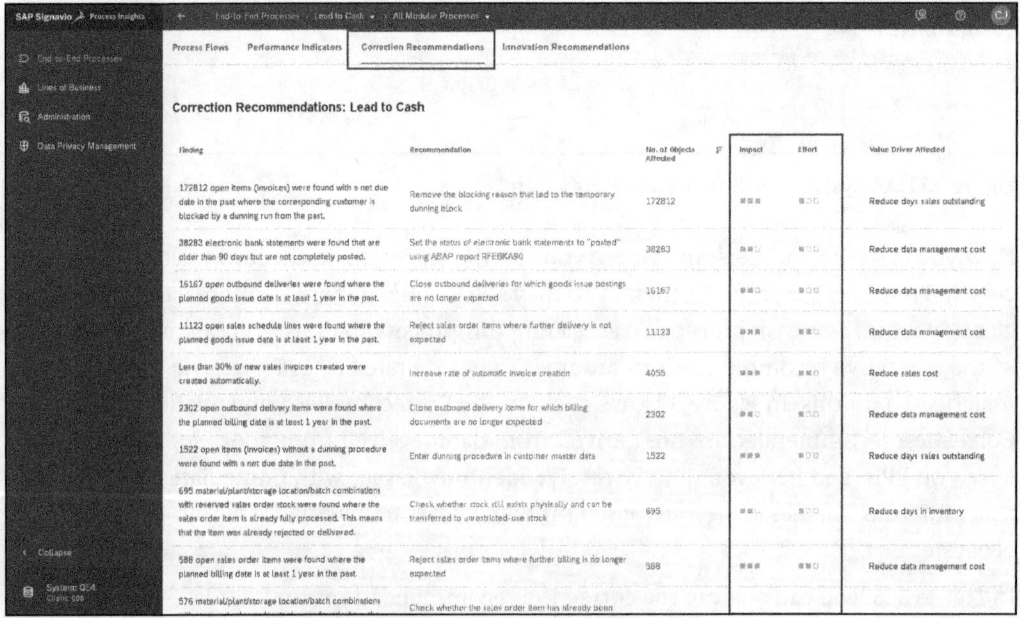

Figure 3.15 Analysis of the Correction Recommendations

In Figure 3.16, first click on the relevant correction recommendation. This relates to automation, has a positive impact on your process flow, and at the same time requires little implementation effort. This sounds like a fantastic way to increase the efficiency of your deliveries.

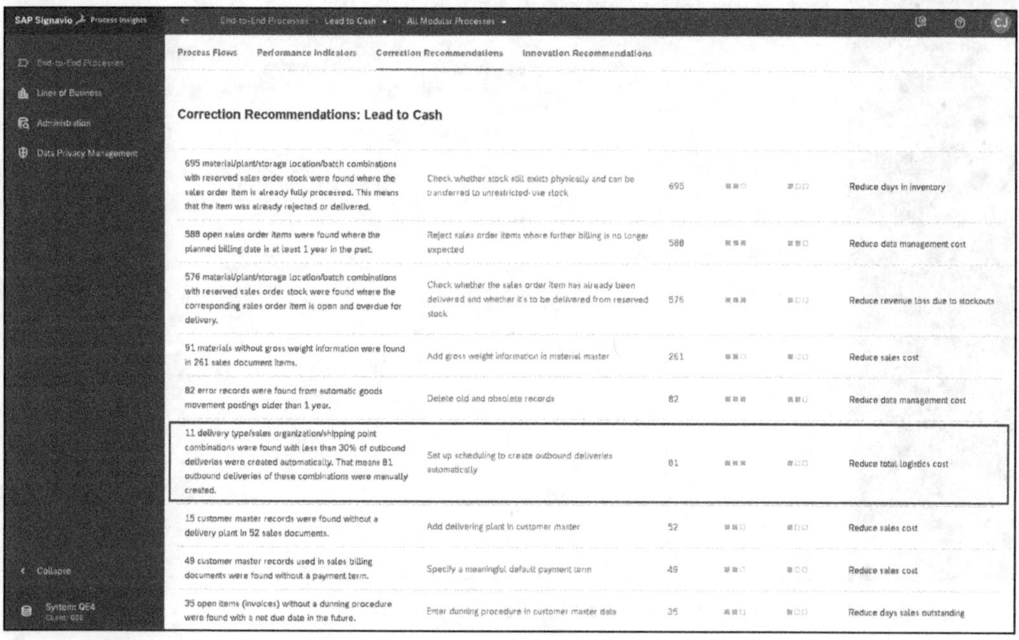

Figure 3.16 Focus on the Relevant Correction Recommendation

The insight from Figure 3.17 correctly states that less than 30% of deliveries were created automatically. On closer inspection, you can see that the potential cause is that we have unused standard automation functions. By creating outbound deliveries automatically, you can reduce overall logistics costs. To see exactly what SAP Signavio Process Insights suggests implementing, click on the **Show Recommended Action** link.

The recommended action in Figure 3.18 immediately explains what exactly needs to be done. This allows you to target your colleagues to implement this improvement.

Correction recommendations are ideal for being able to act quickly based on facts. Furthermore, they are suitable for identifying inefficiencies in individual process steps. With SAP Signavio Process Insights, you get quick insights into your business processes in an uncomplicated way. The improvement recommendations provided are immense as they make available years of SAP consulting knowledge.

All in all, this use case gave you a clear understanding of your lead-to-cash process using the extensive analytics capabilities of SAP Signavio Process Insights. You were able to identify several inefficiencies in the process that can be addressed in the short and long term. Out-of-the-box innovation recommendations help you understand where the next strategic improvement makes sense and generates value. Based on the innovation recommendations, you can create a structured plan to implement long-term strategic improvements with SAP solutions and applications.

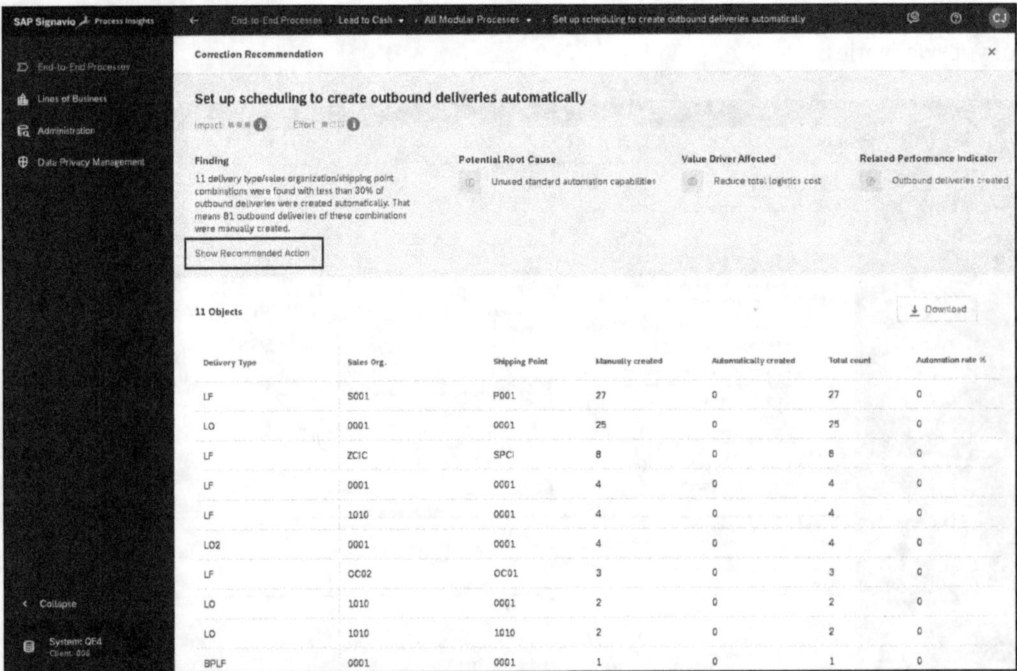

Figure 3.17 Realization of the Correction Recommendation

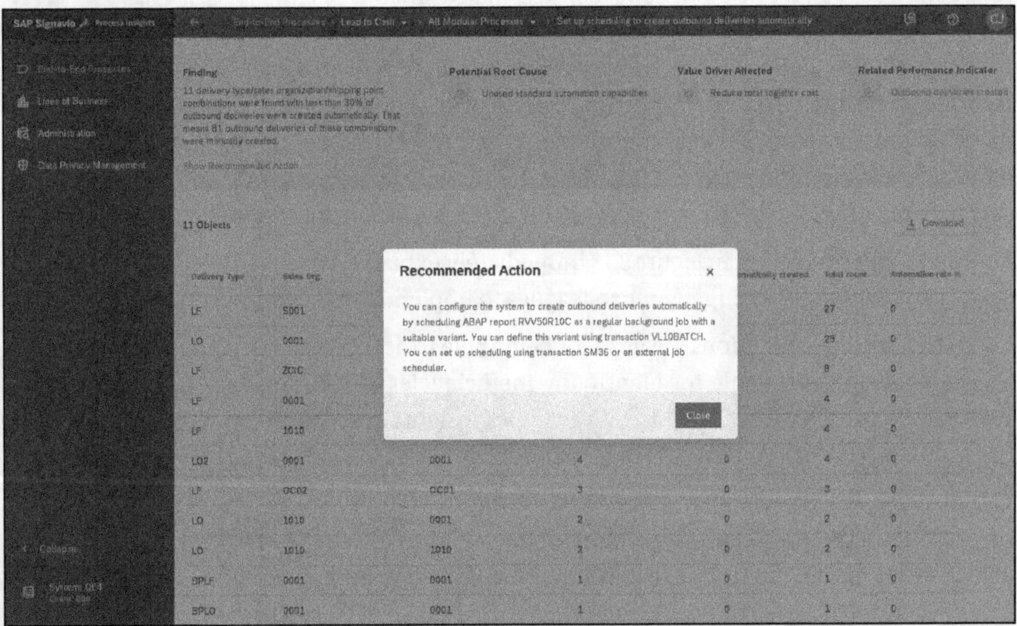

Figure 3.18 Implementation Proposal of the Correction Recommendation

By exploring corrective recommendations, you'll gain a better understanding of your lead-to-cash process as well as a good understanding of where inefficiencies may lie in your end-to-end process. With these insights, you can launch a targeted, fact-based analysis. But it's not just the insights that are comprehensive. All corrective recommendations provide quick solutions, including step-by-step descriptions for effective implementation. As illustrated in our fictitious application example, you can take advantage of standard automation capabilities that may have been previously unknown to you.

As an alternative to end-to-end process scenarios, you can also approach your process analysis per business unit. As shown in Figure 3.19, you can choose between **Asset Management**, **Finance**, **Manufacturing**, **R&D/Engineering**, **Sales**, **Service**, **Sourcing & Procurement**, and **Supply Chain**.

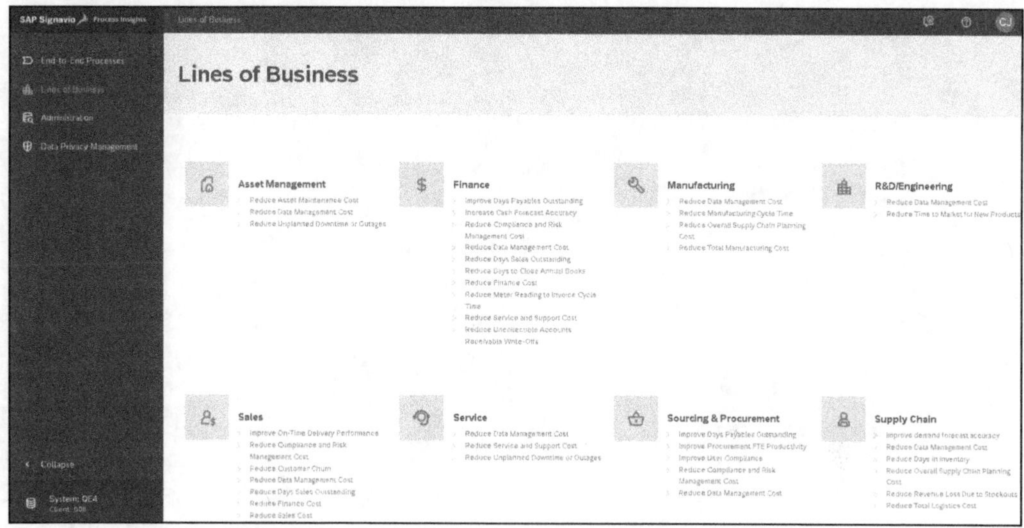

Figure 3.19 Analysis View per Line of Business

3.4 Summary

With SAP Signavio Process Insights, you can secure immediate added value by getting a holistic view of the performance of your end-to-end business processes as well as key business areas. SAP Signavio Process Insights uses advanced analytics to help you dive deep into your business processes and make the right decisions in your business areas. It also helps you deliver greater business value by focusing on the most impactful transformation actions. With granular, easily accessible insights and tailored recommendations, benefits can be achieved faster—within a very short time after implementation. This gives you the opportunity to achieve new levels of optimization in step with improved business resilience, leading to stability, business continuity, and sustained improvement in the customer experience.

Chapter 4
SAP Signavio Process Intelligence

SAP Signavio Process Intelligence helps you to systematically evaluate your business processes. This means that you collect information about all processes that can be analyzed and used for future process improvement.

Process-driven business transformation helps organizations of all industries and sizes compare their status quo, including the as-is versions of their business processes to the to-be state of their processes. It enables organizations to gain an understanding of any variances in their systems, optimize the data collected there, and bring it all together to show the evolution of the process using intuitive dashboards and interfaces. By linking processes and workflows, responsible stakeholders such as line of business management, process owners, or project managers can execute data-driven, end-to-end transformation initiatives based on coherent information. This gives organizations the information, visibility, and quantifiable numbers they need for continued growth and process excellence. All of this is possible with SAP Signavio Process Intelligence (see Figure 4.1). SAP Signavio Process Intelligence allows you to analyze and visualize your target processes in real time and compares your actual processes with target process variants to identify potential for improvement.

[!]

Prerequisites for Use

SAP Signavio Process Intelligence requires the use of SAP Signavio Process Manager (see Chapter 5) and SAP Signavio Process Collaboration Hub (see Chapter 7).

In Section 4.1, we'll first look at the various functions of SAP Signavio Process Intelligence. Section 4.2 shows various benefits of the solution. Subsequently, we'll illustrate the use of SAP Signavio Process Intelligence in practice by means of an example (Section 4.3).

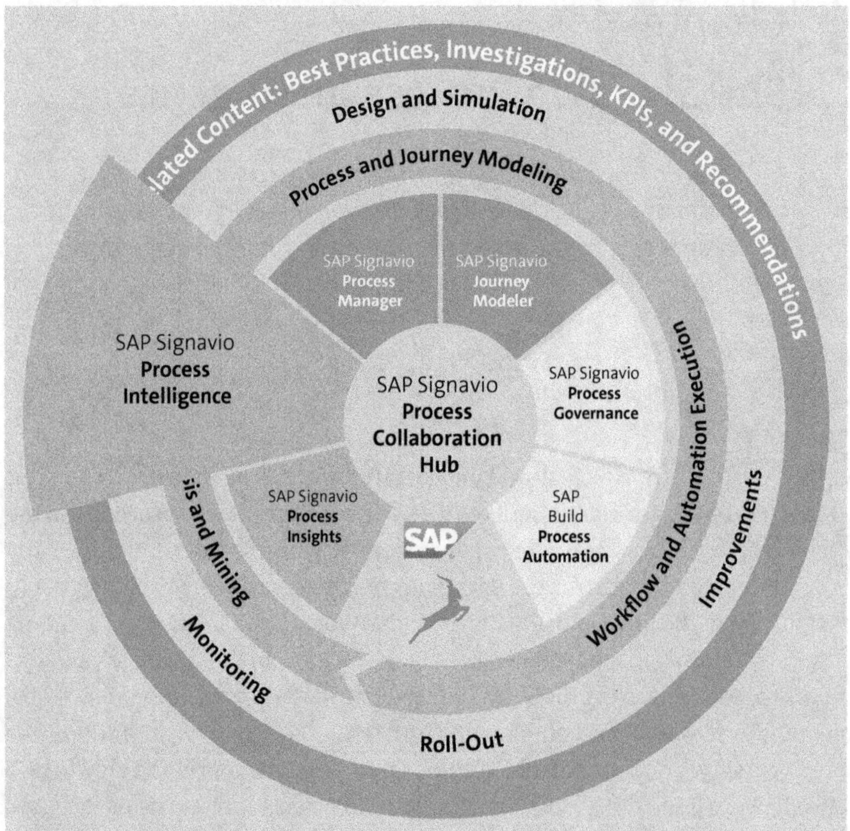

Figure 4.1 SAP Signavio Process Intelligence in Focus

4.1 Functions

SAP Signavio Process Intelligence is an integral part of SAP Signavio Process Transformation Suite. The solution covers the latest process mining capabilities at scale and supports smarter business decisions by providing powerful, fact-based insights into potential risks and continuous improvement opportunities. This allows you to identify the business value hidden in your processes and data. SAP Signavio Process Intelligence enables you to perform detailed analytics for a wide variety of process improvements; create intelligent diagnostics to test bottlenecks and simulations of your business processes; evaluate process change alternatives considering best practices; optimize business processes for robotic process automation (RPA), hyperautomation, artificial intelligence (AI), and machine learning; and derive accurate customer behavior for improved customer journeys.

The following are core functions of SAP Signavio Process Intelligence:

- Collect and prepare process and experience data from a variety of sources.
- Gain insight into actual process execution, highlighting existing inefficiencies and issues and easily monitoring performance at all times.
- Gain insights faster and identify causes of poorly performing processes or low experience metrics.
- Improve process harmonization and compliance by visualizing compliance violations.
- Gain visibility into the combined process and experience reality of your customers by connecting process and experience analytics.
- Strengthen teams and drive collaboration across the organization thanks to insights that can be implemented immediately.

To continuously optimize your processes in today's changing business environment, your company's process mining initiatives must go beyond one-off projects. That's because continuous monitoring of processed data is becoming the norm for successful companies. SAP Signavio Process Intelligence leverages the latest advances in stream processing to handle event data in real time.

[«]

Stream Processing

Stream processing is an alternative to batch processing. The data isn't stored temporarily as in batch processing but is processed and analyzed continuously in real time as soon as it's generated or received. In addition to the digitized processes of Industry 4.0, stream processing is also used, for example, in the big data environment.

Thus, the solution acts as your omnipresent finger on the pulse of your business process landscape. In this context, process analysis with SAP Signavio Process Intelligence is your path to effective data analysis. With its help, you can automatically analyze your data and gain insights for current and future operations. With this tool, you can leverage data that already exists in your database (e.g., in your enterprise resource planning [ERP]/customer relationship management [CRM] systems) to optimize business performance for intelligent operations. SAP Signavio Process Intelligence takes the raw data and translates it into a blueprint that clearly shows how your business operates.

[«]

Wide Range of Applications

With SAP Signavio Process Intelligence, you can visualize your process reality: from detailed process analysis for business transformations (including SAP S/4HANA migrations), to exploring and interacting with business data, to using process mining for fact-based changes to uncover the hidden value in business processes and data.

SAP Signavio Process Intelligence helps you understand not only what your processes look like but also how they actually run in reality. A complete assessment of as-is processes reveals inefficiencies, problems, misbehaviors, and unused processes. In response, for example, you can implement intelligent ERP transformations, optimize resources, significantly improve business metrics, and reduce costs. By combining SAP Signavio Process Intelligence with a customer-centric view of production, marketing, and sales, as well as products and services, customer satisfaction becomes a strategic catalyst: through process mining, you're able to precisely understand and analyze your customers' behavior. You can optimize customer touchpoints and realign processes in a targeted manner. The tool ranges from process recognition with Process Discovery visualization and analysis, to real-time monitoring of your processes based on real data. As even complex processes can be better understood, you benefit from continuous improvement in your organization.

With the SAP Signavio Process Intelligence solution, you can also perform detailed *process analyses* from any location. By gaining detailed insights into process data, you too can optimize your daily operations and make smarter decisions faster. The solution helps you identify root causes of poorly executed processes by detecting and visualizing compliance violations, monitoring process performance, and responding to critical cases and performance bottlenecks. In addition, the tool helps you easily become familiar with the discrepancies in your systems, optimize contained data, and summarize all necessary information. With the help of intuitive dashboards and interfaces, you can easily show your business history. By linking different processes or operations, you can also lead data-driven insights regarding end-to-end processes to holistically support your transformation plans. As a user of SAP Signavio Process Intelligence, you always have access to transparent information and quantifiable figures to establish continuous growth and optimal processes in your organization. With regard to your customer expectations, SAP Signavio Process Intelligence enables you to regularly compare metrics, investigate performance deviations, and make corrections if necessary. As a result, you can use process mining to quickly visualize current system processes, share lessons learned, and gather suggestions for improvement.

Table 4.1 provides an overview of the numerous functions and tools of SAP Signavio Process Intelligence.

Functions	Description
Data onboarding	Upload the process data as files to SAP Signavio Process Intelligence. If you use zipped comma-separated values (CSV) or Microsoft Excel spreadsheet (XLS) files, each zipped file can contain only one file.

Table 4.1 Functions of SAP Signavio Process Intelligence

Functions	Description
Extract, transform, and load (ETL) data pipeline	Get an overview of data sources, integrations, and data models. The pipeline serves as a funnel for converting raw data into process data.
Data extraction	Ingest data flexibly via native connectors for on-premise and cloud systems, and benefit from native data transformation, data modeling and data pipeline management, and integration with SAP Data Intelligence to handle even the most demanding requirements and complex IT landscapes.
Data transformation	Convert your data into event logs. The logs are loaded into the process that is linked to the data model and provides the data for investigations.
Data load management	Upload event logs to processes within SAP Signavio Process Intelligence.
Connectors for different source systems	Use connectors to connect source systems to SAP Signavio Process Intelligence. The tool provides connectors for the following source systems: ▪ AWS Athena ▪ AWS S3 ▪ Elastic ▪ Google BigQuery ▪ Jira ▪ Microsoft SQL Server ▪ MySQL ▪ OData ▪ PostgreSQL ▪ SAP ERP ▪ SAP HANA ▪ ServiceNow ▪ Snowflake ▪ SAP SuccessFactors
Legacy connectors	Use legacy connectors that complement SAP Signavio's ETL functionality to connect to source systems.
CSV event log upload	Upload the process data as zipped CSV files to SAP Signavio Process Intelligence.
API for uploading eXtensible Event Stream (XES) files	Manage the uploads of data with the Data Upload API to SAP Signavio Process Intelligence on a one-time or regular basis. The data must be provided as XES files.

Table 4.1 Functions of SAP Signavio Process Intelligence (Cont.)

Functions	Description
SAP Signavio analytics language (SIGNAL)	Perform customized analytics using SIGNAL, which is an SAP Signavio-specific query language for process analysis.
SIGNAL table or case table	Display any dimensions or key figures in tabular form.
Process metrics	Manage key processes and metrics centrally. Metrics can be added from the central library or created from scratch by users.
Dashboard for process metrics	See core metrics at a glance. SAP Signavio Process Intelligence creates configurable aggregations and visualizations and helps you find weak points in your processes.
Case pattern identification	Identify all existing process variants and analyzes process metrics per variant. This helps you isolate groups of cases that run inefficiently or aren't compliant.
Metrics library	Use metrics in the SAP Signavio Process Intelligence metrics library to monitor your key performance indicators (KPIs). SAP Signavio Process Intelligence includes a set of standard metrics to quickly get started with your analysis. Metrics can be customized using SIGNAL.
Verification of process conformance	View process variant paths, one variant at a time, to show hotspot activity or map each variant to the process's Business Process Model and Notation (BPMN) model.
Process Discovery	Generate an interactive process map from the event log data. By default, the widget displays the most common process paths.
Variant Explorer	View the different process variants that exist in the event log.
Actual versus target process comparison	Find cases that don't conform to the target processes and analyzes the business process data related to the noncompliant behavior. This includes the identification of noncompliant actual processes as well as a detailed visual comparison of process variants based on the actual process data.
Correlation widget	Create a scatter plot, that is, a graphical representation of two numeric variables plotted along two axes, to identify or represent their relationship.
Live insights	Add process performance or experience metrics and KPIs to monitor in BPMN diagrams, widgets, and journey models.

Table 4.1 Functions of SAP Signavio Process Intelligence (Cont.)

Functions	Description
Widget configuration	Create widgets with dynamic content that you can add to your investigations. You can copy, duplicate, delete, or edit them.
Filter	Narrow the scope of analysis flexibly and to multiple levels.
Calculations and formulas	Use calculations and formulas to obtain results based on process data.
Shared investigations	Share investigations in SAP Signavio Process Collaboration Hub with your users.
Volume usage	Display the number of processes, cases, and events in the data store.
Access rights management	Manage access rights for different users and at the data level to restrict permissions to investigations, among other things.

Table 4.1 Functions of SAP Signavio Process Intelligence (Cont.)

4.2 Benefits

SAP Signavio Process Intelligence helps you gain a better understanding of your data to make fact-based changes. By using this solution, you get an improved end-to-end view of your processes—from event logs to user interactions. From a technical perspective, you can manage your data, develop new models, integrate outputs with third-party applications, and set metrics for in-depth analysis to understand actual states, activities, and operations. In this context, you can benefit from out-of-the-box capabilities to reduce time to analysis with ETL connectors and transformation templates for the most common processes, including procure to pay, order to cash, and IT service management (ITSM).

An overview is provided here of the core benefits of SAP Signavio Process Intelligence:

- Delivery of process results in the shortest possible time
- Automated process analysis
- Identification of process weaknesses
- Continuous process improvement
- Process-based reporting and identification of KPIs
- Optimization of process performance

Interaction and analysis of key information has never worked so efficiently, as SAP Signavio Process Intelligence immediately uncovers inefficiencies and bottlenecks in business processes to ensure smooth operations. Process flows can be examined to compare KPIs, perform impact analysis, and learn about variants to streamline business processes. You can dynamically map and connect your process control flows to improve compliance, deviation, and waste. By performing root cause analysis, the solution enables you to investigate delays, skipped activities, and the sequence of events while dramatically reducing cycle times and rework.

SAP Signavio Process Intelligence uncovers process inefficiencies to improve operations. You can also benefit from *widgets* for accelerated performance analysis and sharing of results. These widgets enable simplified use and detailed extractions for immediate insights into performance KPIs, cycle time analysis, and overall indicators of business health. Table 4.2 provides an overview of the numerous SAP Signavio Process Intelligence widgets.

Widget	Description
Activity List	Lists all activities that occur in the loaded process data
Breakdown	Visualizes the division or distribution of your data in charts
Case Table	Displays a table with case information you selected
Correlation	Provides a scatter plot
Diagram	Displays a diagram from your SAP Signavio Process Manager workspace
Distribution	Displays the distribution of the process duration attributes
Over Time	Displays activities in your process over time
Process Conformance	Similar to the Variant Explorer widget, this widget displays the paths of process variants, but with only one variant at a time
Process Discovery	Creates a process model from the event log data by default to visualize the most common process paths
Process Funnel	Displays the traffic through your process
SIGNAL Table	Displays the result of a SIGNAL query as a table
Spreadsheet	Displays a table
Text	Supports viewing and editing rich text content in your investigation
Value	Displays aggregated case data
Variable Importance – Relate	Shows which attributes relate to a selected target attribute

Table 4.2 Available Widgets in SAP Signavio Process Intelligence

Widget	Description
Variant Explorer	Like the Process Conformance widget, this widget displays the paths of process variants, but with one or more variants at a time

Table 4.2 Available Widgets in SAP Signavio Process Intelligence (Cont.)

In addition to widgets, process performance and compliance metrics can be created and combined with process mining to accelerate your business operations.

SAP Signavio Process Intelligence allows you to see more of the information you really need on a single platform! By connecting and extracting data with advanced integration capabilities, it's possible to connect to multiple source systems to explore relevant widgets for instant information. In this way, you can transform your data into an environment where business processes and IT work together to continuously increase efficiency and effectiveness.

SAP Signavio Process Intelligence provides real-time monitoring of your processes and helps you identify the root causes of underperforming processes, locate compliance violations and bottlenecks, and continuously monitor the performance of your processes. With SAP Signavio Process Intelligence, you realize the full potential of real-time data-driven process mining. The immediate knowledge of your business processes ensures end-to-end process orientation throughout your organization. It promotes quick changes and customer-oriented thinking. Using SAP Signavio Process Intelligence, you can now extract relevant information from process data sets even faster. Simple process models come to life in dynamic and responsive *dashboards*. to life. As a result, you're automatically informed, guided, and warned regarding your business processes.

The SAP Signavio Process Intelligence solution represents a revolution for your automated process analysis. Getting inside a company's existing business processes has long been time-consuming and expensive. Highly skilled external personnel were required, and, in some organizations, fact-based insights were supported by an unhealthy dose of intuition. By automatically applying advanced process mining algorithms, SAP Signavio Process Intelligence provides visibility and understanding into your organization's actual business processes. With this tool, you gain insights into how and why your end-to-end business processes are currently running a certain way. Subsequently, you can better align your operations with your processes.

The time and money spent on big data analytics can significantly slow down improvement projects. However, especially sound process mining usually requires large investments because it's not easy to monitor and efficiently execute processes in a constantly changing regulatory environment. SAP Signavio Process Intelligence not only supports transparency and understanding across all business processes but also provides insights on where optimization is needed and how to achieve desired business outcomes. This accuracy of insights supports you in analyzing and optimizing your business processes. SAP Signavio Process Intelligence thus ensures continuous improvement of your processes and KPIs.

Another benefit of this solution comes from the disruptive process mining technology. This is because top managers and executives are under increasing pressure in an ever-changing business landscape and need to achieve better and better results in less time. At the same time, it's becoming more difficult to secure the budget for required investments and consistently demonstrate return on investment (ROI). ROI is a key figure for measuring the actual return in relation to the capital invested. To be successful here, transparency is required as well as a deployment scenario that is based on facts instead of forecasts and estimates. In this context, SAP Signavio Process Intelligence is a key enabler for process-based reporting and KPI identification. Using this tool, you give your analysts a defined framework to identify weaknesses and define your KPIs. You can track and evaluate these directly in SAP Signavio Process Intelligence and consider them in the context of your process models in SAP Signavio Process Manager to make data-driven process decisions with even more confidence. In addition, you can share your insights with all your colleagues using SAP Signavio Process Collaboration Hub. This holistic approach means that collaborative feedback can be shared in one place and results can be presented centrally. The ROI is therefore much more visible.

With SAP Signavio Process Intelligence and the *live insights* function, you're also able to set indicators and thus examine defined thresholds in individual process diagrams in a completely automated way. In this sense, you can add process performance or experience metrics and KPIs to monitor them in BPMN diagrams, widgets, and journey models. In this way, you sharpen your focus on identifying business challenges (e.g., related to poorly performing production equipment). This enables you to launch optimization initiatives in a timely manner because live insights means, among other things, that you narrow down the problem area in a targeted manner by bringing your process management team together and directing their focus directly to the affected processes, systems, and data sources.

As a fully cloud-based process mining solution, SAP Signavio Process Intelligence lets you deliver business results smoothly across the entire software solution. This solution flow makes it even easier to collaborate with colleagues from all areas and leverage collective knowledge to generate more ideas, optimize processes, and reduce resistance to change. Finally, this SAP Signavio solution is also an indispensable tool for the entire internal audit function. It enables particularly strong and coherent initiatives, for example, using Breakdown, Distribution, or Variant Explorer widgets (refer to Table 4.2) when transparency and data- and fact-based analysis are required. Auditors or risk management departments can use SAP Signavio Process Intelligence to assess how processes are executed from a compliance perspective and where specifications aren't followed. Compliance metrics available in SAP Signavio Process Intelligence are particularly suitable for this purpose. Based on the case pattern identification, SAP Signavio Process Intelligence identifies all existing process variants and also analyzes process metrics per variant (see Table 4.1). This helps you isolate groups of cases that are running inefficiently or aren't compliant.

Thus, SAP Signavio Process Intelligence allows you to use the full potential of your business processes because the tool gives you the following options:

- Better data-based and transparent decisions for the implementation of your goals
- Integrated *conformance checks* for viewing hotspot activity, based on a process variant to ensure standard procedures, including fast identification of exceptions
- Complete end-to-end perspectives, performance overviews, and a better understanding of what is happening in your organization over a period of time
- Continuous monitoring and improvement of the exact processes to gain knowledge for root cause analysis
- Data-based facts and continuous process improvement

This results in numerous use cases for SAP Signavio Process Intelligence—from ERP transformation, audit, risk and compliance to increasing customer satisfaction and operational excellence. Use cases related to business transformation and process analytics will be covered in detail in Part III. First, we dedicate ourselves to an introductory example to become aware of the potential of SAP Signavio Process Intelligence.

SAP Signavio Process Intelligence helps you capture and prepare your business data with multiple integration options. You gain complete visibility and understanding of your actual business operations and processes by integrating process data and automatically applying advanced process mining algorithms. The tool allows you to connect and extract online and offline data across different systems and landscapes, including data from experience platforms such as SAP Qualtrics. Raw data is transformed into complete event logs in an environment where your business processes and IT can work together. Using comprehensive data models, you get a holistic overview of your business processes. Namely, with data models in SAP Signavio Process Intelligence, you define how the ETL data pipeline extracts and transforms your process data and where the data is loaded. A data model contains all settings required to run an ETL data pipeline:

- A data source that enables the connection to the source system
- A process into which the extracted and transformed data is loaded, or an integration that defines which data is extracted
- A transformation configuration with a business process and transformation rules for extracted data

Stay flexible and shorten your time-to-analysis with a variety of integration options, including standard connectors and data integration features supported by SAP Data

Intelligence as well as template-based process data transformations for the most common processes and systems. Figure 4.2 provides you with an insight into a data model in the backend of SAP Signavio Process Intelligence. There you can make all settings regarding data source, business process, and data transformation.

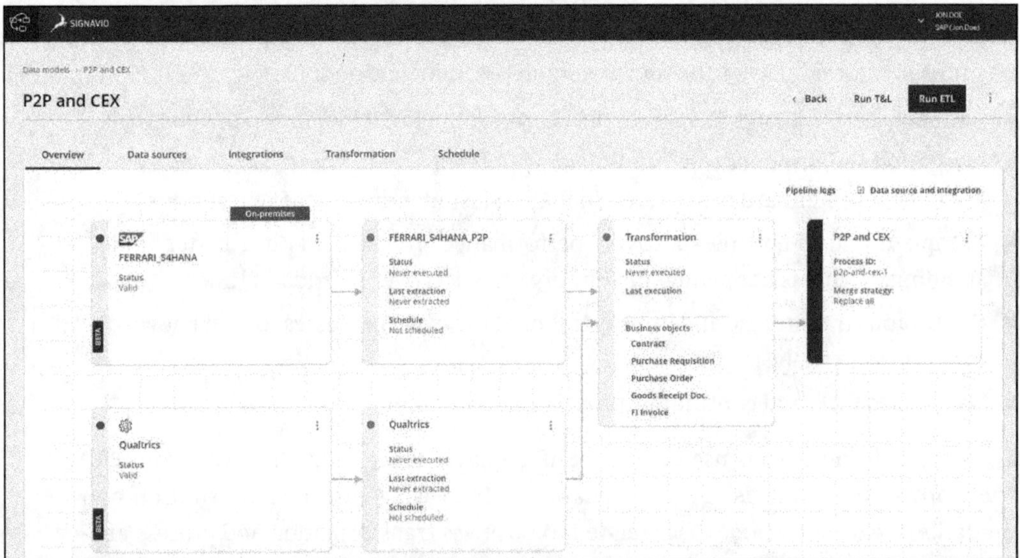

Figure 4.2 SAP Signavio Process Intelligence: Data Model

Based on the data models, you discover the process reality of your business. With this complete insight into your process reality, you can easily identify problems such as long cycle times or bottlenecks, as well as the process variants that affect your overall performance. You can automatically generate new process models from selected variants, saving time in implementing changes and best practices. SAP Signavio Process Intelligence helps you validate your gut feeling with real data points. The tool allows you to fully understand your process dynamics by comparing how processes are implemented in your organization and help you identify process compliance violations. The result is a clearer understanding of how your business actually operates. With this holistic view, you can make informed decisions and immediately begin standardizing, simplifying, and/or redesigning your processes. Figure 4.3 shows an example of a **Process Discovery** that is automatically displayed by default when you start a new investigation in SAP Signavio Process Intelligence.

SAP Signavio Process Intelligence enables you to rapidly drive process analysis and accelerate your overall value creation. You benefit from standard and customized content to quickly start your analysis with SAP Signavio Process Intelligence *accelerators*, which are available for different processes and source systems. Predefined or preconfigured accelerators help you get to your results faster. For example, your data integration can be significantly accelerated using standard process data transformation templates available for different processes and source systems (e.g., procure to pay, order to cash, and SAP S/4HANA). This also significantly reduces the time and effort required to prepare your data.

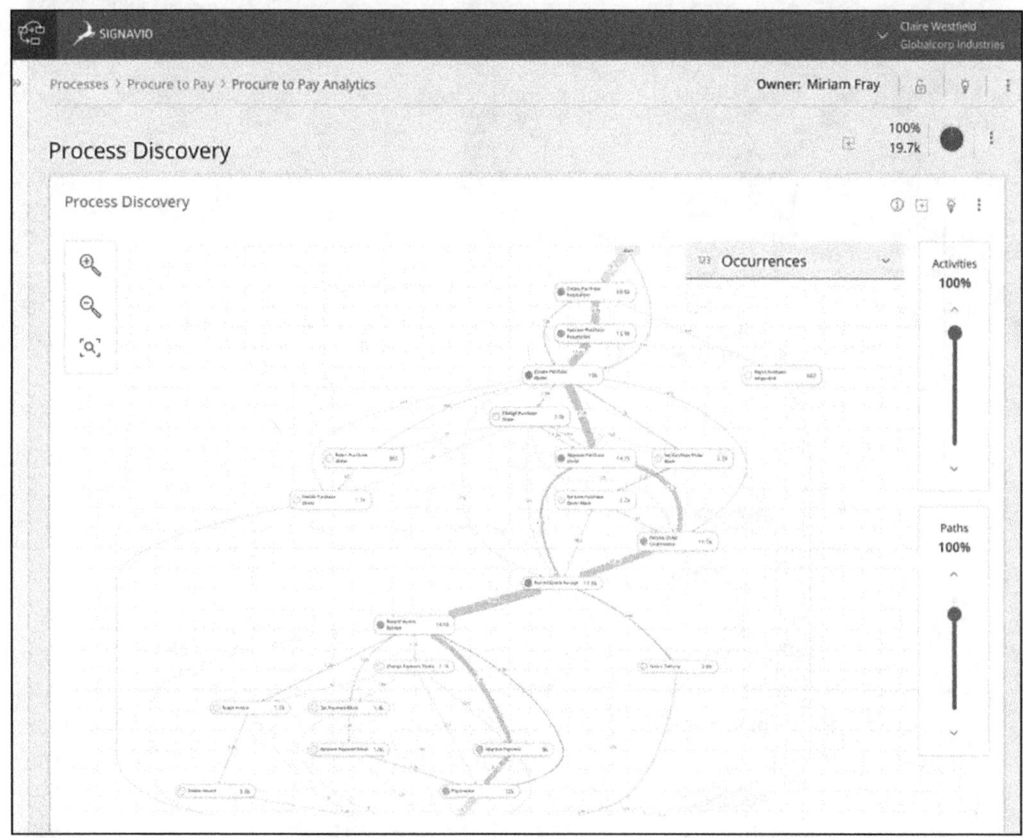

Figure 4.3 SAP Signavio Process Intelligence: Process Discovery

Once you have the data, you save time by using ready-to-use metrics and are fully equipped to start monitoring your KPIs. Metrics are quantitative assessment measures commonly used to evaluate, compare, and track performance or production. SAP Signavio Process Intelligence comes with a set of metrics. The value or values of a metric are determined using a preconfigured SIGNAL query. Some metrics provide results immediately (out of the box), and other metrics contain variables to which you must assign values. In some cases, you may also need to customize the SIGNAL query. To access the metrics, open your process in SAP Signavio Process Intelligence, and click **Metrics** in the sidebar. In Figure 4.4, you get an extract of various metrics available in SAP Signavio Process Intelligence.

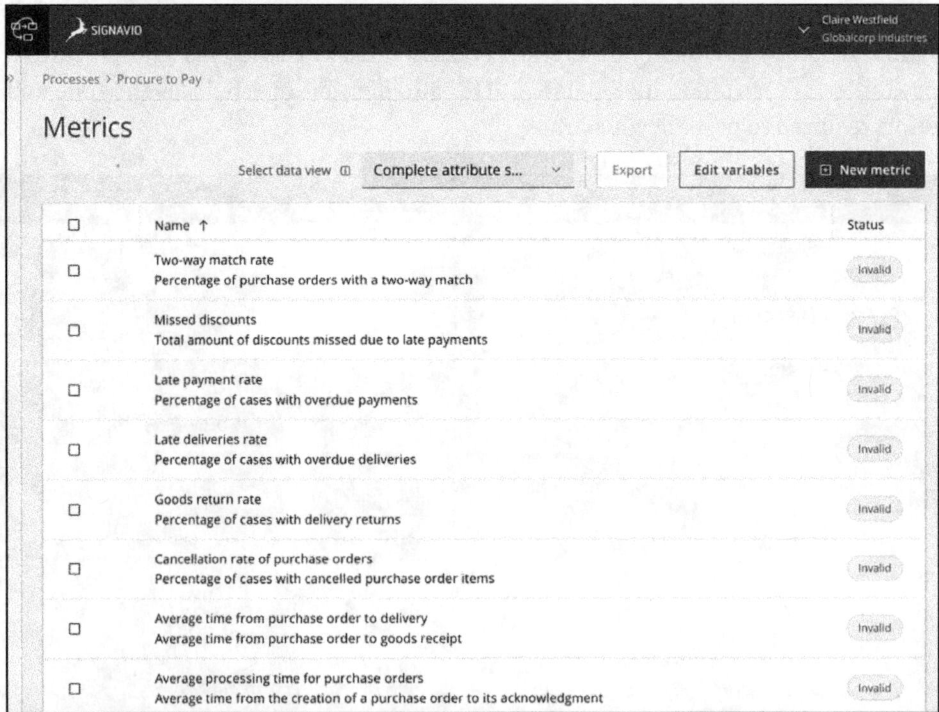

Figure 4.4 SAP Signavio Process Intelligence: Metrics

Now, with SAP Signavio Process Intelligence, you can gain faster insights into your business processes as well as identify new opportunities for process improvement. As a result, you're able to achieve improved business results and provide even more effective and efficient collaborations to your customers, employees and suppliers. SAP Signavio Process Intelligence enables you to analyze and monitor your performance to uncover process inefficiencies and identify opportunities. To do this, you can access customizable out-of-the-box metrics and widgets.

SAP Signavio analytics language (SIGNAL), as mentioned earlier, helps you perform customized in-depth analyses of your data. SIGNAL is an SAP Signavio-specific query language for process analysis based on structured query language (SQL). Like SQL, SIGNAL uses queries to retrieve your data and perform calculations on the data. However, it doesn't allow you to modify or delete process data. The main difference with SQL is in the data model. While with SQL, you usually query data from multiple tables, SIGNAL queries data from only one table that contains nested events. In addition, SIGNAL provides a variety of user-defined functions to work more effectively with this data structure. SIGNAL is optimized for process mining to identify, for example, conformance, lead times, and rework.

SIGNAL also helps you shorten your time-to-insight with built-in formulas and ready-to-use analyses. For example, these uncover hidden correlations or anomalies in your business data. Once you know what may be causing problems or inefficiencies, share your insights with your colleagues directly in SAP Signavio Process Collaboration Hub or in the context of your process models in SAP Signavio Process Manager to make data-driven process decisions with confidence—faster and more efficiently than ever. You can achieve process insights by using the Insights feature because it allows you to uncover problems and inefficiencies in your process in an intuitive way. You can capture personal insights manually as well as generate automated insights. Members of your organization can also access the insights you capture. Insights include their description and a snapshot of the data. The data snapshot visualizes the data at the time of capture. Automated insights are generated by an integrated analytics engine. This engine provides advanced statistics and contextual data, generating automated insights based on algorithms and metrics:

- **Algorithms**
  For example, the algorithms detect correlations or anomalies in the data set; they are integrated and can't be customized.
- **Metrics**
  All metrics associated with your process are used to generate insights.

You can also save your insights and turn them into widgets that are then added to an investigation. You can see how such insights might look in practice in Figure 4.5.

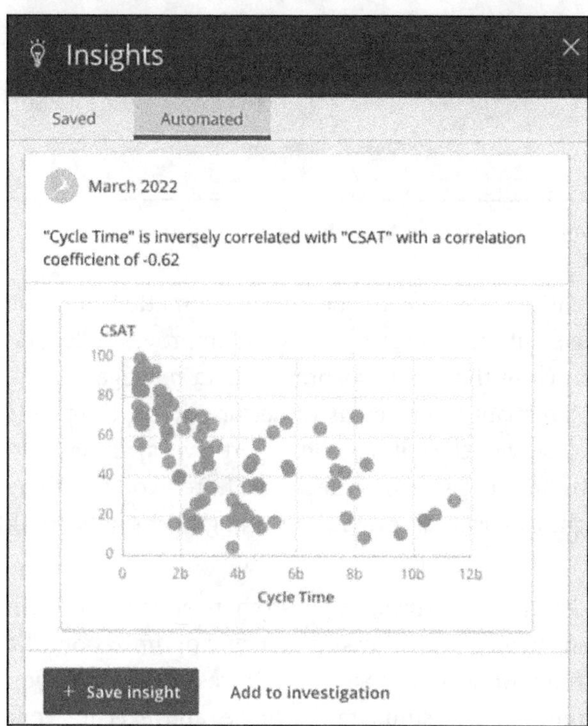

Figure 4.5 SAP Signavio Process Intelligence: Auto Insights

With SAP Signavio Process Intelligence, you also give meaning to every customer experience (CX) in your process analysis. Namely, you're able to bring together process data from your operational systems and experience management solutions. You can use experience-driven process mining capabilities to easily identify inefficiencies and issues that impact your revenue and profit numbers. Examples would be selected process variations that have potentially negative impacts on experience metrics and related dynamics and root causes. Subsequently, you can share your analysis results in the context of your customer journey models in SAP Signavio Journey Modeler, fostering collaboration between your process and experience experts (see Figure 4.6).

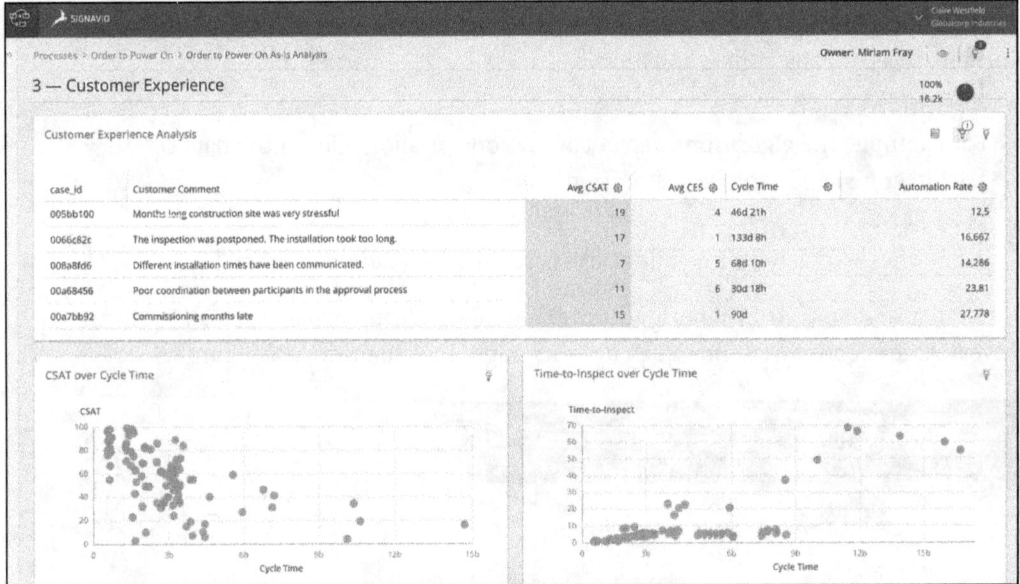

Figure 4.6 SAP Signavio Process Intelligence: Process Analysis

Because people are central to the entire SAP Signavio Process Transformation Suite, the SAP Signavio Process Intelligence solution also strengthens and improves collaboration among your teams. You can close the gap from process data management to insights. At the same time, you can enable your teams to perform in-depth process analysis with a combination of self-service capabilities and expert-level analysis tools. To improve day-to-day operations and make smarter decisions faster, you can share your findings with your colleagues as well as provide meaningful insights to all employees.

As a result, you create a process-driven, data-driven, and customer-centric mindset in your organization with a common language for process design and optimization. The comprehensive transformation framework of SAP Signavio Process Transformation Suite enables you to analyze, mine, model, simulate, optimize, and execute your processes in a single environment. In Figure 4.7, you can see an example where the features of SAP Signavio Process Intelligence are combined with those of other SAP Signavio Process Transformation Suite solutions. This SAP Signavio Journey-to-Process Analytics practice and tool set combines insights and metrics from SAP Signavio Process Intelligence with process models from SAP Signavio Process Manager (see Chapter 5) as well as with journey models from SAP Signavio Journey Modeler (see Chapter 6) and promotes collaborative knowledge sharing via SAP Signavio Process Collaboration Hub (see Chapter 7).

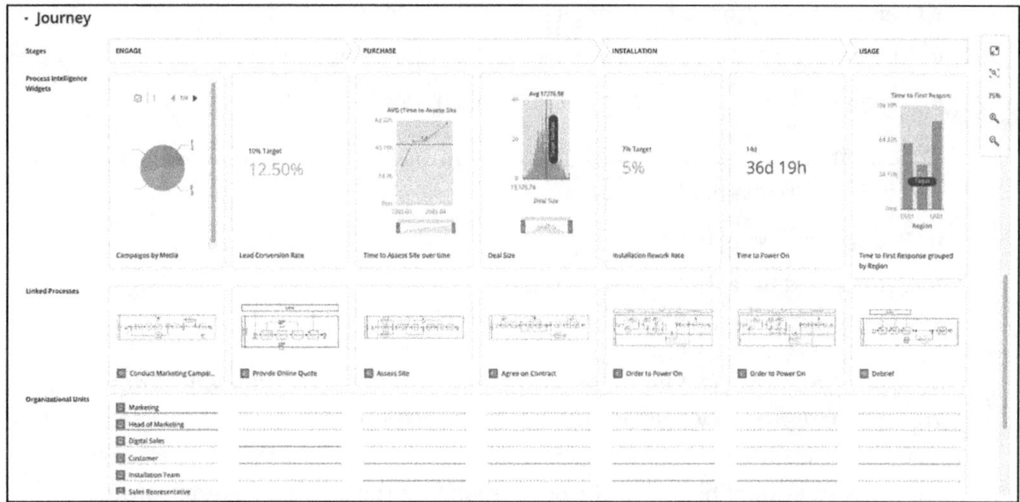

Figure 4.7 SAP Signavio Journey-to-Process Analytics

4.3 Application Example: Transformation of Accounts Payable Accounting

Let's further explore SAP Signavio Process Intelligence by means of a practical application example. In our example, we assume a company where the accounts payable process is expensive and often not compliant. These two problems are closely related for the company. In fact, the processing costs are mainly due to the personnel costs (too many or too expensive accountants) involved in the process execution. On the other hand, the compliance problem is due to payment delays, usually due to too long waiting times in the process execution due to lack of human resources (too few accountants). This resource problem is confirmed by the operational employees of our sample company.

In early 2020, the chief financial officer (CFO) decides to solve both of these problems with a unique automation initiative. The cost of a robot doing programmatic or RPA-based work is much lower than that of an accountant. It's therefore easy to increase the number of robots to reduce waiting times, thus allowing accountants to focus on higher value-added tasks and problems. The go-live date for this project is September 1, 2020. Three months after the end of automation phase 1 (March 2021), the CFO realizes that, despite an improvement in the process, the original goals haven't been met (42.7% automation instead of 50%).

Now, the CFO wants to understand how to execute the new process, thereby achieving his initial goals faster. In doing so, he first looks at the level of automation in accounts payable and comes to the following conclusions:

- Before automation: 25% automated
- Target of phase 1 (as of December 31, 2020): 50% automated
- Target of phase 2 (target: December 31, 2021): 75% automated

The automation process establishes the following key objectives for the process transformation initiative:

- Reduce processing costs
- Reduce lead time through automation
- Ensure process conformity

What is the actual starting point for the transformation initiative just described? A transformation is only effective if the transformation driver knows the actual starting point and the intention of the optimization initiative. In this context, we'll now look at three key indicators regarding the performance of the accounts payable process:

- **Number of invoices over time**
  Are the monthly processing volumes constant or subject to strong fluctuations?
- **Invoice payment term**
  Special clauses agreed between the parties may extend the payment term up to 60 days after the invoice is issued. Thus, this period corresponds to the time between the date of issue of the invoice and the effective date of payment of the invoice (knowing that the date of issue of the invoice is day 1).
- **Processing costs**
  This indicator is the sum of processing costs. In our example, we distinguish between two processing types: automatic billing and manual billing. According to our SAP system, these two processing types have average processing costs of EUR 6 and EUR 18 per settled invoice, respectively.

The data shown in Figure 4.8 illustrates the relevance of the mentioned volume. The monthly volume of invoice processing is approximately 35,801 invoices for the month of March alone. Moreover, these figures are constant each month with an average value of 35,000 (see the bar chart in Figure 4.8). The perceived and observed performance problems are therefore not related to an increase in the number of invoices to be processed but result from structural and organizational problems.

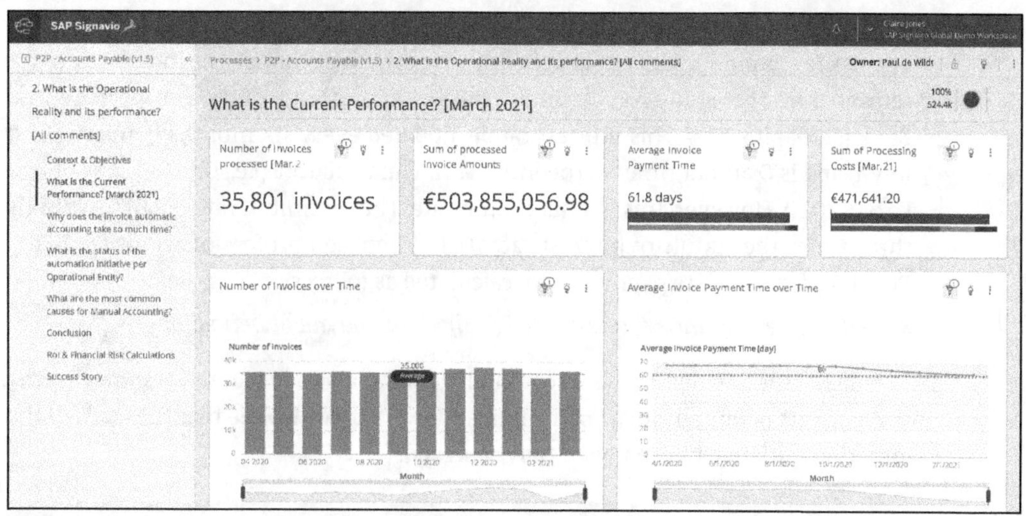

Figure 4.8 Actual Performance of the Accounts Payable Department

Despite the improvement since the automation initiative began, the CFO notes that the average invoice payment time is more than 60 days, and therefore contractual agreements with suppliers aren't being met. This results in the company having to pay overdue payment fines and government fines. It also recognizes that current processing costs are significantly higher than the targets set when the automation initiative was launched (9% higher than target 1 and 44% higher than target 2, according to our SAP system).

In summary, noncompliance with contractual agreements between the company and its suppliers regarding payment deadlines leads to two types of financial risk:

- Suppliers can impose penalties on the company for late payment.
- During an audit by state financial services, the company risks a fine for noncompliance with state laws.

Accordingly, we take a closer look at the following key indicators of our accounts payable process with SAP Signavio Process Intelligence:

- **Rate of overdue payments**
  The rate of overdue payments is calculated as follows:
  Number of overdue payments – Number of payments × 100
- **Average payment term for overdue payments**
  The average payment term for overdue payments is calculated as follows:
  AVG (average) [(Invoice payment date – Invoice issue date) – 60 days] [Where (where) (invoice payment date – invoice issue date) > Maximum contractual payment period (60 days)]

- **Risk for reminder fees**
  The risk for reminder fees can also be calculated. In this context, the interest rates for reminder fees generally correspond to the key interest rate of the European Central Bank (ECB), plus 10%. This interest rate is updated every six months (January 1 and July 1) and is 0 at that time, so the interest rate may be 10% (i.e., 0.00 + 10) as of January 1, 2020. However, this is a maximum rate. The minimum rate is 2.61%, which is three times the statutory interest rate (0.87% between professionals for Q1 2020). Penalties for risky late payments are calculated as follows:
  SUM [(Invoice amount × Interest rate [10%]) × (Late payment period [day] – 365)

The results of the SAP Signavio Process Intelligence analysis in Figure 4.9 show the first positive effects of our automation initiative. However, we also see that the goals of the automation initiative haven't been fully achieved.

[»]

Navigation in SAP Signavio Process Intelligence

With SAP Signavio Process Intelligence, you can use *process investigations*, as in our application example for the accounts payable process, which show various process analyses and statistics. By simply scrolling, you can access the different parts of your investigation.

As you can see from the **Payment Overdue Rate** key figure in Figure 4.9, 18.4% of payments of supplier invoices are still not made in accordance with contractual and legal obligations. This leads to high financial risks, as you can see from the **Risky Payment Overdue Penalties** key figure. This is EUR 245,000, resulting in penalties of EUR 2.9 million per year. We must not forget that if the company is audited by the state financial services, it additionally risks a fine for noncompliance with state laws.

Figure 4.9 Overview of Potential Financial Risks

The system also allows us to check the status of the already started automation initiative. So, in the next step, the CFO looks at the results of the automation initiative and wants to visualize them using SAP Signavio. Here, it's important to check whether the goals of phase 1 have been achieved and to find out where the company stands with regard to the previously set goals.

To optimize cash management, the company waits until the payment deadline (60th day from invoicing) to quickly check supplier invoices. For this reason, during our study, we chose the **Account the invoice** task just before payment as the endpoint. This means that the **Invoice accounting date** is the end time of the **Account the invoice** task.

Let's take a look at the key indicators:

- **Automatic settlement rate**
  The automatic settlement rate corresponds to the proportion of settled supplier invoices without the help of accountants (i.e. without deviations or with resolution by robots) to the total number of settled invoices in the period. This is calculated as follows:

  COUNT, that is, the number (All cases with the task Settle invoice and without the task Take over invoice manually) – COUNT (All cases with the task Settle invoice) × 100

- **Invoice settlement time**
  This indicator measures the time between the first supplier invoice check and the posting of this invoice. Therefore, this indicator is useful to compare automatic and manual processing. The invoice posting time is calculated as follows:

  AVG [Invoice posting date – MIN (Invoice verification date)]

  MIN by itself returns the smallest value within an argument list.

In Figure 4.10, you can see how SAP Signavio Process Intelligence visualizes the first positive effects of the automation initiative regarding our accounts payable process. We can also get to this section again by simply scrolling down. It's clearly visible that the goals of the initiative haven't yet been achieved, as the **Automatic Accounting Rate** is only 42.7% compared to the initial target of 50% and the final target of 75%. The **Average Invoice Accounting Time** is still too high. It's almost 50% of the statutory invoice payment time. This time is even more important for manual billing, but the automatic billing time attracts our attention first. True, this time should have an average value of no more than 1 day. However, as you can see in the **Invoice Accounting Time** per **Accounting Type** tile in the **Automatic** column, it has an average value of more than 3 days.

To find out how this weak result comes about and why invoice automation takes so much time, the company needs to ask itself the following questions:

- Is there a problem with the execution of the robots?
- Is there a problem in the configuration of the robots?
- Is the number of robots used too low?

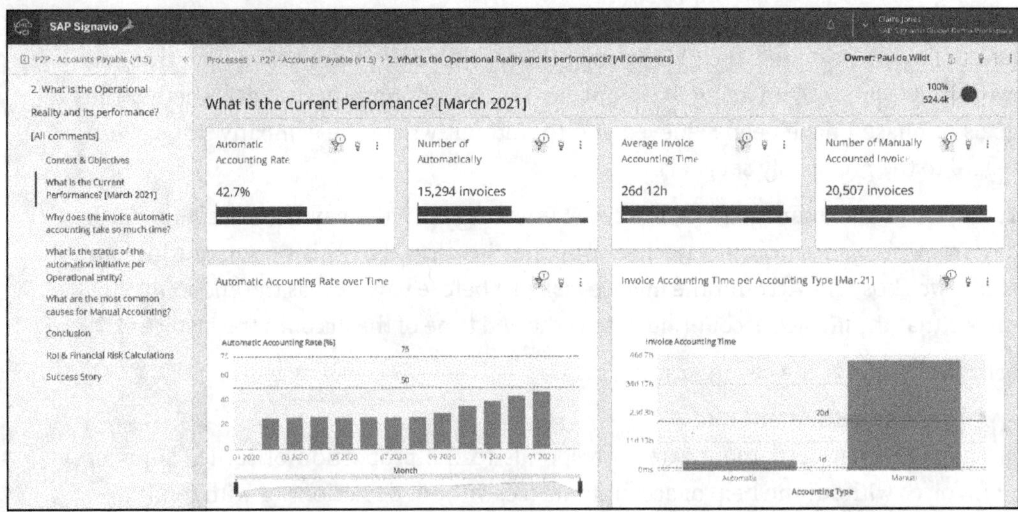

Figure 4.10 Status of the Automation Initiative

If we scroll further down in our SAP Signavio Process Intelligence investigation, we reach the screen shown in Figure 4.11 displaying the possible theoretical paths of automated accounting (phase 1 of automation).

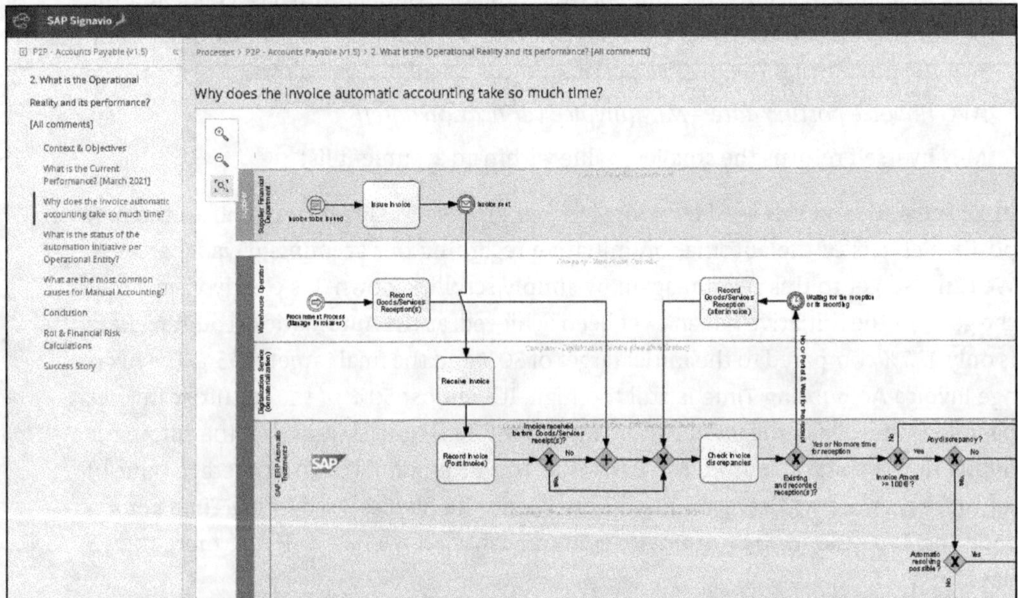

Figure 4.11 Possible Processes of Automatic Accounting

To understand the operational reality of automated accounting, the CFO decides to create a benchmark test between actual processing on more than one day and processing on less than one day. During this process, SAP Signavio Process Intelligence enables visualization by events and then by cycle time (refer to Process Discovery in Table 4.2).

If you've selected the two Process Discoveries in Figure 4.12, you'll notice that the only real difference between these two processing flows is that the **Record Goods Reception (after Inv.)** task isn't executed at the same time as the processing flows.

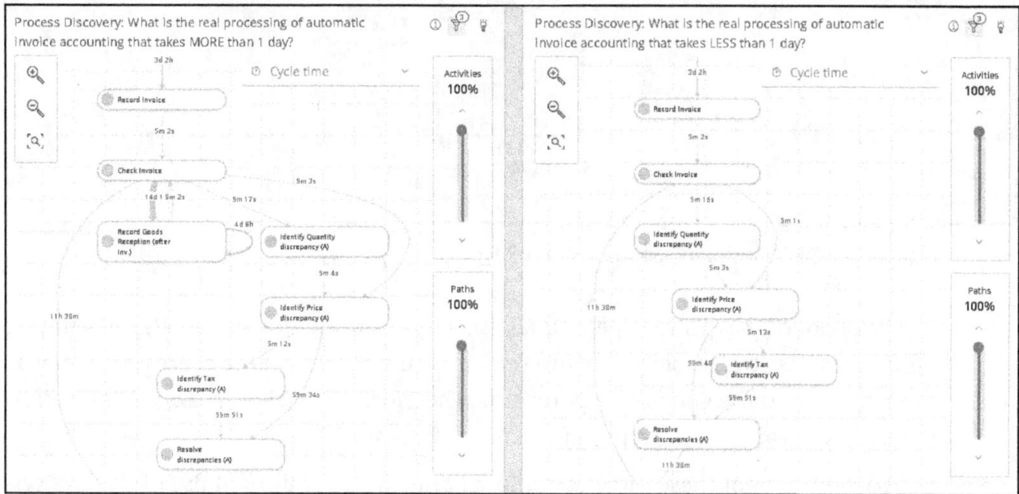

Figure 4.12 Distinction between Two Processing Sequences

In fact, for more than 1 day, the automated accounting system always makes a record of the goods receipt after receiving the supplier invoices, and this record is made on average about 14 days after the first check of the supplier invoice, according to our Process Discovery in the left half of Figure 4.12. This calculation is done using the transactional data from our source system.

To determine whether this problem is due to delivery delays or employee behavior, the CFO needs to take a closer look at these goods receipt records.

If you scroll down, the distribution of these records over time indicates that this problem is due to the poor behavior of three operational units (05, 08, and 12) (see Figure 4.13). These three units use the ERP system differently and have all goods receipts recorded on two specified days at the end of each month.

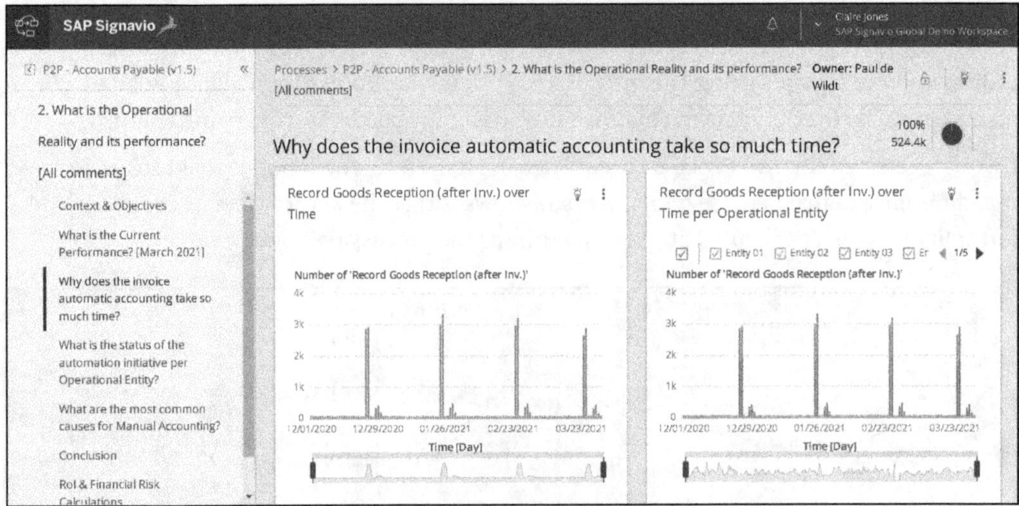

Figure 4.13 Distribution of Goods Receipt Records

The behavior of these three operational units significantly slows down the accounting process because the resolution of discrepancies and the rest of the process can only be executed after these goods receipt records. Therefore, this behavior affects the performance of each accounting type (automatic and manual).

The bad behavior of these three operational units with the deviant use of the ERP system contributes to more than 50% of the overdue payments and leads to high financial risks: EUR 2.3 million in risk penalties per year (EUR 195,000 × 12 = EUR 2.3 million). You can see this information in the SAP Signavio Process Intelligence analysis (see Figure 4.14).

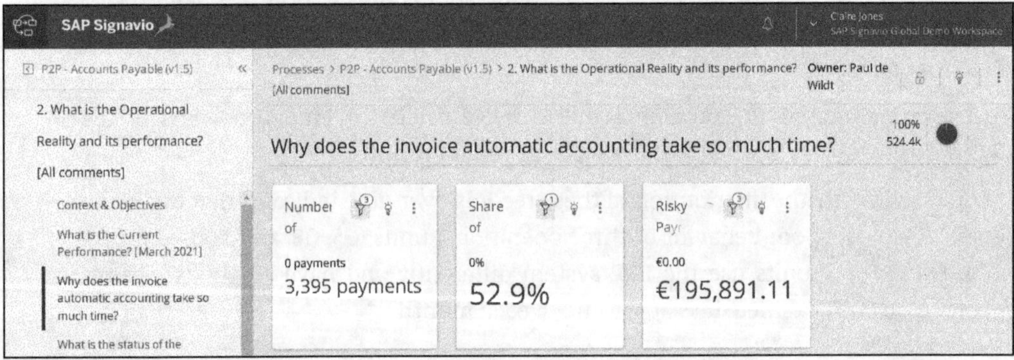

Figure 4.14 Impact on Overdue Payments

Given the preceding observation, the CFO is subsequently interested in the status of the automation initiative per operating unit. As shown in Figure 4.15, operating units 5 and 11 have 50% lower automation rates than the other units. These low automation rates have a direct impact on accounting time. In fact, these two operating units are among those that take 50% to 85% more time to settle an invoice.

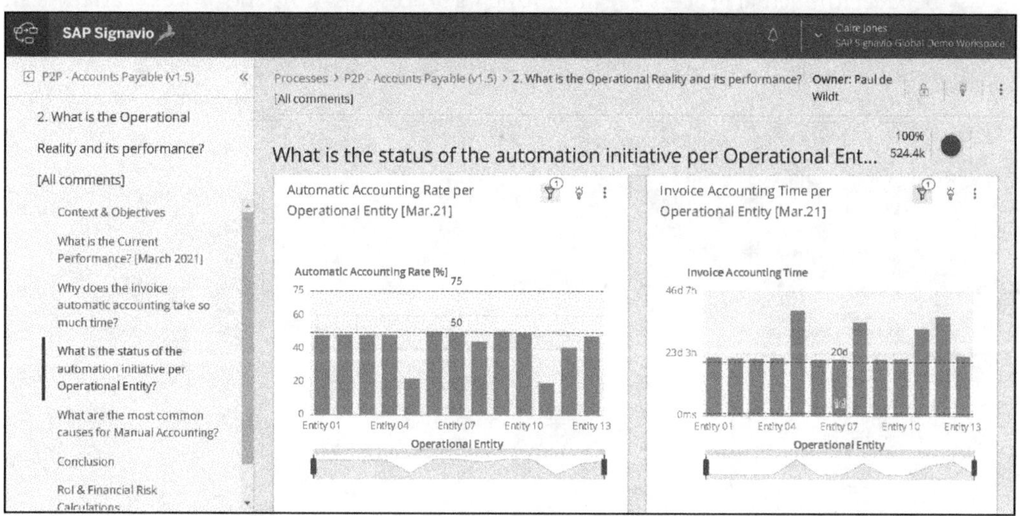

Figure 4.15 Status of Automation Initiative per Operational Unit

To quickly understand the reasons for these low rates for automated billing, the CFO uses a Process Discovery widget that compares the actual processing of operating units 5 and 11 with that of the other units (see Figure 4.16). To do this, he simply clicks **Add widget** and selects **Process Discovery** as the widget type.

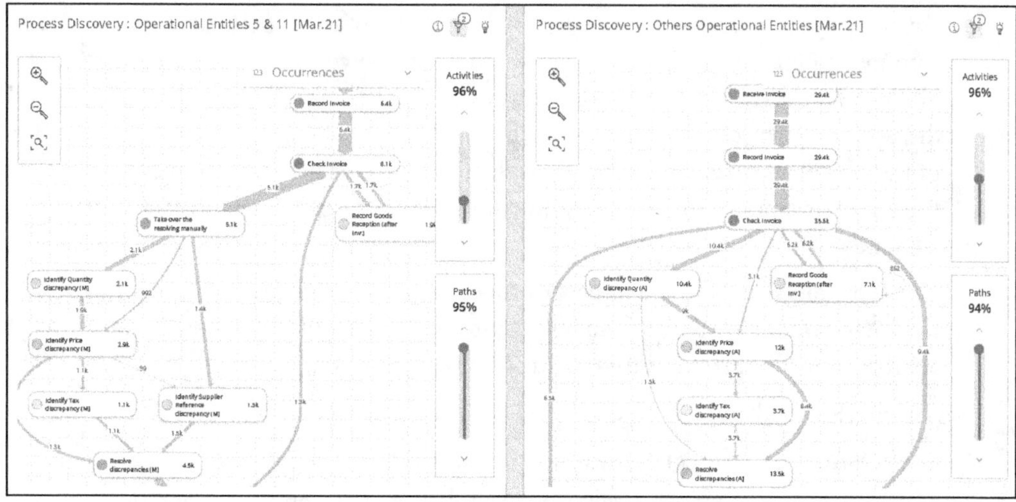

Figure 4.16 Process Discovery of the Operating Units in Comparison

At the same time, the process funnel (refer to Process Funnel in Table 4.2) is checked. The actual process flow is compared with the conceptual process flow: What is the actual process performed by operational units 5 and 11 and their conceptual breakpoints (visualization in Figure 4.17 with the main variant to show the main breakpoints between the actual process execution and its process design)? The process variants can be deepened in the right screen area using the **Variants** slider. The illustrated process funnel can also be added and configured via **Add widgets**.

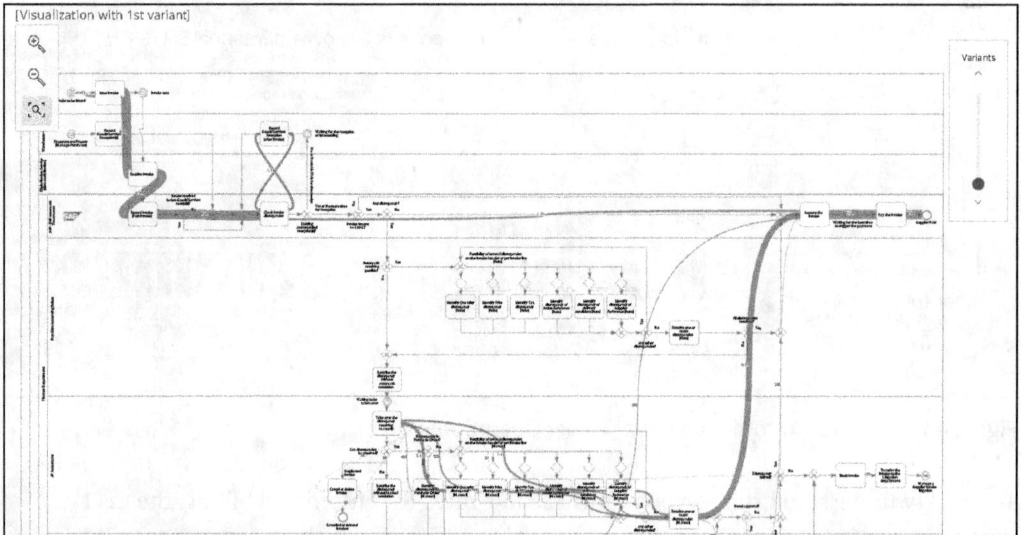

Figure 4.17 Process Funnel Analysis

Once the real processing performed by operational units 5 and 11 through a Process Discovery widget is visualized, it can be seen that the tasks aren't performed by the robots. Automatic billing is performed directly from invoice verification without correction. This is confirmed by visualizing the first variant of the process throughput. We note a disruption in execution compared to the recommendations due to the nonexecution of the robotized tasks. In fact, there have been some delays in the implementation of the automation initiative, so operating units 5 and 11 aren't yet benefiting from these advantages.

The question is, what is the impact of this delay in the company's automation initiative? What would the automated billing rate be without this provisioning delay? What are the additional processing costs due to this provisioning delay?

As a result, we take a closer look at the following indicators:

- The automatic billing rate without the operational units 5 and 11
- The comparison between our current processing costs and the theoretical processing costs if there had been no deployment delay

Of course, the company would be far from the final targets of the automation initiative, but it would still have achieved 95% of the phase 1 targets without the deployment delay (47.6% – 50% × 100 = 95%). This deployment delay, as shown in Figure 4.18, has a direct impact on processing costs. This delay resulted in additional processing costs for the company of EUR 18,651 in March, or more than EUR 55,000 since December (EUR 18,651 × 3 = EUR 55,953). This delay in provisioning should be resolved as soon as possible.

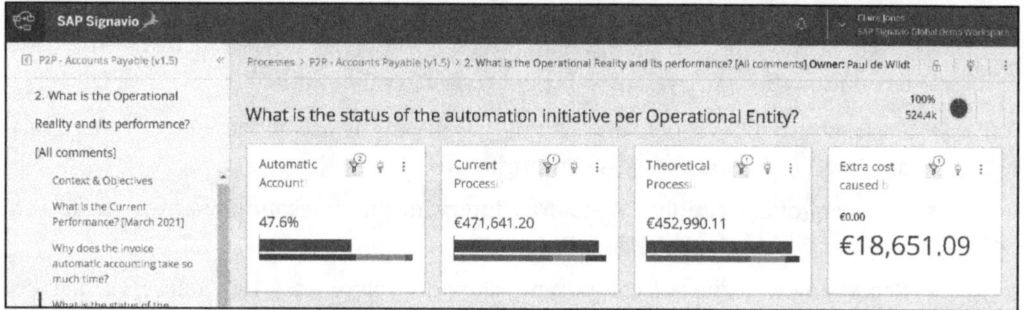

Figure 4.18 Analysis of Processing Costs

To prioritize future automation developments, the CFO would like to find out what the most frequent manual interventions are by the company's accountants. Figure 4.19 shows different variants within the process in which manual interventions occur.

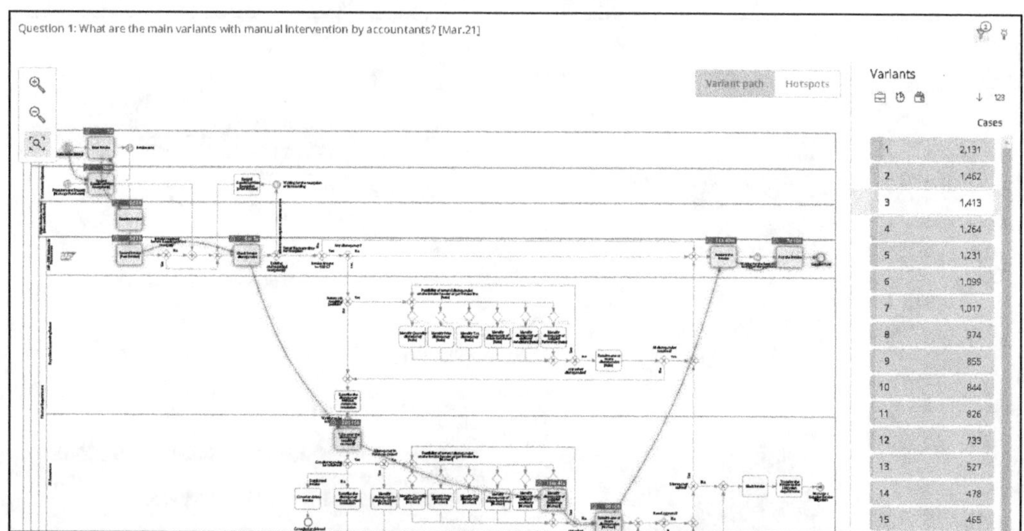

Figure 4.19 Main Variants with Manual Intervention by the Accountants

One in four invoices had changes made manually in the form of supplier reference corrections (10.4% + 7.1% + 6.9% = 24.4%). This type of correction wasn't prioritized in phase 1 of the first automation initiative and was therefore not automated, possibly because its scope was underestimated. Today, the goal is to reduce and automate the number of this type of correction. This point confirms the feelings of the operational staff about the disruptions of this process.

In the next step, the CFO wants to check whether the right types of corrections were prioritized in the first phase of automation and also assess how efficient the automated corrections are in the first place. To answer these questions, he first ranks the correction types per volume and per execution mode (automatic or manual). Then, he focuses on the types of automated correction by using filters to exclude the cases where automation isn't yet available or even currently impossible:

- Processing of operational units 5 and 11: Delay in the use of phase 1 robots (e.g., variant 8 of the previous process conformity)
- Approval of corrections: Mandatory intervention of accountants (e.g., variant 5, 7 and 9 of the previous process conformity)
- Processing with the correction types: Supplier reference, article reference, delivery status, and order correction; correction type not automated in phase 1 (e.g., variant 4 and 6 of the previous process conformity)

The classification of correction types according to the number of corrections confirms that the three types of corrections automated in phase 1 are performed most frequently. These are shown in Figure 4.20 on the right-hand side: **Price**, **Quantity**, and **Tax**.

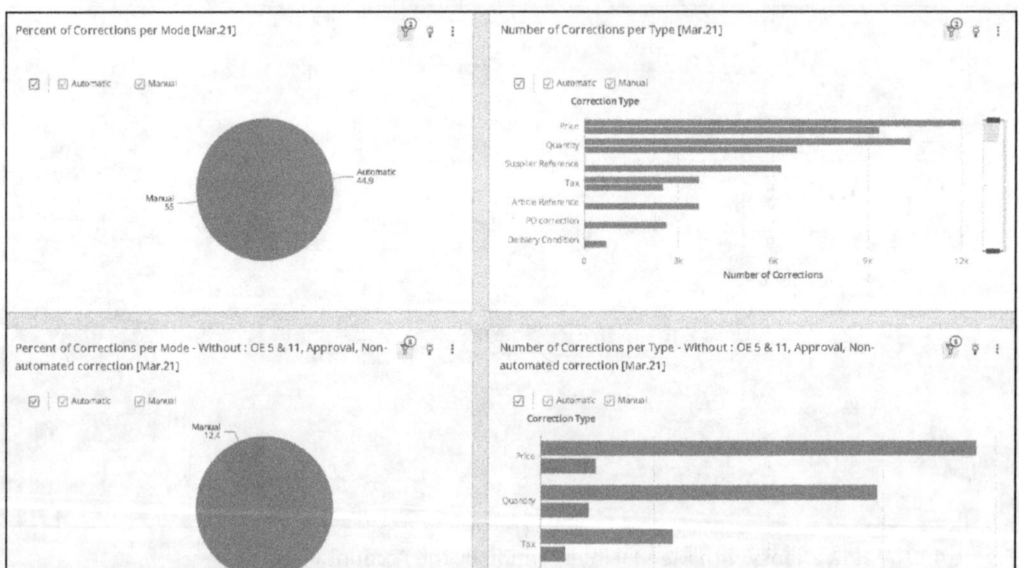

Figure 4.20 Distribution of Correction Types per Execution Mode

If we look at the distribution of correction types per execution mode (automatic or manual) in the pie charts in Figure 4.20, at first glance, we might wonder about the success of this first automation phase. But if we exclude the cases where automation isn't yet available (cases of multiple corrections with correction types not yet automated) or impossible (need approval), the percentage of corrections executed automatically is very high at 87.5% (lower pie chart).

The results confirm that the priority isn't to further optimize the already automated correction types. On the other hand, the priority is to automate the correction types that aren't yet automated (especially the supplier reference) and to optimize the cases of multiple corrections. This leads to the following question: What are the most frequent manual corrections of the supplier references?

If we examine in detail the most common manual corrections of supplier references in our example, we can see that many new supplier references have the same labels and that these supplier references are only distinguished by the supplier numbers (see Figure 4.21). For example, you can see that there are two supplier references with the label **ECONOCOM FRANCE**, one with the number **67301364824** and one with the number **47414967984**, making it difficult to identify and select the correct supplier reference. In fact, when buyers enter a supplier order in the information system, they have great difficulty identifying the correct supplier reference and choose by default the first reference proposed for this group of suppliers. This behavior has a direct impact on accounts payable accounting. In fact, a correction is required to match the vendor information of the purchase order and the vendor invoice. Automating this correction isn't the only solution, nor is the optimal one. An update of the supplier reference labels in the supplier directory should be planned to limit this type of malfunction. Furthermore, after this update of the supplier directory, a complementary training for buyers is recommended.

Previous Supplier Group	Previous Supplier Reference	New Supplier Group	New Supplier Reference	Number of Corrections	Percentage of this correction [%]
ECONOCOM	(65326966777) ECONOCOM	ECONOCOM	(67301364824) ECONOCOM FRANCE	236	3.7
			(47414967984) ECONOCOM FRANCE	183	2.9
EIFFAGE	(21388773772) EIFFAGE	EIFFAGE	(34391905486) EIFFAGE CONSTRUCTION	183	2.9
			(14402096267) EIFFAGE TRAVAUX PUBLICS	162	2.5
			(08388727240) EIFFAGE ENERGIE	152	2.4
			(20500704820) EIFFAGE CONSTRUCTIONS	118	1.8
			(01398762211) EIFFAGE TRAVAUX PUBLICS	108	1.7
			(48329009559) EIFFAGE ENERGIE	95	1.5
			(33340023225) EIFFAGE ENERGIE	89	1.4
	(00333916386) EIFFAGE METAL	EIFFAGE	(00333916385) EIFFAGE CONSTRUCTION METALLIQUE	163	2.5
INEO INDUSTRIES	(63409881083) INEO ACTIVITE NUCLEAIRE ET CENTRALE	INEO INDUSTRIES	(16409899077) INEO ACT. NUCLEAIRE ET CENTRALES	152	2.4
			(89916409078) INEO ACTIVITE NUCLEAIRE CENTRALES	85	1.3
			(77164098990) INEO ACTIVITE NUCLEAIRE ET CENTRALE	70	1.1
CEGELEC	(01537934028) CEGELEC SAS	CEGELEC	(79582750188) CEGELEC SDEM	103	1.6
			(68537934309) CEGELEC CEM	100	1.5
			(80537934507) CEGELEC NDT PSC	90	1.4
			(44537933913) CEGELEC NDT-PES	85	1.3
			(37537908238) CEGELEC SAS	84	1.3

Figure 4.21 Listing of the Most Frequent Manual Corrections of Supplier References

The Variant Explorer widget in Figure 4.22 also helps to visualize various variants in the course of supplier references. With the Variant Explorer, you can display one or more paths of process variants. Similar to the Process Funnel widget, the Variant Explorer can also be added and configured via **Add widgets**.

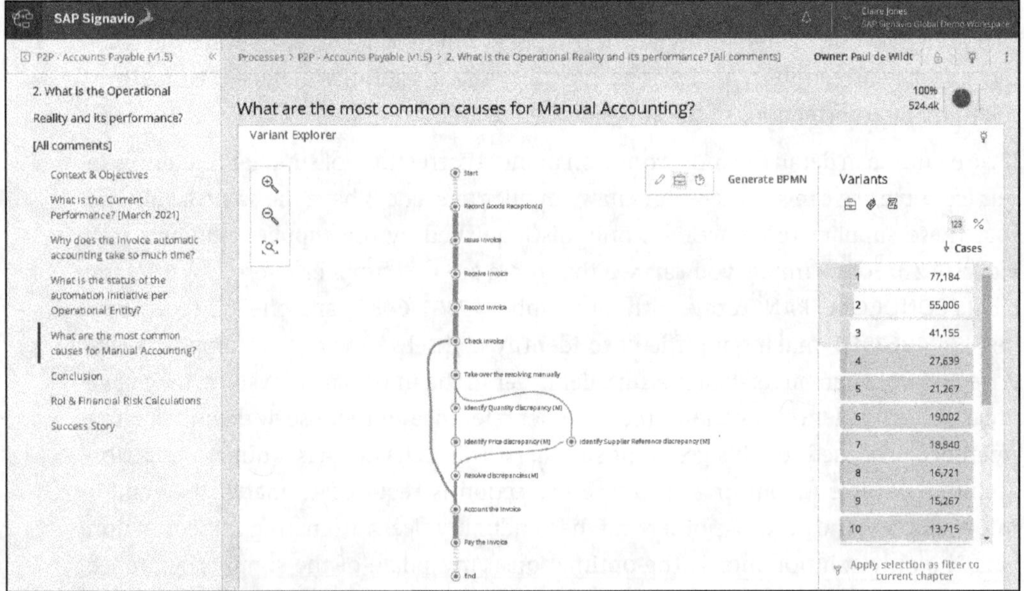

Figure 4.22 Variant Explorer

Finally, following are the main objectives of the CFO in our application example as a reminder:

- Reduce processing costs.
- Reduce lead time through automation.
- Ensure process conformity.

To achieve these goals, an initiative was launched to automate the accounts payable process. Initial results have been somewhat lower than expected. Initial performance measurements confirmed high global processing costs and a compliance issue related to long payment delays. The payment delays are the result of a bottleneck in the transfer of invoices due to a lack of human resources (accountants). Therefore, the automation initiative is the right solution to achieve the three main objectives. To achieve the automation goals under good conditions, the following points must be taken into account:

- **Project management**
  Any delay in the implementation project will result in additional processing costs that will undermine the year 1 performance commitments (ROI) of this automation initiative.

- **Acceleration of the automation initiative**
  Feedback from operational staff on process execution disruptions is validated by this operational reality analysis. Anticipating phase 2 of the automation initiative provides the best ROI for future transformation efforts.
- **Organization**
  All operational contributors to the process should be reminded of efficient processing procedures and best practices. For example, goods-in registration prevents robots from blocking tasks and thus favors timely payment.
- **Master data management**
  Master data such as supplier references, price, delivery status, and so on must be consolidated. A robot can't guess what the right data is when it has multiple options in context.

Of course, the application example explained is only an excerpt from many different use cases that you can analyze in depth with SAP Signavio Process Intelligence.

4.4 Summary

This chapter serves as an introduction to the SAP Signavio Process Intelligence solution. You've learned what SAP Signavio Process Intelligence is and what the solution can do. In addition, the individual functions and benefits were briefly presented as well as highlighted from a practical point of view using an application example.

In summary, with SAP Signavio Process Intelligence, you can fully visualize your as-is process from detailed process analysis for business transformations (including SAP S/4HANA migrations), to exploring and interacting with business data, to using process mining for fact-based changes to uncover the potential of your business processes and data. SAP Signavio Process Intelligence is a true game changer for business process management. The solution equips you with next-generation end-to-end capabilities to make your business processes more efficient and effective: from precise improvement of the current state to root cause analysis and continuous monitoring. Based on transparent process data, SAP Process Intelligence enables improved decision-making about process flow. Consequently, compliance auditing can be used to ensure that standard procedures are executed as originally intended and enables quick identification of exceptions. Finally, thanks to the derivation of complete end-to-end perspectives and performance overviews, you can better understand what is happening in your organization in real time to ensure sustainable operational health.

Chapter 5
SAP Signavio Process Manager

SAP Signavio Process Manager provides you with a comprehensive understanding of your entire business processes. The solution allows you to document, model, compare, simulate, and outline possible inter-relationships.

In Part I of this book, you already learned that you could rely on *SAP Signavio Process Manager* to adapt business processes or completely redesign them, depending on your needs.

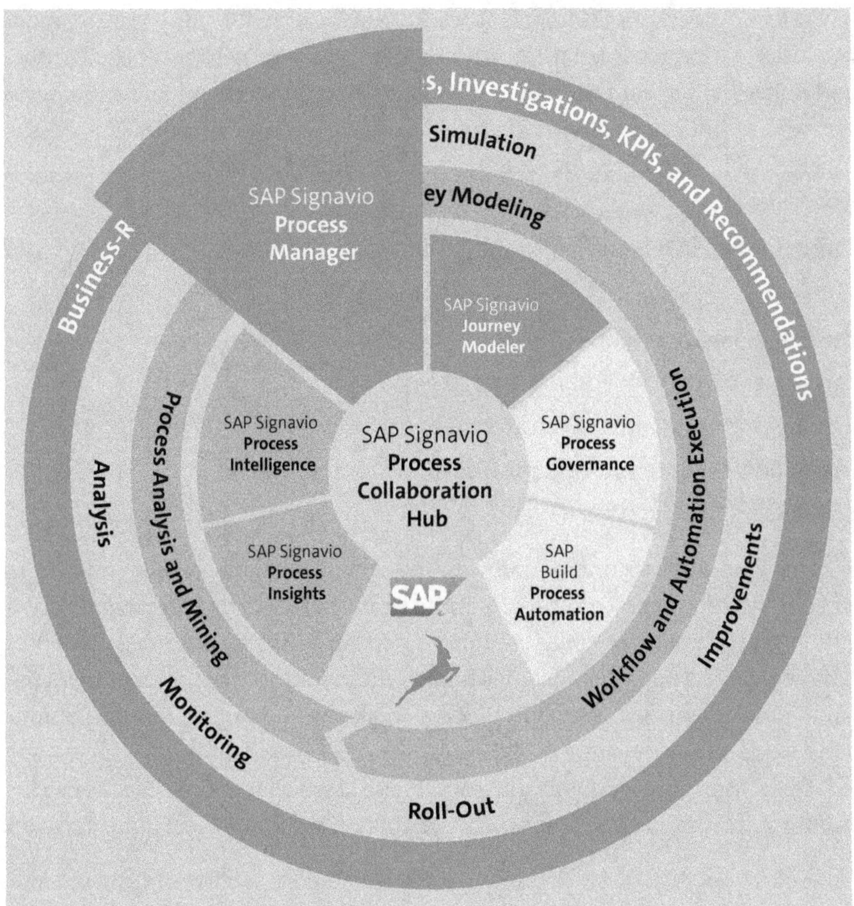

Figure 5.1 SAP Signavio Process Manager in Focus

This tool helps you react quickly to unpredictable business and regulatory changes, collaboratively scale your company's process repository, and simulate business processes to define alternative business scenarios. SAP Signavio Process Manager is a web-based platform and provides the ideal environment for collaborative work on processes thanks to its modern and user-friendly interface.

In this chapter, we'll give you a deeper dive into SAP Signavio Process Manager and show you how to harness the power of your business processes to succeed in the new normal and beyond. In Figure 5.1, you can see how the solution is located in SAP Signavio Process Transformation Suite.

In Section 5.1, we'll first look at the various functions of SAP Signavio Process Manager. Section 5.2 shows the different benefits of the solution. Subsequently, we illustrate the use of SAP Signavio Process Manager in practice by means of an example (Section 5.3).

5.1 Functions

SAP Signavio Process Manager is an intuitive, fully web-based business process management solution for professional process modeling. It allows you to get started immediately and engage your team members in collaborative process modeling, analysis, simulation, and optimization. With this tool, you can put your organization on the path to operational excellence and quantify and forecast the return on investment (ROI) of your changes.

The core functions of SAP Signavio Process Manager are as follows:

- Model processes together.
- Automatically check modeling rules.
- Align processes better with corporate strategy.
- Comprehensively implement a transformation process.
- Record and document the current process infrastructure.
- Make processes more sustainable.

This SAP Signavio tool is a technologically advanced process management solution to keep your complex day-to-day business running smoothly. With SAP Signavio Process Manager, you get the ability to create and monitor your business processes. Any process landscape is far-reaching and includes finance, human resources, marketing, sales, and possibly production. SAP Signavio Process Manager allows you to bundle your processes, including your customer journeys, in one central location. This makes processes across your entire organization accessible from that central location. This supports enterprise-wide transparency and enables efficient process and decision modeling across your organization. SAP Signavio Process Manager helps you connect your business processes in a way that improves the customer journey and enhances the customer experience (CX). You can see how all these positive changes are brought about by the tool and with which functions this is done later in Table 5.3.

SAP Signavio Process Manager is ideally suited for beginners as well as for experts in process management. After a short introduction to the system by trained staff, employees can immediately start process modeling. In addition, your employees are able to share, comment, and collaborate on their process knowledge at any time using the embedded sharing feature. In addition to sharing, the export, save, and print diagram functions are additional collaboration options for modelers. Diagrams can be transferred between different workspaces in SAP Signavio and between SAP Signavio and other modeling software (via Business Process Model and Notation [BPMN] 2.0 compliant XML), saved locally, and sent via email (as PDF, SVG, or PNG) or printed. Advanced users can further collaborate by embedding diagrams in websites or Microsoft SharePoint-based systems. This new form of collaboration puts an end to misunderstood or unclear processes, and enterprise-wide process transparency becomes the norm throughout your organization.

SAP Signavio Process Manager enables fast and efficient processes by supporting continuous process optimization through *process simulation*, including multiple scenarios for operational comparisons. You can use this feature to run your process step-by-step under changed factors after modeling, and thus see exactly how process improvements can be achieved in your organization. The solution supports multiple modeling languages such as the previously mentioned *Business Process Model and Notation* (BPMN) 2.0, the *event-driven process chain* (EPC), *Decision Model and Notation* (DMN) 1.1, and so on, as well as standardized process models through the use of modeling conventions. Table 5.1 shows the most common modeling notations supported by SAP Signavio Process Manager.

Modeling Notation	Description
Business Process Model and Notation (BPMN) 2.0	BPMN is an industry standard for modeling business processes. The industry standard is published by the Object Management Group (OMG) and supported by various vendors and consultants. In SAP Signavio Process Manager, BPMN 2.0 is supported with all diagram types, modeling elements, and attributes. This version of the standard enables the representation of activities, control flows, data flows, organizational dependencies, and system dependencies.

Table 5.1 Supported Modeling Notations (Source: SAP)

Modeling Notation	Description
Customer journey maps	Customer journey maps depict the customer's perspective on your process landscape. Customer journey maps are intuitive diagrams that focus on the CX rather than internal processes. They help you understand how customers perceive your products and services in the context of everyday life and how key decisions that lead to a purchase or churn, for example, are motivated.
Process maps	You can use process maps to create overviews of your process landscape. Each modeling element in a process map represents a process or group of processes for a specific business unit. You can link elements chronologically and show hierarchical relationships between processes and process groups.
Navigation maps	Use navigation maps to create graphical views of your processes. Similar to process maps, you create higher-level perspectives for your process landscape. Compared to process maps or other modeling languages, navigation maps are more flexible and not bound to a strict standard. You can even upload your own images as modeling elements.
ArchiMate	With SAP Signavio Process Manager, you can model enterprise architecture diagrams in ArchiMate notation. This is an open modeling language used to describe, analyze, and visualize enterprise architectures within business units and between business units.
Event-driven process chain (EPC)	EPCs are used to model business processes and to capture and visualize processes. However, unlike BPMN, they aren't executable. EPCs usually focus on the lower levels of the process hierarchy (process flows). While EPCs were popular in some European countries in the late 1990s and early 2000s, they are currently being replaced by BPMN because BPMN is more attractive to both business users and technical experts.
Decision Model and Notation (DMN)	DMN can be used to describe and model decisions that are frequently made in an organization. DMN isn't suitable for modeling strategic decisions. It includes the decision diagram and the decision table. The decision diagram is the graphical representation of the decision rule. For each input, the corresponding decision is read from the decision table. DMN and BPMN diagrams can be linked together so that processes can be viewed separately from decisions. This makes the process clear and the decision comprehensible.

Table 5.1 Supported Modeling Notations (Source: SAP) (Cont.)

Modeling Notation	Description
Case Management Model and Notation (CMMN)	CMMN is a notation developed to enable more flexibility in the business process landscape. To complement your existing process landscape, you can seamlessly integrate CMMN with BPMN and DMN diagrams. For example, in SAP Signavio Process Manager, you can model the standard framework of the corresponding process in a BPMN diagram and then link a BPMN subprocess to a CMMN diagram that defines a flexible flow. In addition, you can change a task in a BPMN diagram to a subprocess linked to a CMMN model to describe a set of flexible actions in more detail.
Organizational charts	Organizational charts represent the internal structure of a company. By representing a company in this way, the internal hierarchy and the relationships between the different roles within an organization are shown.
Choreography diagram	Choreography diagrams depict the complex collaboration between process participants. They illustrate how information is exchanged and how the participants coordinate their activities. Choreography diagrams are part of the BPMN standard but are rarely used. In the context of this work, it's also possible to visualize choreography diagrams using BPMN process diagrams.
Conversation charts	Conversation diagrams illustrate the interactions between process participants. They make it possible to capture the relationships between the participants at a glance. Conversation diagrams are a component of the BPMN standard but are rarely used. Therefore, BPMN process diagrams are also suitable for modeling interactions between process participants.
Unified modeling language (UML) diagrams	As part of the UML standard, use case diagrams can be used to represent actions that can be performed jointly by systems and users. Similarly, UML class diagrams can be created to visualize the properties, methods, and relationships of system classes. They are often used to describe object-oriented programming code and are also part of the UML standard.

Table 5.1 Supported Modeling Notations (Source: SAP) (Cont.)

BPMN 2.0 is the current industry standard for modeling business processes and is especially maintained in SAP Signavio Process Manager. Table 5.2 lists the core elements of this notation.

Destination	BPMN Element	Description
Model activities	Task	Tasks are activities that represent individual task steps of the business process.
Collapsed subprocesses	Represent several tasks combined	Collapsed subprocesses represent several tasks combined.
Consider events	Start event	Start events begin new process instances.
Intermediate event	Represent states or milestones in the process	Intermediate events represent states/ milestones in the process.
Final event	Represent the final states in the process	End events represent the end states in the process.
Connect objects	Association	Associations connect data objects, IT systems, or text annotations with other elements.
Frequency flow	Sequence flows connect events, gateways, tasks, and subtasks	Sequence flows connect events, gateways, tasks, and subtasks.
Message flow	Represent communication and interaction between pools	Message flows represent communication and interaction between pools.
Illustrate artifacts	Data object	Data objects represent documents or files that are used to perform activities.
IT system	Represent specific systems that are used to perform activities	IT systems represent specific systems used to carry out activities.
Show gateways	Exclusive gateway (XOR)	XORs are used when several conditions are mutually exclusive and only one selection is possible (e.g., yes or no).
Inclusive gateway (OR)	Used when one or more conditions are possible	ORs are used when one or more conditions are possible. The merge waits for all incoming paths.
Parallel gateway (AND)	Activate all outgoing paths simultaneously	ANDs activate all outgoing paths at the same time. The merge waits for all paths.

Table 5.2 Overview of BPMN 2.0

Destination	BPMN Element	Description
Event-based gateway	Used when the subsequent process flow depends on an event	Event-based gateways are used when the subsequent process flow depends on an event. The process continues with the event that occurs first.
Represent roles	Swimlane	Areas of responsibility can be represented with pools and lanes. Each pool represents an organization, lanes represent responsible workgroups or employees, and the end result is a swimlane representation of the roles involved.

Table 5.2 Overview of BPMN 2.0 (Cont.)

Thanks to cloud-based process management, there is no need for maintenance, administration, or storage space on your part. Thus, SAP Signavio Process Manager enables rapid process optimization and change implementation, continuous monitoring and action on individual cases, and a holistic examination of the overall process state of your organization. Finally, this SAP Signavio solution promotes process and decision modeling for all employees in the organization. You don't need any prior knowledge to use the solution.

Table 5.3 provides an overview of the numerous application features of SAP Signavio Process Manager.

Feature	Description
Process modeling	Model and document your business processes: ■ Modeling notations: BPMN 2.0, customer journey maps, process maps, navigation maps, ArchiMate, EPK, DMN, CMMN, organization charts, choreography diagrams, conversation diagrams, and UML diagrams ■ Graphical modeling via drag and drop ■ Suggestion system for modeling ■ Diagram import (SAP Signavio Archive [SGX], XML Process Definition Language [XPDL], BPMN 2.0 XML, ARIS Markup Language [AML]) ■ Diagram export (SGX, BPMN 2.0 XML, Portable Network Graphic [PNG], Scalable Vector Graphic [SVG], PDF) ■ Syntax check for BPMN 2.0 and EPK

Table 5.3 Functions of SAP Signavio Process Manager (Source: SAP)

Feature	Description
Central diagram repository	Manage process models in one place: ■ Folder structures ■ Process hierarchies (subprocess structures) ■ Private modeling area (**My Documents**) ■ Recycle bin function ■ Central object repository with autocompletion ■ Version management ■ Graphical version comparison or model comparison ■ Full-text search over models (including element attributes and comments) ■ Possibility to link external documentation
Central object management repository (dictionary)	Reuse individual modeling elements: ■ Management of specific modeling elements ■ Consistent terminology and elements in your organization-specific modeling environment ■ Default dictionary categories
Custom dictionary categories	Create dictionary categories that meet your specific business needs: ■ Defining attributes for dictionary entries ■ Adding your own categories to existing dictionary categories
QuickModel	Create BPMN 2.0 process models quickly and easily with the QuickModel feature: ■ Tabular capture of BPMN 2.0 process models ■ Intuitive operation ■ Revision of existing diagrams
Modeling conventions	Perform modeling checks and adjust process models: ■ Configuration of the modeling language scope (BPMN subsets) ■ Definition of custom/additional attributes (meta-model customization) ■ Process structure (including syntax and semantics check) ■ Layout conventions

Table 5.3 Functions of SAP Signavio Process Manager (Source: SAP) (Cont.)

Feature	Description
Customer journey maps	Map the customer perspective on your organization's products and services: ▪ Define customer journey maps that map the customer's perspective of your organization's products and services ▪ Integrate customer journey maps into the business process landscape ▪ Perform collaborative modeling of customer journey maps
Multilingualism	Model processes in multiple languages: ▪ Deposit of multiple languages in one model ▪ Multilingual process portal in SAP Signavio Process Collaboration Hub
Document management	Upload files and manage them: ▪ Upload screenshots, work instructions, etc. directly to the diagram repository ▪ Uniform rights management analogous to the diagrams ▪ Direct display of selected file types (e.g., PNGs) in SAP Signavio Process Collaboration Hub ▪ Linking documents in diagrams ▪ Shared memory for all modeling users
Access rights management and security settings	Manage access rights for different user groups and personalize security settings: ▪ Manage users and user groups ▪ Fine-grained rights definition (folder/model level) ▪ Restriction of functions for user groups possible ▪ Rights definition for the dictionary ▪ Password policy definition ▪ Restriction to IP ranges, additional authentication via certificate is configurable
Configurable modeling conventions	Review process models and adjust modeling conventions: ▪ Definition of own rules and regulations ▪ Selection, prioritization, and parameterization of rules ▪ Naming conventions (German, English) ▪ Automatic verification when saving diagrams ▪ Reporting for batch validation of charts

Table 5.3 Functions of SAP Signavio Process Manager (Source: SAP) (Cont.)

Feature	Description
Attribute visualization	Visualize attributes with symbols and colors: ■ Visualization of attribute values with different icons and colors ■ Definition of own rules
Configurable process manual	Customize process documentation with templates: ■ Definition of own process manual templates via drag and drop ■ Flexible design of the process manual (content and layout)
Process cost analysis	Analyze process models for cost and resource audits and improvements: ■ Deposit of time and cost attributes on models ■ Plausibility check for attribute definitions ■ Evaluation as cost accounting as well as resource requirements calculation in Microsoft Excel ■ Calculation also taking into account for subprocesses
Process reports	Create, save, and share reports in a variety of formats: ■ Process manual function (PDF, Microsoft Word) ■ Microsoft Excel reports (responsibility assignment according to the responsible, accountable, consulted, and informed [RACI] matrix, responsibility transfers, document usage, IT system assignment, process profile with element details, and process model metrics)
BPMN simulation	Use step-by-step process simulation: ■ Animated visualization of process flows ■ Step-by-step traceability of the process flow In addition, use the full process simulation: ■ Step-through simulation with cost and time attributes ■ Automatic simulation (one operation vs. multiple operations) ■ Definition of distribution curves/arrival rates ■ Resource modeling, including availability ■ Replay mode with interactive visualization ■ Drilldown function for simulation result details ■ Comparison of different simulation scenarios

Table 5.3 Functions of SAP Signavio Process Manager (Source: SAP) (Cont.)

Feature	Description
Risk management	Manage risks and controls in the process landscape: ■ Visualization of controlled and uncontrolled risks in process diagrams ■ Preparation of risk management reports ■ Definition of individual data structures for risks and controls ■ Central dictionary for managing risks and controls ■ Planning of regular risk management reviews with SAP Signavio Process Governance
Approval workflows	Define approval workflows for process models and manage them together with SAP Signavio Process Governance: ■ Freely definable workflows for the release of process diagrams ■ Involvement of various stakeholders ■ Automatic publication of diagrams possible for releases in the workflow ■ Monitoring of current and executed releases
Intelligent folders (smart folders)	Smart folders allow you to save search results from queries: ■ Definition of dynamic folders via search queries ■ Reevaluation of the search query each time the smart folder is opened ■ Use of report function directly in the smart folder
Decision modeling in DMN	Model business decisions in DMN: ■ Modeling decisions with DMN ■ Graphical modeling via drag and drop ■ Suggestion system for modeling ■ Definition of the structure of decisions ■ Clear presentation of the decision logic in tabular form ■ Definition of input and output values for decisions ■ Automatic check for consistency Generate test cases: ■ Generation of Drools test cases as an extension to the export of rules ■ Test cases based on input value ranges and decision tables ■ Enables automatic testing of modeled and exported decision diagrams

Table 5.3 Functions of SAP Signavio Process Manager (Source: SAP) (Cont.)

Feature	Description
Decision modeling in DMN (Cont.)	Verify decisions: ▪ Automatic verification for completeness and uniqueness ▪ Review of individual decision tables ▪ Consistency checks in complex structures between linked decision nodes and diagrams Simulate decisions: ▪ Interactive simulation of decisions based on appropriate inputs ▪ Calculation of result sets for partially specified input set ▪ Graphical display of applicable rules within decision tables ▪ Simulation of subdecisions Test decisions: ▪ Definition of expected outputs for specific inputs ▪ Executes multiple test cases for one diagram Export decision logic to third-party systems, or run decision models directly in the tool: ▪ Export of decisions in the Drools Rule Language (DRL) ▪ Capability to execute exported decisions directly in Drools
Enterprise architecture	Model enterprise architecture diagrams: ▪ Modeling enterprise architecture diagrams with ArchiMate 3.0 ▪ Direct link between processes and enterprise architecture diagrams ▪ Activation of additional languages: UML class diagrams, UML use case diagrams
HTTP REST API	Access APIs of SAP Signavio Business Transformation Suite from applications outside SAP Signavio: ▪ Access to the entire SAP Signavio repository via HTTP GET/PUT/POST/DELETE ▪ Access to models via XML, JavaScript Object Notation (JSON), SVG, PNG ▪ Easy-to-understand JSON format ▪ Full integration into rights management ▪ Realization of numerous integration scenarios

Table 5.3 Functions of SAP Signavio Process Manager (Source: SAP) (Cont.)

Feature	Description
Business process model connector for SAP Signavio solutions	Enable the connection to SAP Solution Manager: ■ Import from SAP Solution Manager ■ Export to SAP Solution Manager ■ Synchronize changes
Mashup API	Visualize your own data in process diagrams as a mashup: ■ Simple JavaScript API for numerous typical visualization scenarios ■ Access to model data via JSON ■ Compatible with all major web browsers ■ Realization of numerous web-based integration scenarios
ITIL/ISO 9000 library	Enables standardized quality management alignment: ■ Collection of best practice processes ■ Based on international standards, a large number of BPMN 2.0 diagrams available via the Information Technology Infrastructure Library (ITIL) ■ Customizable for an organization's specific IT service management (ITSM) requirements

Table 5.3 Functions of SAP Signavio Process Manager (Source: SAP) (Cont.)

5.2 Benefits

SAP Signavio Process Manager offers numerous benefits and opportunities for your company. The solution enables you to gain detailed insights into all of your business areas, increase productivity and thereby reduce costs, better connect processes to the overall strategy, implement a strategic business plan on a large scale, and quickly respond to regulatory changes. All of these useful capabilities make your business even more competitive.

The core potentials of SAP Signavio Process Manager are as follows:

- Accelerated process optimization
- Rapid implementation of changes
- Process modeling for all without prior knowledge
- Continuous process monitoring
- Process simulation for more efficiency
- Understanding the effects of process changes
- Optimal coordination between business units and IT

The solution provides an effective way to improve your operational activities through simplified model creation. Business processes can be easily created, and new activities can be added, edited, and then reviewed before the process model is published. You maintain full control and improve process output with holistic editing capabilities, no matter how many thousands of processes you have.

Shared input and feedback raise the standard of your organization's operations. Your employees can share their expertise and get the support they need in a timely manner. What does that look like exactly? People involved can add comments to task messages and respond in real time to colleagues from other departments and locations. They can also ask for feedback and have conversations that lead to faster and better results. In this way, you can eliminate silos and continually improve performance.

To prepare for alternative business scenarios, you can simulate your processes with SAP Signavio Process Manager. This creates a what-if environment to quantify and predict the potential ROI of changes. You can get on with the tasks that matter most to you by putting the process simulation in motion by setting and monitoring case scenarios. In this simple way, you can uncover hidden opportunities and find the best alternatives for complex projects.

SAP Signavio Process Manager supports you in implementing your business transformation. The result is that everyone in your organization is up to date and informed, not only about the "what" of their work but also about the "why" and "how." Process models created with SAP Signavio Process Manager are an effective way to visually represent your business processes and clarify how your organization works together. However, only when these models are connected to other factors do they unleash their true potential. As part of the SAP Signavio solution portfolio, SAP Signavio Process Manager allows you to view your process models from different angles in the context of customer journey, risk and decision management, resource planning, IT implementation, and many more. As a result, SAP Signavio Process Manager comprises the foundation to bring your processes to life. The tool also helps you broaden your view of your process landscape. It encourages you to view your process management no longer as a project, but as a dynamic initiative. Instead of simple documentation, SAP Signavio Process Manager provides you with valuable insights that enable you to plan and profitably manage your day-to-day business even under changing conditions. In this context, the tool also promotes easier management of day-to-day operations, regardless of department or location.

With SAP Signavio Process Intelligence, you can additionally build an operational *process mining cockpit.* with SAP Signavio Process Intelligence. This process mining cockpit connects your process data with your process models from SAP Signavio Process Manager and thus transforms static process models into dynamic and responsive dashboards. In this process, you also can place indicators and define specific thresholds. You can easily decide which processes and activities you want to monitor more closely or understand better within these processes. In this way, you're informed, guided, and warned of any process bottlenecks at an early stage and in a holistic manner with regard to your process management.

All business processes of your entire company can be modeled together with SAP Signavio Process Manager. You benefit from easier and more comprehensive process modeling for process experts as well as for business users who don't have BPMN knowledge.

SAP Signavio and BPMN

SAP Signavio has been committed to BPMN for years, is involved in the standardization process, and promotes BPMN in both industry and academia.

Model processes use drag-and-drop capabilities for elements, including for automatic positioning of elements and for reusing elements from a central repository. You can compare process models and model revisions, track potential process changes, and sustainably improve your process excellence. In addition, you can simplify process models for different roles through the use of overlays, as well as showing and hiding relevant attributes. These attribute overlays are attribute visualizations that are added by modelers when they create a diagram in SAP Signavio Process Manager. Which overlays are visible to a user group is also defined in SAP Signavio Process Manager. If a diagram contains attribute overlays, the number of available overlays and the number of visible overlay categories are displayed. In addition, SAP Signavio Process Manager together with SAP Signavio Process Collaboration Hub (see Chapter 7) promote collaborative process modeling. Collaborative process modeling is a collective activity in which team members discuss, design, and document business processes together. You can provide feedback and comments on entire business processes or specific (sub)tasks and share process diagrams immediately to gather feedback across the organization. Figure 5.2 shows an example of how individual process steps can be commented on directly in the graphical editor.

In addition, you benefit from highly configurable modeling notations (see Table 5.1) based on your specific business requirements. With the configurable modeling conventions of SAP Signavio Process Manager, it's thus possible to define your own rules and have them checked during modeling to see whether these rules have actually been followed.

Consider Custom Modeling Conventions

SAP Signavio Process Manager with its integrated process editor not only offers a professional platform for standard modeling but also the possibility to individually adapt modeling conventions to the respective company.

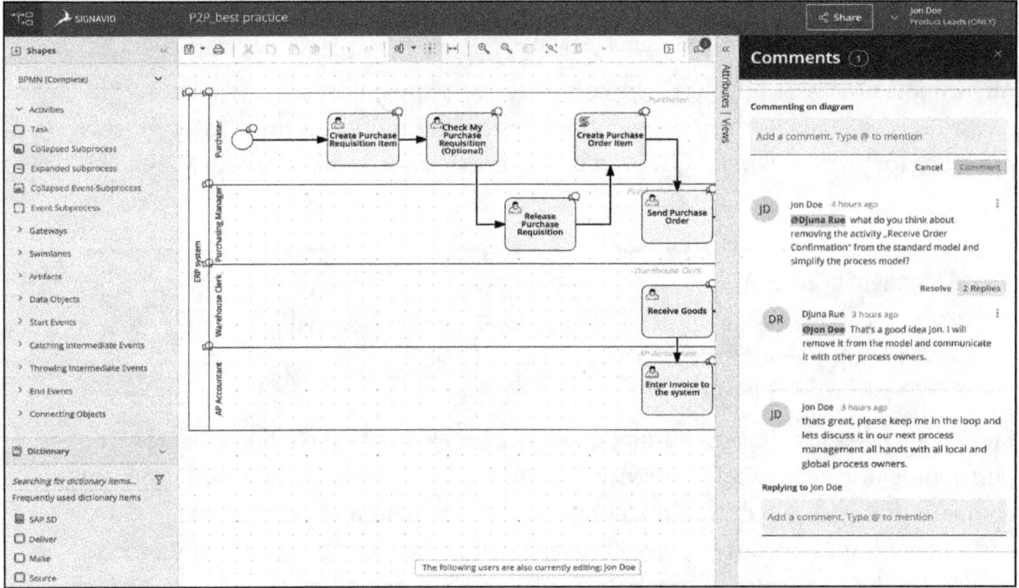

Figure 5.2 SAP Signavio Process Manager: Collaborative Process Modeling

You can create high-level perspectives on process hierarchies in your organization with value chains and ensure compliance with modeling conventions. A concise list of all modeling syntax suggestions and errors is available to help improve the quality and usability of your process models. You can validate modeling rules with warnings, hints, and errors in real time to improve the quality and understandability of models. In addition, your process owners can design processes based on best practices and modeling standards. Custom modeling conventions are defined once by the workspace administrator and automatically validated each time the workspace is saved. Figure 5.3 shows what such a configuration can look like in practice. In this step, you can activate different modeling conventions to be checked in your workspace. Apart from preconfigured conventions, you can define your custom conventions using a set of rules, such as regarding architecture or notation, which can be checked automatically.

With the *QuickModel* function, SAP Signavio Process Manager also allows you to create and edit BPMN diagrams. This allows you to record business processes in a decentralized way and, for example, also assign responsible roles. QuickModel is particularly practical for employees without BPMN knowledge or if you want to quickly model processes in an uncomplicated manner. All process steps are recorded in tabular form, and operation is intuitive and simple—similar to Microsoft Excel. The QuickModel feature offers full integration with the existing process repository, promotes automatic generation of BPMN diagrams, and allows you to revise existing process diagrams. This allows you to obtain process flows that follow modeling notation standards in the shortest possible time.

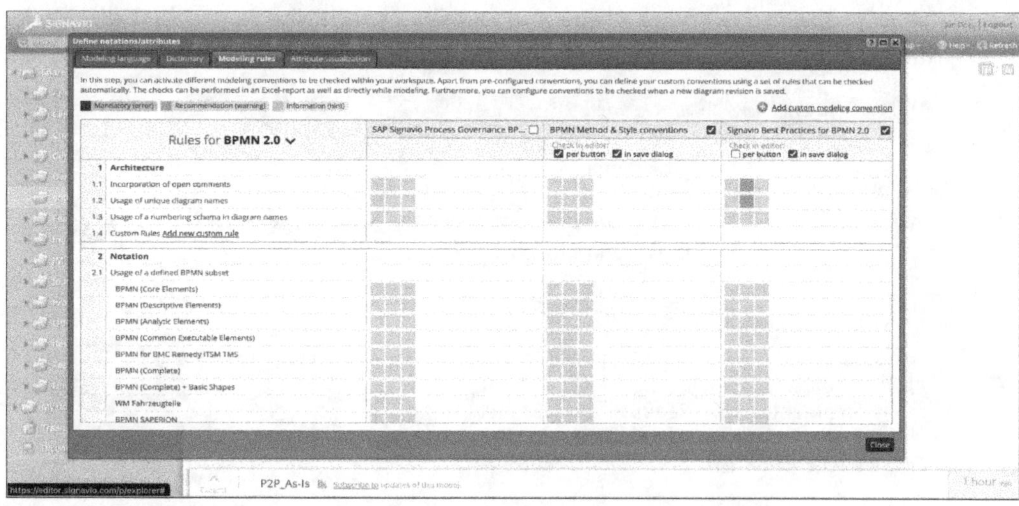

Figure 5.3 SAP Signavio Process Manager: Configurable Modeling Conventions

Figure 5.4 illustrates the QuickModel user interface (UI). In the upper-left area of the screen, you specify the name of the business process you want to model, the organization (your company), and the start and end events of your process. In the upper-right screen area, you can add a relevant process description. In the middle screen area, you list the individual process activities, including responsibilities and involved IT interfaces and documents in tabular form. You can see the result at the bottom of the screen: an automatically generated process model for further use.

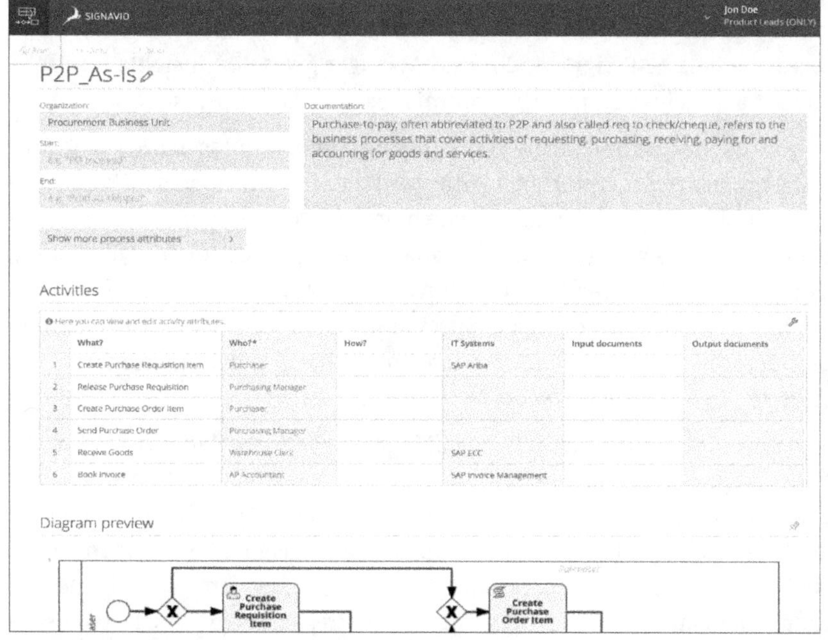

Figure 5.4 SAP Signavio Process Manager: QuickModel

Table 5.4 gives an overview of the main attributes for tasks within the QuickModel function.

Main Attribute	Description
What?	Task name
Who?	Participant or role assigned to the task (a separate lane is created in the diagram for each role)
How?	Task description
IT systems	IT system used in the execution of the task
Input documents	Required data artifact
Output documents	Created or changed data artifact
Execution costs	Costs necessary for the execution of the task
Cost center	Department to which the task costs are allocated
Execution time	Time needed to perform the task

Table 5.4 Main Attributes for Tasks in QuickModel

With SAP Signavio Process Manager, you can analyze your process hierarchy offline and in formats your process analysts already know. A variety of reports on process governance, license, and IT system usage; RACI overviews; and much more provide you with comprehensive insights regarding your business processes. A *risk and control report* provides you with an overview of any risks and related controls identified in selected process models. SAP Signavio Process Manager gives you insights into relevant process risks and helps you identify control measures and process requirements.

Using SAP Signavio Process Manager, you compare simulation scenarios to better understand the impact of potential changes and further improvements on your process results. The tool also helps you decide which process changes you want to implement. In addition, SAP Signavio Process Manager lets you calculate cycle times, resources, and costs to get detailed information about expected process performance and bottlenecks in simulation scenarios. Specifying process simulation parameters such as cost, duration, frequency, and resources to define simulation scenarios ultimately enables you to optimize your complete business transformation.

To get an overview of the user activities in SAP Signavio Process Manager workspace, you can refer to the *governance report*. This allows you to view aggregated metrics (e.g., the number of unpublished diagrams). Based on these metrics, you can draw conclusions about the success of your process modeling initiative. Figure 5.5 shows that each tile displays a different usage metric. Clicking on it opens a new browser tab and displays the results list of an advanced search filter that corresponds to the selected usage metric to show detailed information.

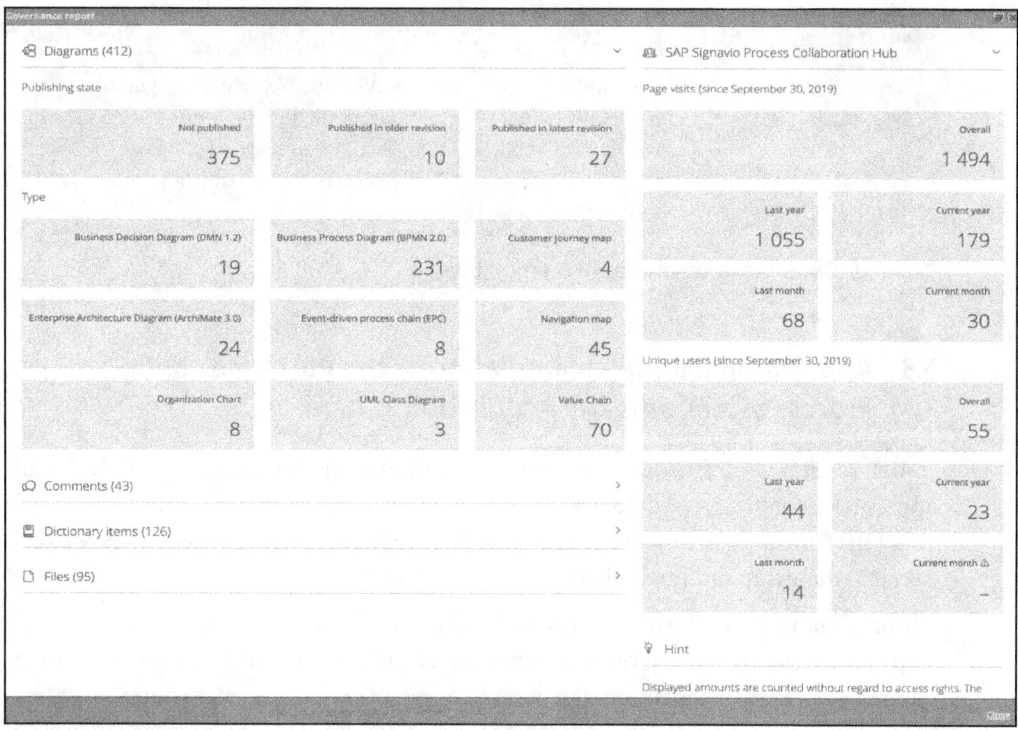

Figure 5.5 SAP Signavio Process Manager: Governance Report

Table 5.5 lists all available usage metrics of the governance report.

Data	Description
Diagrams	The total number of diagrams in your workspace is displayed in parentheses. The number of diagrams, grouped by publication status and type (e.g., process map, organization chart), is displayed on each corresponding title.
Comments	The total number of existing comments is displayed in brackets. The number of comments, grouped by comment status, is displayed on the respective title.
Dictionary entries	The total number of existing dictionary entries is displayed in parentheses. The number of dictionary entries, grouped by publication status and category type (e.g., events and requirements), is displayed on the respective title.
Files	The total number of existing files is displayed in parentheses. The number of files grouped by publishing status and type (e.g., PDF and JPG) is displayed on each corresponding title.

Table 5.5 Governance Report Usage Metrics

Data	Description
SAP Signavio Process Collaboration Hub	The number of page views represents a single user opening any published object (diagram, file, or dictionary item) in SAP Signavio Process Collaboration Hub. Each time someone opens one of these items in SAP Signavio Process Collaboration Hub, it's counted as a view and displayed in the report.

Table 5.5 Governance Report Usage Metrics (Cont.)

5.3 Application Example: Simulation of the Purchase-to-Pay Process after Company Acquisition

In this section, we'll introduce you to the solution using a practical example. In this application example, we model a process model with SAP Signavio Process Manager using the QuickModel function. Subsequently, we simulate the process model with certain parameters to understand potential bottlenecks.

In our scenario, we're the process owner for the *purchase-to-pay* process. Our purchase-to-pay process is currently running smoothly. Now our company has acquired a new company. After the acquisition, the parent company's shared service center must handle the subsidiary's purchase requests. As a result, we simulate the increased volume of purchase requests to understand potential bottlenecks. In this context, we also get a use case of how our purchase-to-pay process can be improved using innovative technologies.

To do this, we first model a purchase-to-pay process with SAP Signavio Process Manager, as shown in Figure 5.6.

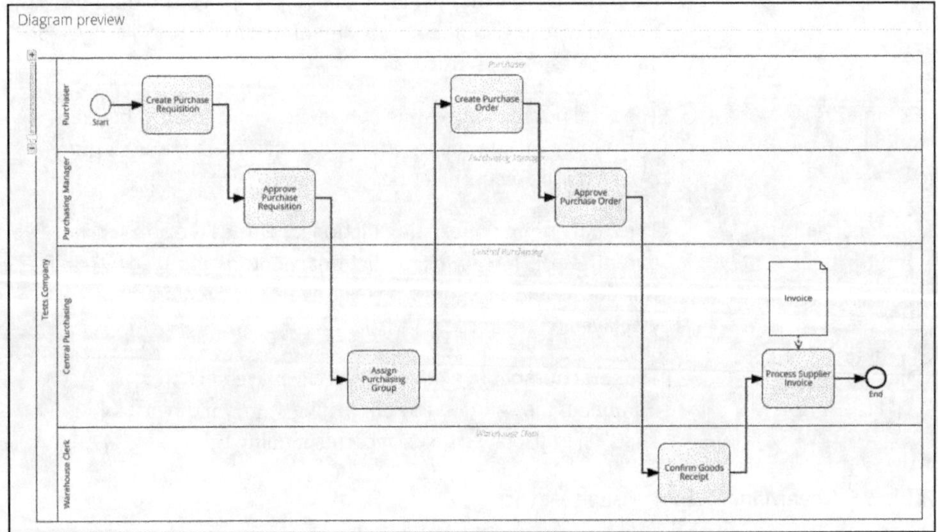

Figure 5.6 Process Model Preview for the Purchase-to-Pay Process

We create the model using the QuickModel function. How exactly we proceed will be explained in the course of the application example.

After logging in to SAP Signavio Process Manager, select the **QuickModel** option via the **New** tab (see Figure 5.7).

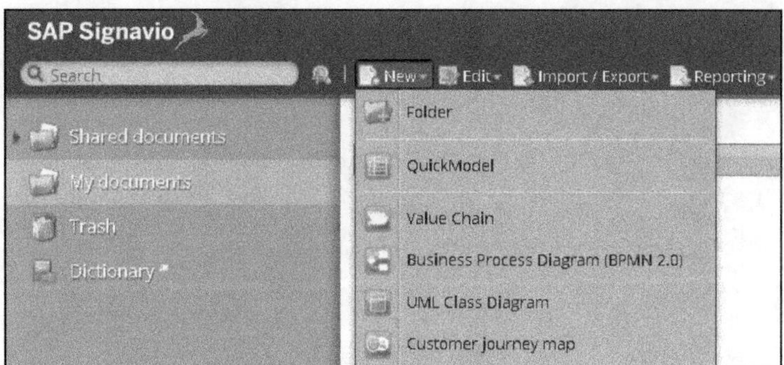

Figure 5.7 Opening the QuickModel Function

When the QuickModel function opens, you can see the initial screen of the function, as shown in Figure 5.8. Here, you can create a new process model by specifying the essential parameters such as description, start event, and end event, as well as the individual activities of your business process.

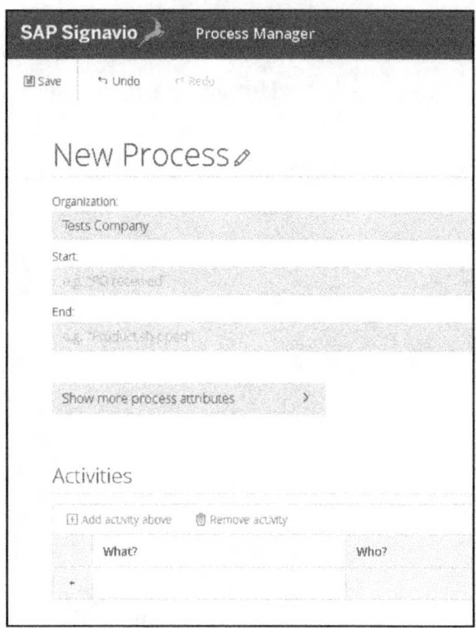

Figure 5.8 Creating a New Process Model in QuickModel Mode

First, in Figure 5.9, rename the process model to "P2P-Process". Enter "Start" as the **Start** event and "End" as the **End** event.

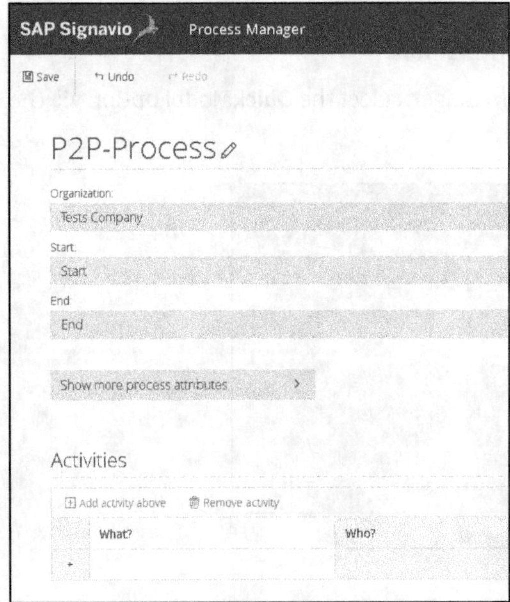

Figure 5.9 Renaming the Process Model

Now it's a matter of entering the individual activities of the process into the model. This is easily done via the tabular view in the **Activities** area (see Figure 5.10) by entering the activities in the **What?** column. You can use the plus button to add new rows or press the Enter key after entering the text to automatically move to the next row. If you press the Enter key for the second time, you can edit and enter the process activities in the rows directly below.

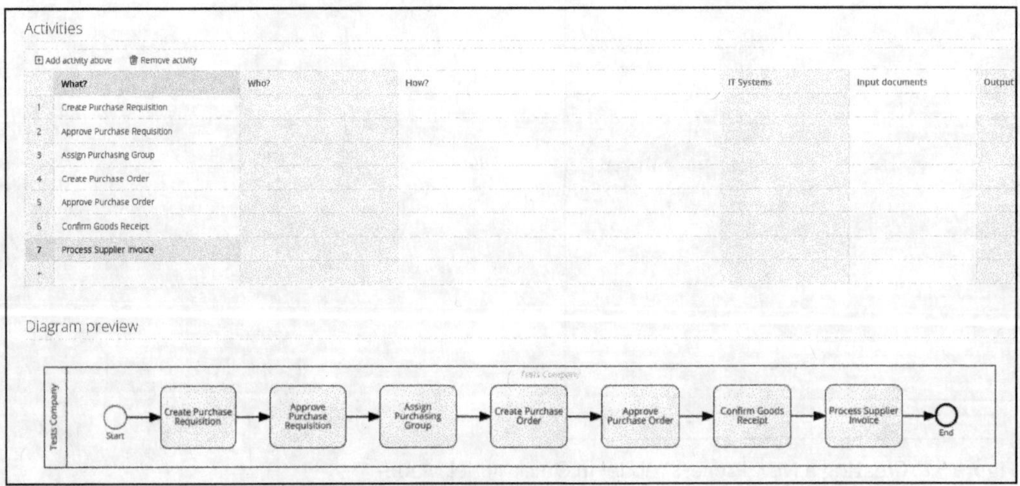

Figure 5.10 Adding Process Activities

You now insert the following activities:

- **Create Purchase Requisition**
- **Approve Purchase Requisition**
- **Assign Purchasing Group**
- **Create Purchase Order**
- **Approve Purchase Order**
- **Confirm Goods Receipt**
- **Process Supplier Invoice**

You can store the responsibilities for the respective process activities via the **Who?** column in a similar way by entering the role that is responsible for the subprocess. The role names are freely selectable. As shown in Figure 5.11, you assign the following responsibilities specifically for your purchase-to-pay process:

- **Purchaser**
- **Purchasing Manager**
- **Central Purchasing**
- **Warehouse Clerk**

Activities

Add activity above | Remove activity

	What?	Who?	How?
1	Create Purchase Requisition	Purchaser	
2	Approve Purchase Requisition	Purchasing Manager	
3	Assign Purchasing Group	Central Purchasing	
4	Create Purchase Order	Purchaser	
5	Approve Purchase Order	Purchasing Manager	
6	Confirm Goods Receipt	Warehouse Clerk	
7	Process Supplier Invoice	Central Purchasing	
+			

Figure 5.11 Adding Roles

Next, in the previously added process step **Process Supplier Invoice**, add the entry **Invoice** to the **Input documents** column (see Figure 5.12). Input documents are data artifacts that are required for the execution of the process step. To be able to process the supplier invoice in our case, you need an invoice as an input document.

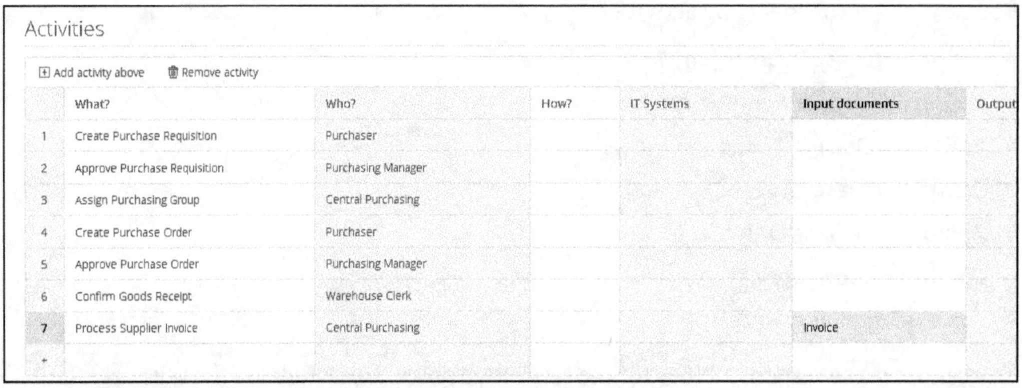

Activities

⊞ Add activity above 🗑 Remove activity

	What?	Who?	How?	IT Systems	Input documents	Output
1	Create Purchase Requisition	Purchaser				
2	Approve Purchase Requisition	Purchasing Manager				
3	Assign Purchasing Group	Central Purchasing				
4	Create Purchase Order	Purchaser				
5	Approve Purchase Order	Purchasing Manager				
6	Confirm Goods Receipt	Warehouse Clerk				
7	Process Supplier Invoice	Central Purchasing			Invoice	
+						

Figure 5.12 Adding Input Documents

Based on the role information added in the **Who?** column next to the process activity, as shown in Figure 5.13 in the **Diagram preview** area, a swimlane diagram is added that visualizes the process you entered.

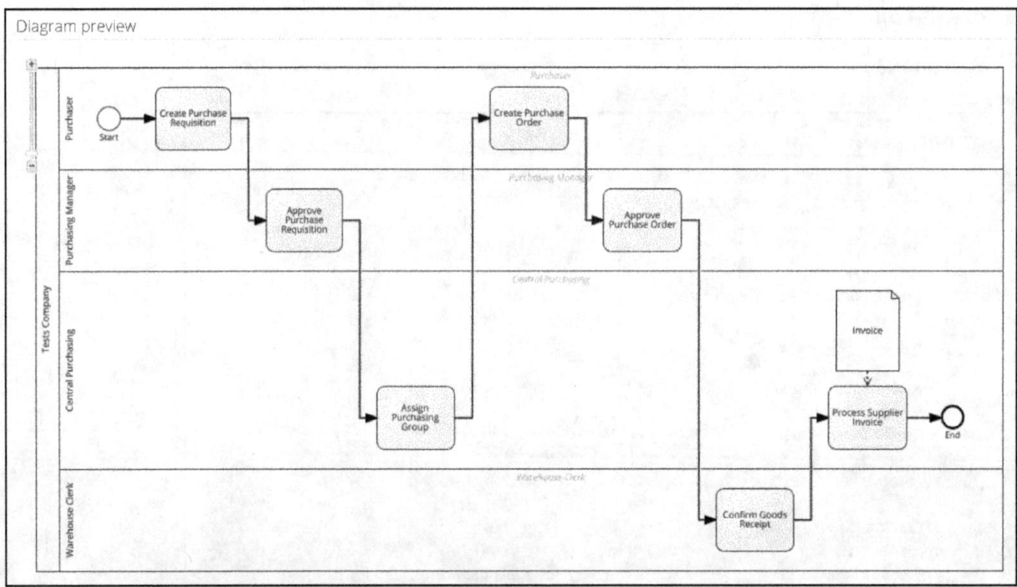

Figure 5.13 Diagram Preview for the Created Process

When satisfied with the result, save your work by clicking the **Save** button (see Figure 5.14).

Now you can view the newly created process model in the process documents. To do so, click on the icon in the menu bar to access the navigation menu. In the dropdown menu, select the **Graphical Editor** menu item.

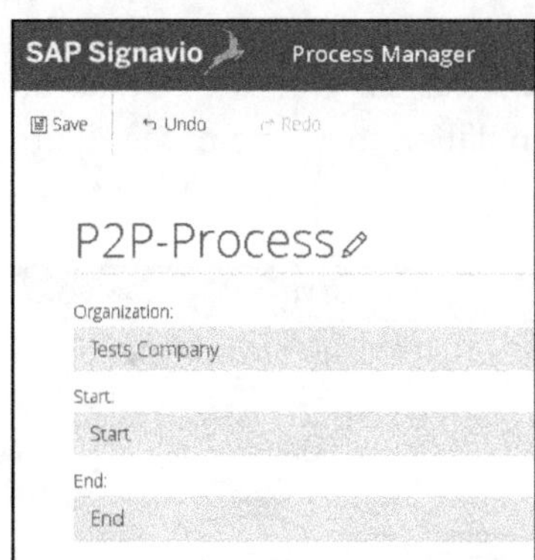

Figure 5.14 Saving the Newly Created Process Model

You'll be redirected to the graphical editor UI of SAP Signavio Process Manager by modeling or modifying the process as needed (see Figure 5.15). On the left side of the editor, a bar displays various BPMN elements for refining the process model, such as **Task** or **Collapsed Subprocess**. You can simply drag and drop these elements in the graphical editor and create your own processes or edit existing processes.

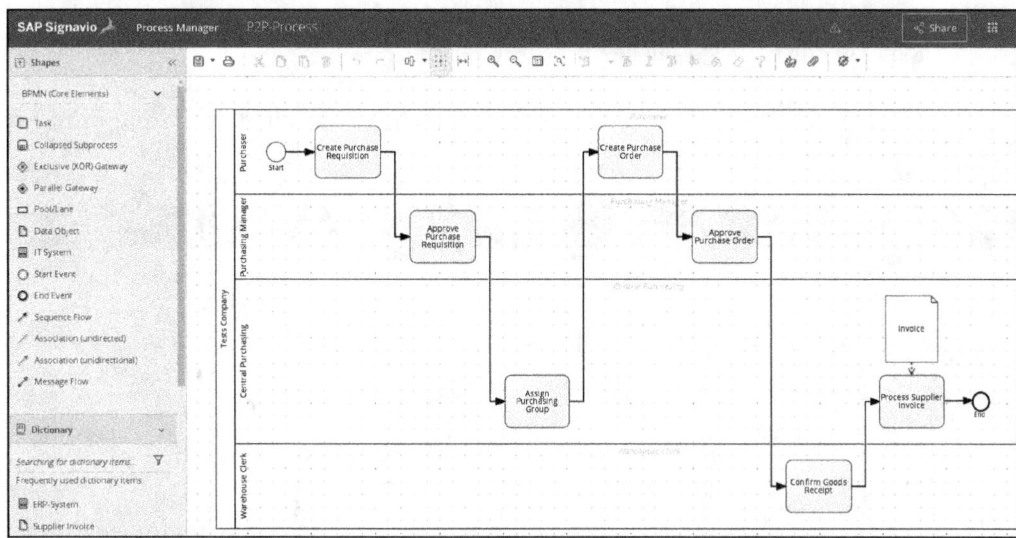

Figure 5.15 Graphical Editor in SAP Signavio Process Manager

> **Graphical Editor**
>
> The graphical editor is a drag-and-drop environment for creating or modifying process flows. It provides another way to model business processes.

For our application example, we won't change the process we created earlier with the QuickModel function any further but continue to use it as is. We're even more interested in simulating our process model to be able to make adjustments, if necessary, that will prevent bottlenecks in the future. To do this, click on the navigation menu at the top right of the screen, and select the **Simulation** menu item (see Figure 5.16).

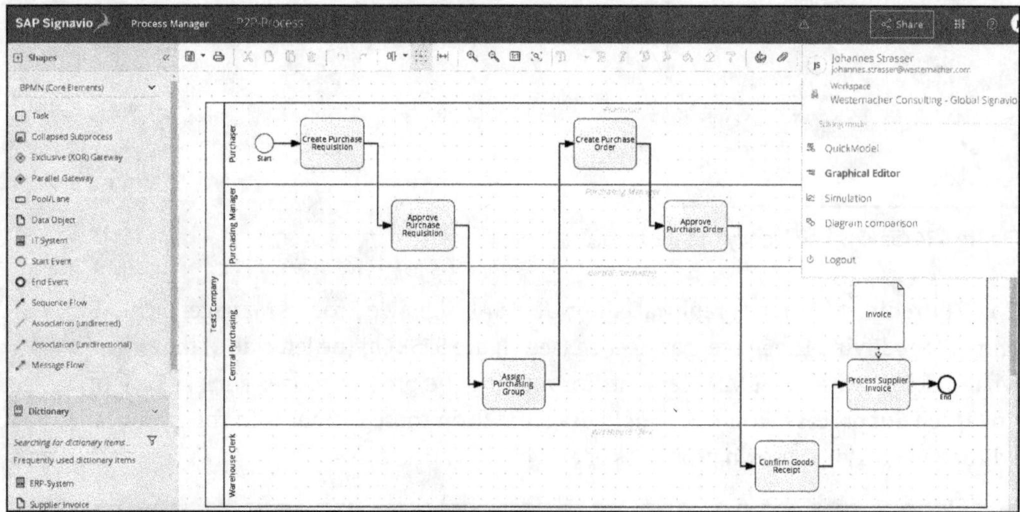

Figure 5.16 Open Process Simulation

As you can see in Figure 5.17, you'll be taken directly to the initial screen of the process simulation.

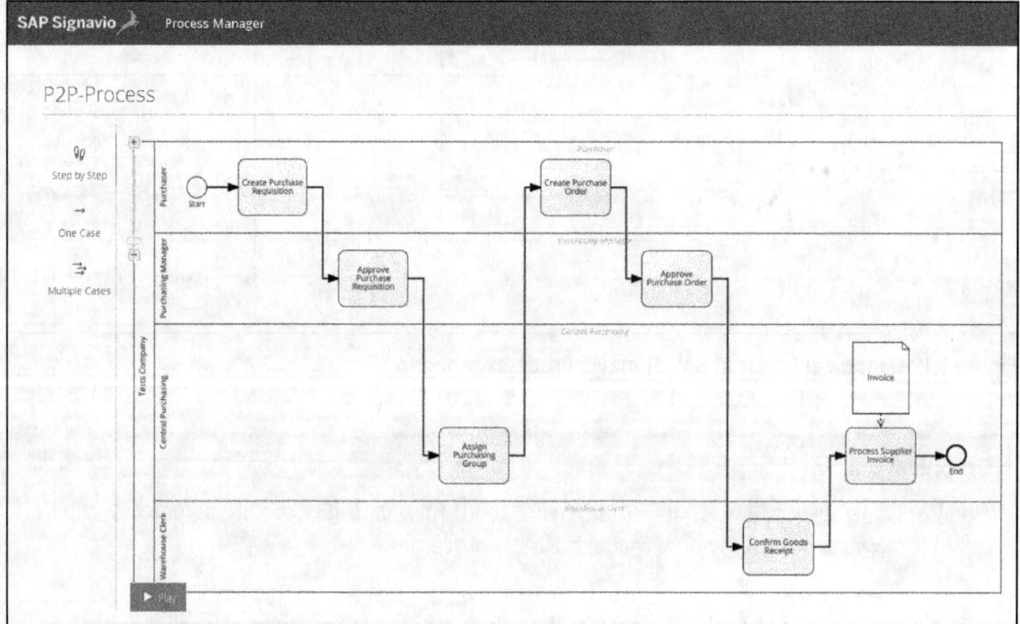

Figure 5.17 Initial Screen for Process Simulation

To view the entire process in the UI, you need to click a few times on the zoom out icon located to the left of the process diagram in the upper half of the screen until you see the complete process in the view.

In the lower part of the screen, you can define the scenario to be simulated. First, click on the **Costs** button, which you can see in Figure 5.18. Here, you can configure the execution costs for each task in the scenario. For example, you set the cost per execution to a flat rate of 10 EUR per activity.

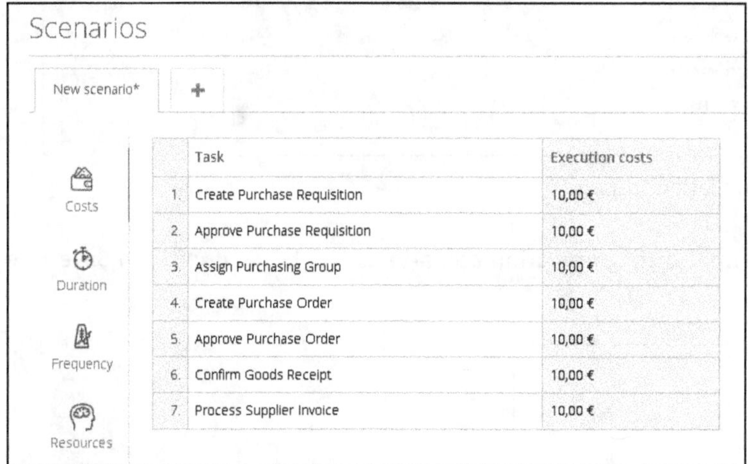

Figure 5.18 Enter Costs per Execution

In the next step, click the **Duration** button, and, based on experience, set a flat 15 minutes as the execution time per activity (see Figure 5.19).

Figure 5.19 Enter Execution Time per Activity

Save your simulation scenario by clicking the **Save scenario** button (see Figure 5.20).

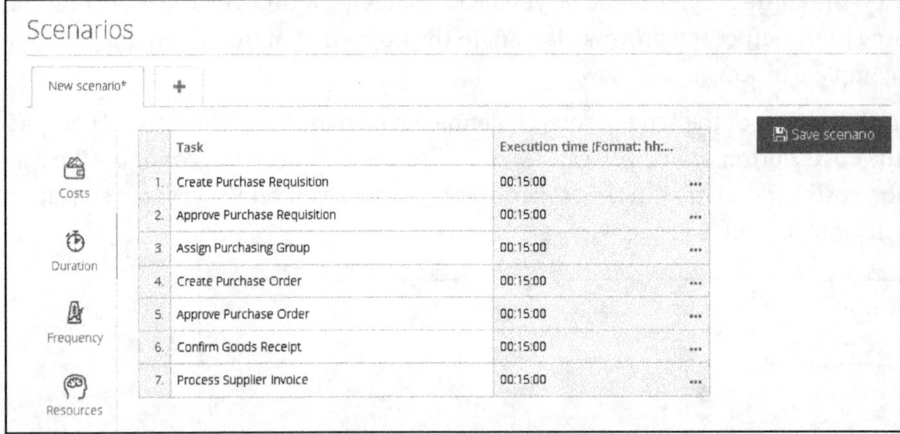

Figure 5.20 Save Simulation Scenario

Now define the name of the simulation scenario as "AS-IS", and click on **Save** (see Figure 5.21).

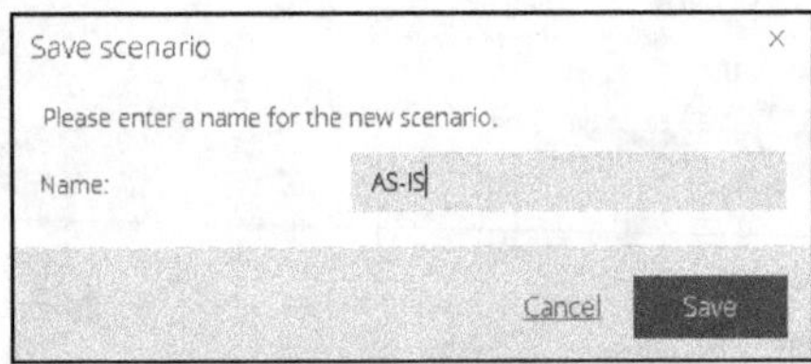

Figure 5.21 Name Simulation Scenario

In the **Frequency** area, you can still determine the activity frequency (see Figure 5.22). At the moment, the number of purchase requests is 20 per week (Monday to Friday). For now, leave the frequency at that, and simulate the process with this value.

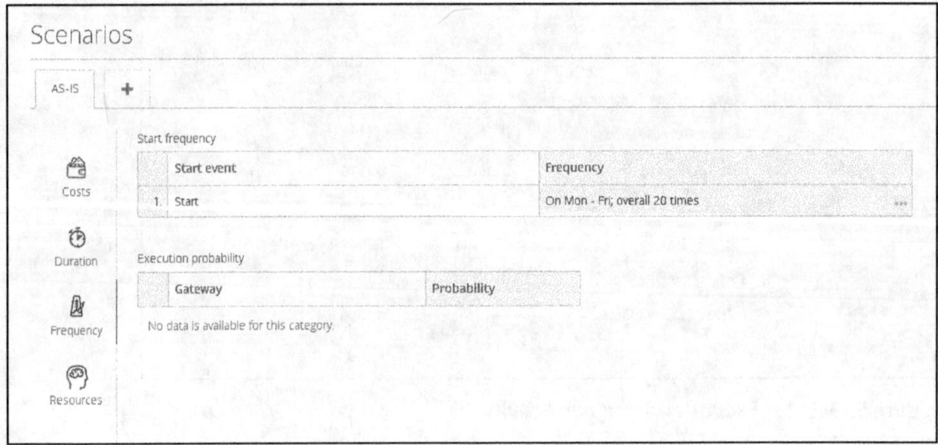

Figure 5.22 Creating the Activity Frequency

Now, you want to simulate all the purchase requisitions created in an entire week to check whether the process has bottlenecks or not. As shown in Figure 5.23, click on the **Multiple Cases** button on the left side of the screen, and then click on the **Start** button for your actual scenario.

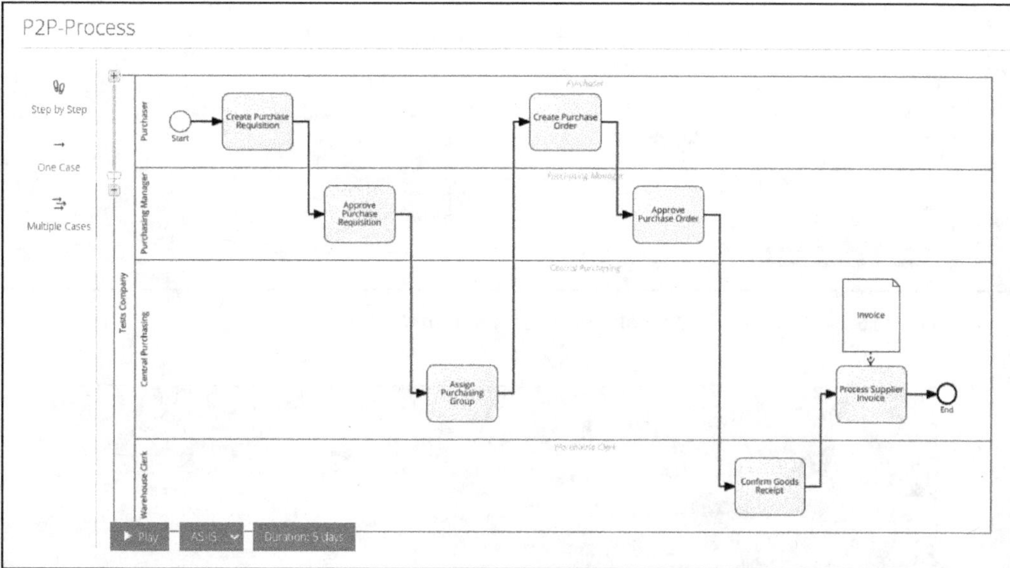

Figure 5.23 Starting the Process Simulation

[!]

> **Simulate Multiple Cases**
>
> A single simulation case involves running an entire cycle from start to finish for a single purchase requisition. To simulate all purchase requisitions, you must select the **Multiple Cases** area.

According to the insights from Figure 5.24, both the inputs at the beginning and the outputs at the end of the process are 20 each. This indicates that there are no bottlenecks, and the process is running smoothly. In addition, the system indicates that there are no concerns with the entry, as shown under **Bottlenecks** on the right side of the screen.

However, due to the acquisition, the parent company must now also process the purchase requisitions of the new subsidiary. You'll therefore simulate this scenario with further orders to be processed.

In Figure 5.25, click on the plus button under **Scenarios** in the lower half of the screen next to **AS-IS.** Give the new scenario the name "TO-BE", and select the **AS-IS** process as a template. Save the new simulation scenario.

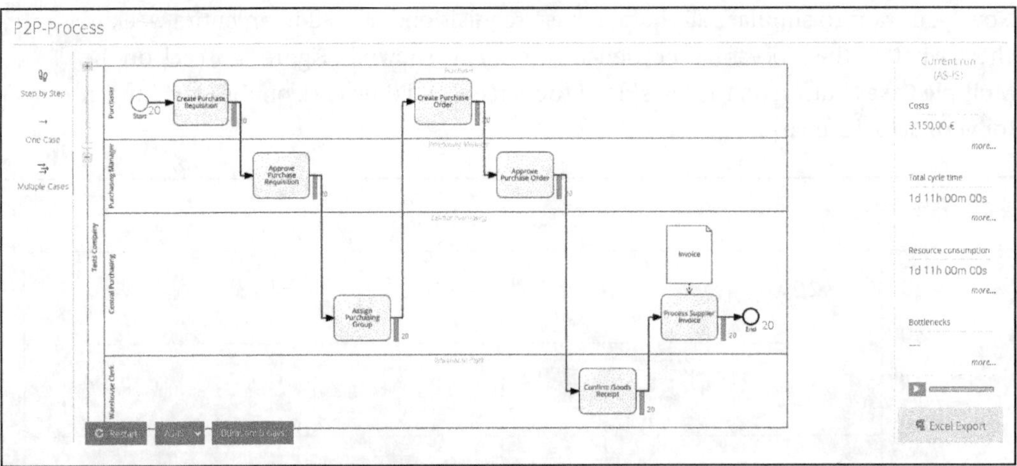

Figure 5.24 Result of the Actual Process Simulation

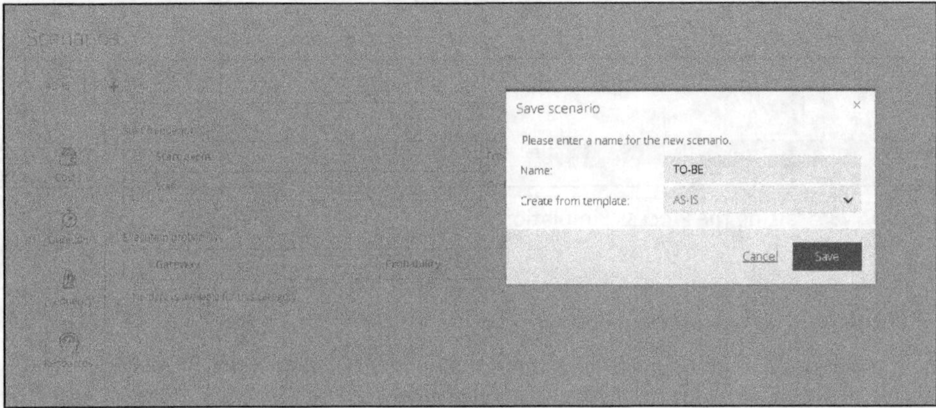

Figure 5.25 Creating the Target Simulation Scenario

Now, you need to increase the activity frequency in the target simulation scenario. To do this, change the value in the **Frequency** column from **4** per day to **20** per day. Figure 5.26 illustrates this process.

After you've saved the new scenario with the updated frequency, you can select it in the simulation run. To do this, in Figure 5.27, select the simulation scenario **TO-BE** from the dropdown menu, and start it accordingly.

You'll notice that the process has a bottleneck at the **Create Purchase Order** step. The bottleneck is indicated in the process diagram by blue stacks of dots, and the **Purchaser** entry is also shown under the bottlenecks on the right side of the metrics. You can also see that only 58 requisitions complete the entire cycle. The count is displayed at the end step.

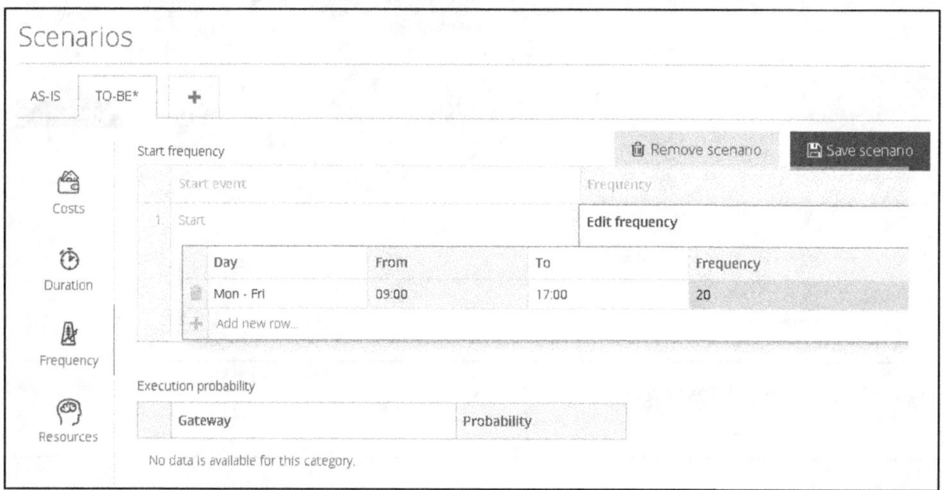

Figure 5.26 Update Activity Frequency

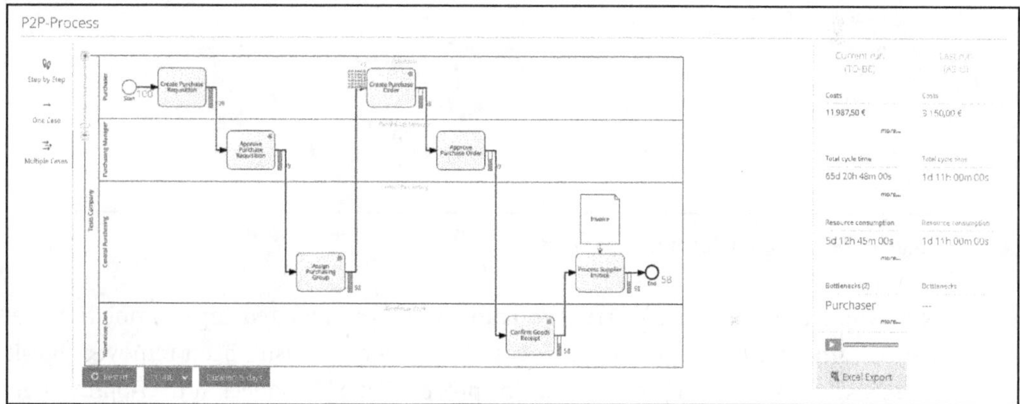

Figure 5.27 Result of the Target Process Simulation

We're now looking to increase resources for central purchasing and the buyer because requests have increased and reflect more load in these areas.

In this context, change the resources for central purchasing and the purchaser to two resources per role, and save this operation (see Figure 5.28). In Figure 5.29, the increased resources regarding the roles of **Purchaser** and **Central Purchasing** are displayed.

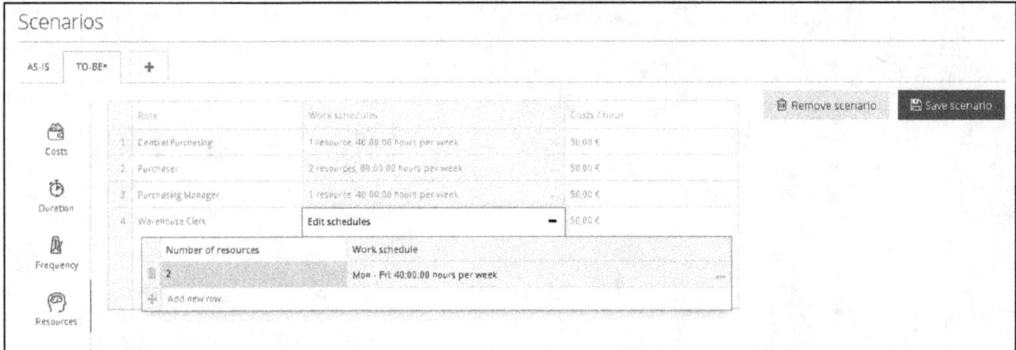

Figure 5.28 Changed Resources per Role

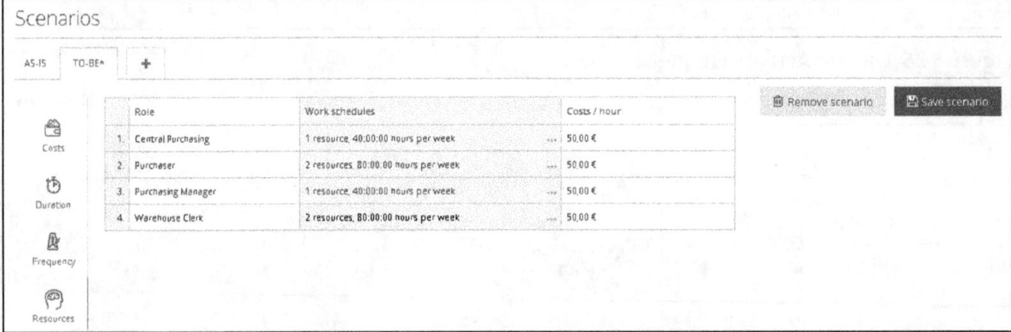

Figure 5.29 Updated Resources per Role

Now, you can see in Figure 5.30 that you can restart your updated target simulation scenario. Even after adding more resources, the bottleneck hasn't been removed, but it merely arises in another process step. The process has a bottleneck at the **Approve Purchase Order** step now with blue point stacks. This is also shown under **Bottleneck** in the **Metrics** section on the right side of the screen.

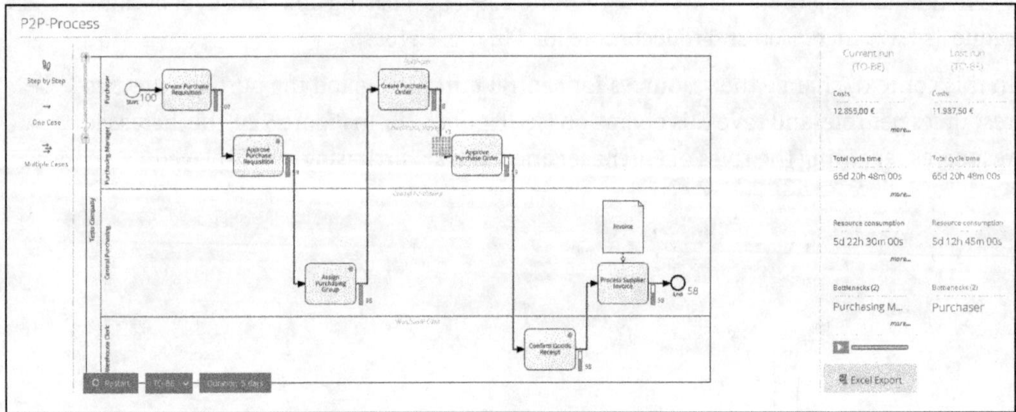

Figure 5.30 Result of the Updated Target Process Simulation

[«]

Simulation Result Metrics

After you run the simulation for multiple cases, the result metrics for the current run of the simulation are available in the right pane of the screen. In addition to the bottlenecks, the metrics **Costs**, **Total cycle time**, and **Resource consumption** are also available. By clicking on the corresponding tile for the metric simulation, you can access detailed information about the metric.

From this, we can conclude that under the new conditions, we're lacking one employee or one employee in the role of **Purchasing Manager**. With regard to our TO-BE simulation, the total costs, the closed cases, and the total cycle time are also listed on the right of the figure. Out of a total of 100 cases, only 58 cases were closed.

In this application example, we've designed the purchase-to-pay process and manually added cost, effort, and time for simulation. However, it's important to understand what the real, operational process data looks like within the company and how it can be coupled to the process model. So, this use case offers scope for much more. For example, it's worth revisiting the process model and trying to address bottlenecks with innovative technologies. For this purpose, you can subsequently use SAP Signavio Process Intelligence (see Chapter 4).

5.4 Summary

SAP Signavio Process Manager is a cloud-based software solution that supports your company in documenting, modeling, and simulating your business processes. You can rely on SAP Signavio Process Manager to adapt your business processes or to completely redesign them. This tool helps you react quickly to unpredictable business and regulatory changes, to scale your company's process repository in the long term, and to simulate business processes to define alternative business scenarios.

You can use this tool to gain better insight into inefficiencies and improve your performance in a big way. Putting the customer first—that is, the ability to enhance the CX by connecting business processes into an improved customer journey—is one of the biggest benefits of SAP Signavio Process Manager. The support for enterprise-wide process visibility and the ability to enable process and decision modeling for all areas definitely increases continuous process excellence across the board. Noncompliant processes can be identified and then corrected as needed. The tool also facilitates editing and collaboration as process models can be shared quickly. You can align your business and technical requirements and manage processes and decisions through a single platform. Say goodbye to process chaos!

Chapter 6
SAP Signavio Journey Modeler

SAP Signavio Journey Modeler enables you to gain an external view of how stakeholders actually experience individual areas of your company. The tool focuses on modeling journeys, aggregates relevant information, and provides you with models that can be used to improve various business processes.

When we think of the most important factors in today's digitized world, we usually think of agility and transformation.

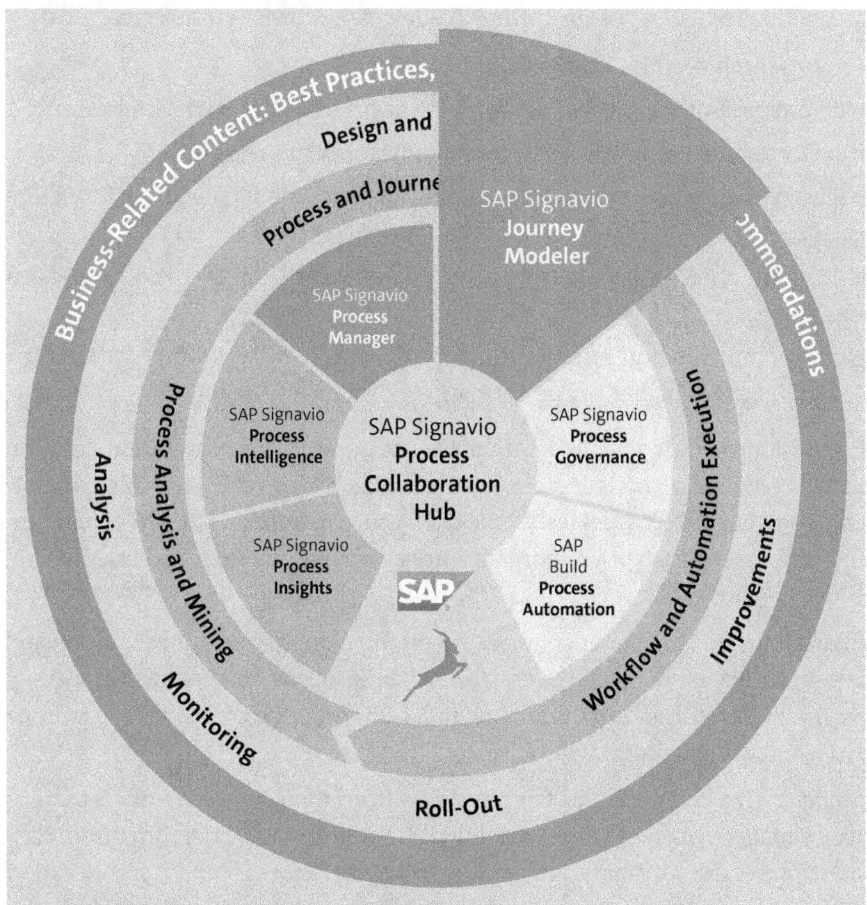

Figure 6.1 SAP Signavio Journey Modeler in Focus

Agility is a given, transformation is a business imperative for survival, and neither is sufficient to achieve a sustainable competitive advantage without placing special emphasis on the customer experience (CX). Tomorrow's businesses are therefore shifting their attention from a product- and efficiency-centric approach to customer centricity to compete in a crowded and fast-moving digital landscape. *SAP Signavio Journey Modeler* gives you the tools to understand, improve, and transform the way your customers interact with your business activities to create enduring customer excellence (see Figure 6.1); however, it offers much more than continuous CX improvement. Let's now dive deeper into the topic.

In Section 6.1, we'll first look at the various features of SAP Signavio Journey Modeler. Section 6.2 shows different benefits of the solution. Subsequently, we'll illustrate the use of SAP Signavio Journey Modeler in practice by means of an example (Section 6.3).

6.1 Functions

The core functions of SAP Signavio Journey Modeler are shown here at a glance:

- Gain a centralized, real-time view of the customer journey.
- Simplify process flows and collaboration for transparent work environments.
- Align customer and employee interaction points for better results.
- Leverage experience data and process mining analytics for detailed transformation views.
- Share insights via SAP Signavio Process Collaboration Hub as a central process source.
- Increase recommendation rate (Net Promoter Score [NPS]), customer satisfaction, and the CX index for higher key performance indicators (KPIs) and value drivers.

SAP Signavio Journey Modeler is another helpful process transformation solution included in SAP Signavio's holistic suite. With this tool, you can operationalize your CX by linking your business processes to the way your customers actually experience them, and subsequently adapt your organizational systems, metrics, and roles to support the best possible CX.

SAP Signavio Journey Modeler enables you to do three important things: quickly and easily design journey models, effectively manage these journey models within SAP Signavio Business Transformation Suite, and, most importantly, connect your journey models to your business processes.

Journey models give you a view of your business from the perspective of customers, employees, suppliers, or other parties dealing with your business. They differ from classic business processes, which are mapped with Business Process Model and Notation (BPMN). While business processes model business processes using flowcharts from the inside perspective out, journey models are designed to represent the experiences of your customers from the outside perspective. Using SAP Signavio Journey Modeler, you can bring together operational excellence from the inside out and CX practices from the outside in, creating a solid connection between journeys, processes, and data to widely delight your customers. To make this possible, all business processes already modeled in SAP Signavio Process Manager can be linked to the journey models created in SAP Signavio Journey Modeler. For example, the process model can be linked to SAP Signavio Journey Modeler on a specific process step or task level. In addition, linked processes can be opened directly in SAP Signavio Process Collaboration Hub.

Especially in the course of SAP Signavio Journey Modeler, the term *customer journey* is used again and again. A customer journey is literally the customer's journey from the first contact with the company to the purchase of a product or the use of a service. With SAP Signavio Journey Modeler, the customer journey can be captured in a graphical or tabular structure, in which the individual phases of the journey and the touchpoints between customers and companies are listed and described in detail. Touchpoints are points of direct interaction with your company. In addition, you can visualize customer sentiment for each stage of the journey using smiley-face emojis, link process maps, and add data visualizations as widgets. Such widgets can be embedded directly from SAP Signavio Process Intelligence. By integrating operational and external data with customer journeys, you can discover opportunities for improvement and ensure that you achieve customer excellence. Customer excellence is about combining the internal and external perspectives: Which of the business processes have touchpoints with the customer, and, more importantly, how do customers perceive these touchpoints? Such insights lead to various potential for your company.

Table 6.1 provides an overview of the numerous application features of SAP Signavio Journey Modeler.

Feature	Description
Standard journey modeling	Create journey models in a tabular structure. Use phases to represent all stages of a journey in a timeline approach. Add sections to include customer opinions or IT systems. There is no limit to the number of these sections. Add multiple columns to each section for a phase. The smallest element you can edit is the map, which is contained in a table cell. What you can add to a map depends on the section type.
Linking BPMN processes with journeys	Link one or more processes to a journey model. You can link complete processes or select individual process elements.
Connection of external business intelligence (BI) tools	Add external widgets using link addresses or code snippets to visualize data, for example. Currently, Tableau and Looker Studio are supported.

Table 6.1 Functions of SAP Signavio Journey Modeler

Feature	Description
Linking of widgets from process analyses	Add widgets from SAP Signavio Process Intelligence to illustrate the results of the investigation.
Automatic recognition of IT systems and organizational units based on linked processes	A section in the system automatically displays IT systems that have been detected in the linked process models. The first element lists all IT systems that are detected in a linked process. If a column contains a linked process, the IT systems in this process are marked with a solid line.

Table 6.1 Functions of SAP Signavio Journey Modeler (Cont.)

6.2 Benefits

With SAP Signavio Journey Modeler, you have a direct way to operationalize new CXs. SAP Signavio Modeler enables you to translate CXs as they interact with specific business processes into quantifiable and manageable information, so you can quickly adapt to changing customer expectations and delight customers on an ongoing basis. This also leads to smarter decision-making about which customers to serve, when, and how.

The core potentials of SAP Signavio Journey Modeler are as follows:

- Design perfect customer journeys.
- Visualize the CX.
- Better understand the customer journey.
- Connecting data with experiences.
- Display journeys, processes, and data on one platform.
- Increase the CX.
- Improve customer retention and loyalty.

The solution helps you develop a centralized, real-time view of the CX by linking journeys to business processes, related IT applications, integration points, and data flows. You can align your entire organization around critical customer outcomes by leveraging journey analytics and sentiment analysis. In addition, you can better understand the interdependencies between customer sentiment, moments of truth, and underlying process flows.

Paradigm Shift in Customer Experience

Today, the CX is more important than ever. Whereas the focus used to be on product and service sales and productivity, there is a paradigm shift in companies: millions of people can see at a glance how a product or customer service is rated. In today's world, hardly anyone can afford a bad rating, an inferior product, or unfriendly service. Accordingly, the contemporary focus is particularly on the CX and the positive experiences along the customer journey.

Another interesting opportunity lies behind the already existing process landscape. Using the process mining capabilities of SAP Signavio Process Intelligence, you can identify critical customer interaction points. Customer data and process mining analytics help you understand the root causes of your customers' frustration or satisfaction and then adjust or redesign business processes to effectively increase customer satisfaction, retention, and loyalty over the long term.

You can also break down silos between CX and process teams and engage in common goals and definitions of journey modeling. In this way, you can minimize operational complexity and manage variation in process flows to create a consistent CX across the enterprise. Corresponding insights can be shared and collected across the organization via SAP Signavio Process Collaboration Hub to further improve your CX.

With SAP Signavio Journey Modeler, you link business processes and experiences. This is the first step toward customer excellence because it gives you a complete 360-degree outside-in and inside-out view of your business processes. It helps you better understand, improve, and transform the way your customers interact with your business, and turn experience and process mining into operational reality. In the process, you can leverage your journey models by embedding widgets, extracting process mining analytics, and connecting processes to customer sentiment to gain a deeper understanding of how your business works. In Figure 6.2, you can see an example of a created customer journey for a company's e-commerce division. Clicking on the comment icon (two speech bubbles) opens a comment area where you and your colleagues can add comments to the journey.

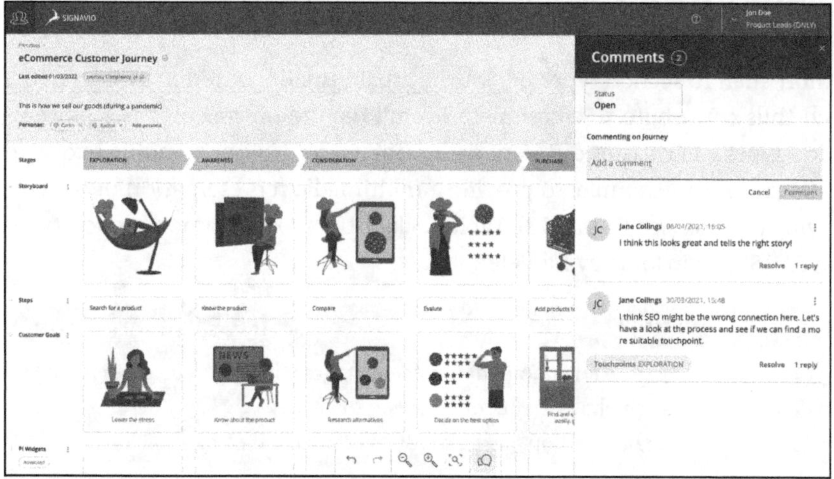

Figure 6.2 Collaborative Journey Modeling

The tool also helps you better understand your customer journey: Using the journey models, together with the modeled business processes, you visualize all end-to-end experiences of your customers, employees, and/or suppliers. With the advanced customization options, you can refine and adapt your journey models.

Journey models are created in a table structure. Table 6.2 shows the elements relevant for the table structure.

Element	Description
Phases (stages)	The top row of the tabular journey model represents the phases of a journey, and phases always represent the first row of your journey model. Phases are required for each journey. The different phases represent points on the timeline. Each phase can contain multiple columns, commonly referred to as steps. You can't remove phases from your model, and you can't have more than one row of phases. The row with phases is fixed, so it remains visible even if you scroll down in longer journey models.
Sections	The rows of the table represent the different stages of the journey. The horizontal sections below the phases add different types of information in layers for each step of a journey. Use the top rows to tell the user story, the middle rows to map touchpoints or channels, and the bottom rows to show the company's perspective and related processes. For example, you use the **Sentiment** section to model changing customer sentiment, or the **IT Systems** section to show which systems are relevant to a particular step. You can add any type of section as often as you like next to another section or at the end of the journey.
Columns	A column spans all sections. A phase can contain several columns.
Cells	The smallest element you can edit is the map contained in a table cell. What you can add to a card depends on the type of section.

Table 6.2 Elements of the Table Structure

There are more than 10 section types per journey that visualize different aspects of the journey and thus contribute to a complete overall picture of the customer journey. These section types each accept different data formats, so you can combine text with images or diagrams, for example, to provide a multimedia representation of the customer's buying process (see Figure 6.3). Table 6.3 lists the most common section types available in SAP Signavio Journey Modeler.

Section Type	Description
Phases (stages)	■ Create an unlimited number of stages within the customer journey. ■ Each phase may contain one or more steps. ■ Drag and drop makes it easy to reorder and edit.

Table 6.3 Section Types in SAP Signavio Journey Modeler

Section Type	Description
Pictures (images)	■ Browse the image library by file name and get an image preview. ■ Images are supported in PNG, JPNG, GIF, BMP, and SVG formats. ■ Images up to 10 MB are supported. ■ Use the bulk upload and the drag-and-drop upload.
Text	■ Use text paragraphs to define pain points, business goals, customer thoughts, moments of truth, etc. ■ The text function supports basic formatting: bold, italic, bullets.
Touchpoints	■ Use the predefined icon library for each touchpoint. ■ Link any single dictionary entry as a touchpoint.
Mood (sentiments)	■ A five-point scale shows customer sentiment throughout the customer journey. ■ Symbols and colors automatically adjust to the score. ■ The mood is easily customizable via drag and drop.
Linked processes	■ Link processes modeled in SAP Signavio Process Manager in advance. ■ Link the entire process, or auto-populate this section with information in linked processes. ■ View the linked process in the **Pages** panel, and open it directly in SAP Signavio Process Collaboration Hub.
IT systems and organizational units	■ Auto-fill this section with information from your linked processes. ■ See how different IT systems and organizations are involved during the journey. ■ Manually add additional IT systems or organizational units as needed.
External widgets	■ Embed third-party widgets directly into a step. ■ Looker Studio and Tableau are supported.
SAP Signavio Process Intelligence widgets	■ Embed widgets directly from SAP Signavio Process Intelligence. ■ Display interactive widgets in the side panel and access your investigation directly.
Linked journey models (linked journeys)	■ Link directly to other journey models as a subjourney or reference. ■ Easily switch between connected journeys.

Table 6.3 Section Types in SAP Signavio Journey Modeler (Cont.)

Figure 6.3 illustrates the **Stages**, **Storyboard**, **Steps**, **Customer sentiment**, and **Process** sections.

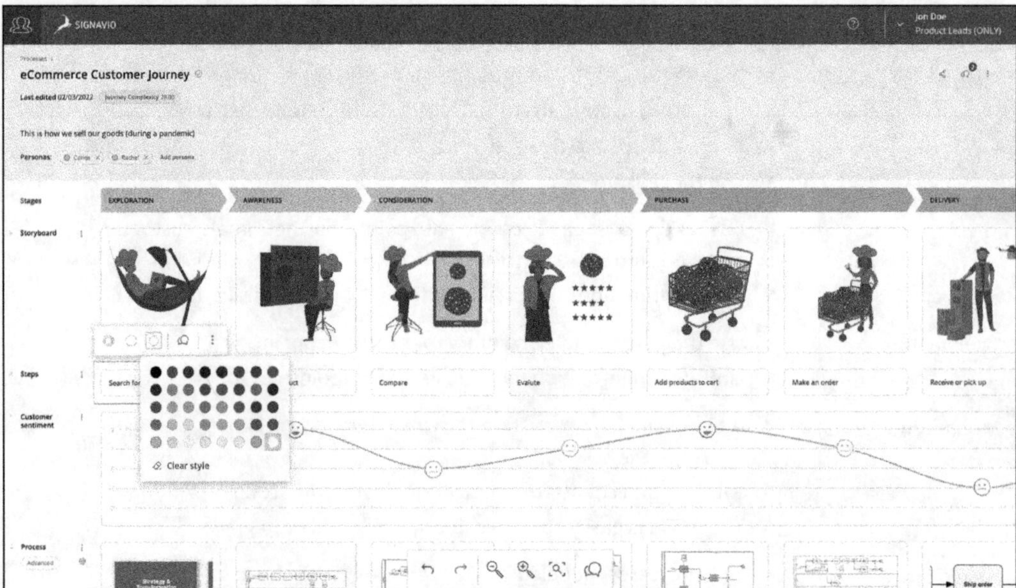

Figure 6.3 Tabular Journey Modeling

All of these customer journeys are based on standards and don't require expert knowledge. By using tabular journeys, you not only follow the industry standard but also ensure access to all business processes related to a specific customer journey. This allows you to identify, among other things, bottlenecks, duplications, and all unnecessary steps that have a negative impact on your customer relationships. With SAP Signavio Journey Modeler, you create highly transparent environments so that all employees can identify improvement potentials, simplify processes, and suggest innovations of individual business areas.

In addition, you benefit from a true suite approach by linking operational or CX data from the SAP Signavio Process Intelligence solution. It's worth mentioning here that you're not necessarily bound to SAP Signavio solutions. You can also connect SAP Signavio Journey Modeler to external systems such as Tableau and Looker Studio to view the data associated with each customer journey.

In Figure 6.4, you can see a customer journey for e-commerce. The tabular structure of the journey modeling takes into account the data shown in Table 6.3, including external widgets, customer sentiment, and linked process models. Based on the example of customer sentiment, we can see that the customer sentiment at the **DELIVERY** stage is negative and therefore offers room for optimization. We can subsequently take a closer look at the data from relevant metrics as well as the linked processes to better understand the background of the negative CX.

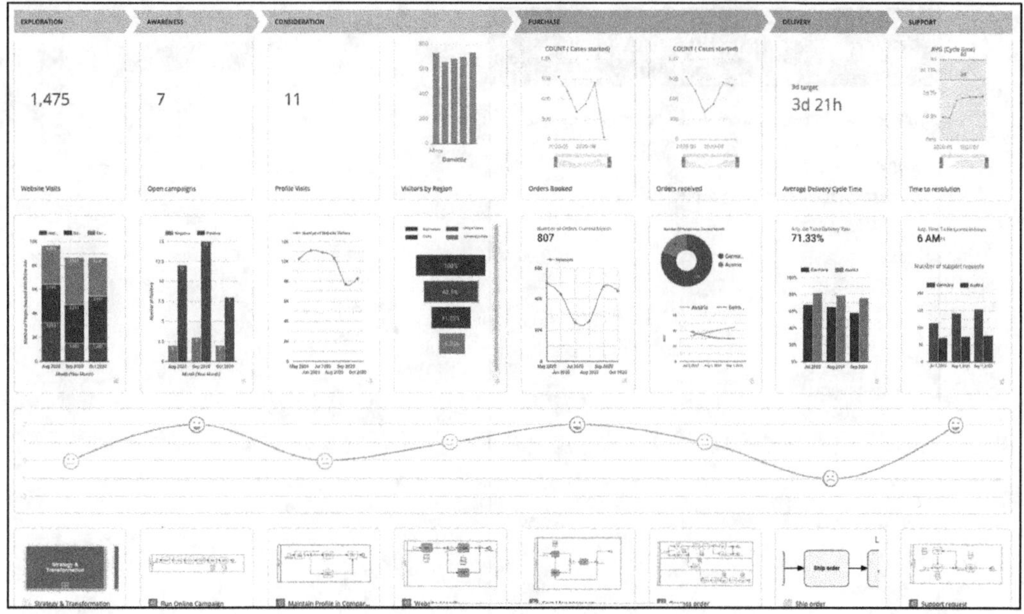

Figure 6.4 Data-Driven Journey Modeling

The perspective of a person interacting with your company is expressed in a journey model. The journey complexity score provides you with an additional data point for evaluating your journey models. The *journey complexity score* gives you an idea of how difficult it is to complete the journey. Typically, the CX is improved by reducing the complexity of journeys and making the underlying processes more efficient. Journey complexity scores can help you measure this progress. With SAP Signavio Journey Modeler, you can automatically create a journey complexity score based on IT systems, gateways, and handoffs for process complexity as well as stages, touchpoints, organizational units, and linked journey models. The journey complexity score provides insight into how many departments and people in your organization are involved in the journey and interact with each other until the journey is complete. A complexity score is also provided for processes. This score shows how efficient a process is and how many departments and people are involved in a particular process. You can visualize the journey complexity score directly within a journey and stay informed about journey and process complexity in real time. As you link new processes and journeys, the complexity score is automatically updated (see Figure 6.5).

The complexity score is displayed next to the journey model metadata in the right pane of the screen; the process complexity score is displayed in the details pane for all linked processes.

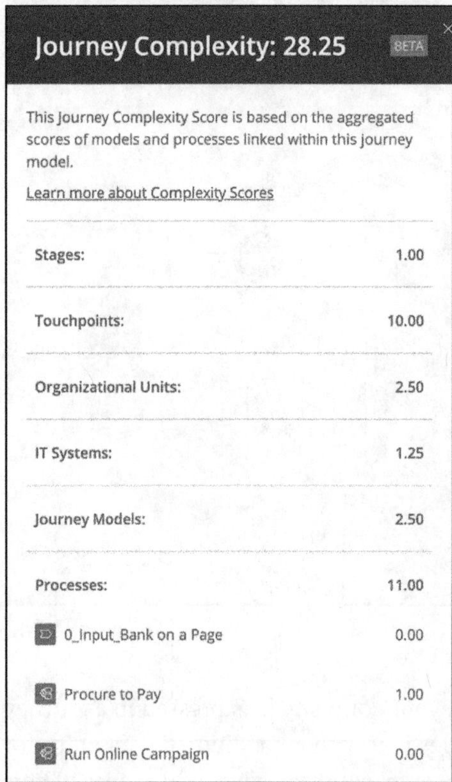

Figure 6.5 Journey Complexity Score

[»]

Determine Complexity Score

The complexity score for a journey model aggregates the complexity scores of all linked processes and journey models. It also takes into account properties such as the number of phases, the number of contact points, and the added organizational units and IT systems. The complexity score for a process takes into account how many decisions, handoffs, and parallel procedures are included. It also takes into account the number of IT systems involved, documents required, and linked processes. A more complex process has a higher process complexity score. Linking more complex processes to a journey model consequently increases the complexity of the entire journey. The numbers displayed for journey sections are ultimately aggregate scores, but not the number of specific elements in the journey model.

The great potential of SAP Signavio Journey Modeler is also that you can efficiently manage customer journeys, business processes, and data together with SAP Signavio Process Collaboration Hub on a single platform. With SAP Signavio Process Collaboration Hub, content is created, captured, and shared by any user at any time and across journey, customer, or transformation initiatives. As briefly touched on earlier, you can leverage the power of process mining by connecting SAP Signavio Process Intelligence process mining with customer journeys for in-depth analysis. In doing so, you can consider your operational, transactional, and outside-in data as part of your transformation initiatives. Finally, SAP Signavio Journey Modeler allows you to link all touchpoints with transactional and performance data and, as a result, monitor the evolution in real time. This in turn helps you see where your operation interacts with customers and how, providing you with a holistic view to improve your CX. Figure 6.6 illustrates process models, IT systems, and organizational units linked to the customer journey. In the example of the linked IT systems, we can see which IT systems are used in which process step by means of a solid line. More details on the linked content can be found in Table 6.1 earlier.

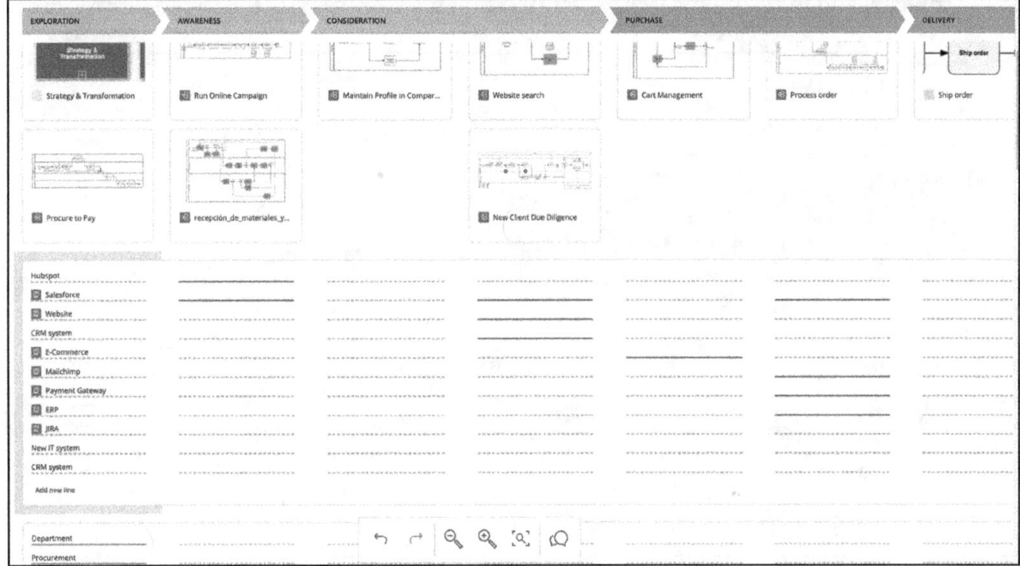

Figure 6.6 Linked IT Systems

6.3 Application Example: Working with a Customer Journey Map

As you already know from the previous chapters, we would like to introduce SAP Signavio Journey Modeler to you by means of an application example. This example is based on a customer journey map that can be modeled with SAP Signavio Process Manager (see Chapter 5).

What Is a Customer Journey Map?

A *customer journey map* shows a business process from the perspective of a customer. This often involves only a few steps, where the customer's main interest is in the outcome of the process rather than the effort put into the activities to get to the end result. A customer can be a person outside the organization (e.g., an end customer) or someone inside the organization (e.g., an employee). As a result, a customer journey map is a visual representation of all touchpoints throughout the customer journey with your organization from discovery to purchase to engagement.

In Figure 6.7, you can see a customer journey map for an order-to-cash process as it can be represented in the system. There we see Linda, who is a buyer of goods in your company. Her goal is to procure the best raw materials in a short time and at competitive rates from inventory suppliers as well as new suppliers. She researches and requests a quote from our sample company where Alan is employed as a sales representative. He prepares a quote for Linda. His goal in doing so is to ensure that the order is processed quickly and in accordance with quality standards within the company to satisfy Linda.

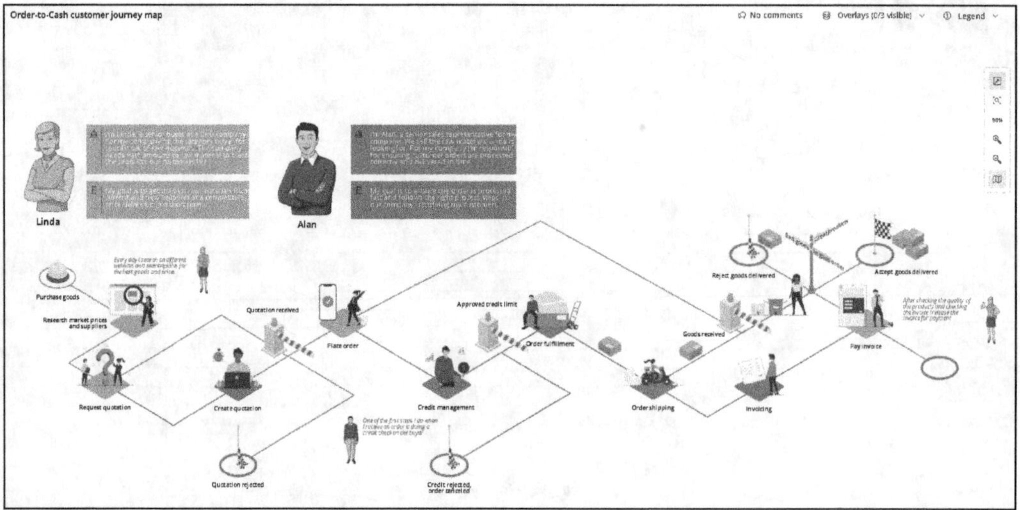

Figure 6.7 Example Customer Journey Map

There is no predefined format for the development of customer journey maps. However, most customer journey maps contain the elements listed in Table 6.4.

Element	Description
Persona (or actor)	A customer or user underlying the journey map.
Scenario	The desired action or goal that the persona is trying to achieve.
Actions	The steps that the persona and other personas go through during the journey.
Motivations	Thoughts or emotions about why the persona did something specific.

Table 6.4 Elements for Journey Maps

Element	Description
Moments of truth	The moments of truth when the persona forms a positive or negative opinion.
Opportunities	Improvement potentials of the journey. Opportunities make the journey map actionable.
Handover points/ responsibility	The moment/point at which interactions move from one team or person to another. This emphasizes accountability.
Metrics	How success is evaluated so teams can assess whether changes have positively impacted the experience.

Table 6.4 Elements for Journey Maps (Cont.)

Linda, the customer, researches a product and requests a quote from Alan, the seller, from a company that sells the product. Once the offer is made by Alan, there is what is called a *moment of truth*: Will Linda accept the offer or not? This moment is shown in Figure 6.7 by the barrier labeled **Quotation received.** If the offer is accepted, the order is placed; otherwise, the process ends. Once the order has been placed, Linda can only wait for the goods to arrive, while the sales organization must now get moving: first, the customer's creditworthiness has to be checked, and then the goods are delivered as quickly as possible and in good condition to meet Linda's expectations.

In this regard, some internal process steps need to be completed. For example, in Figure 6.8, click on **Credit management** in step four to open the side panel and display the link to the credit management process. This makes it apparent which processes are linked to a particular step on the customer journey map. The performance of the process and the outcome impact the customer journey and therefore Linda's experience. For example, if the credit management process produces a negative result, the order is canceled, which certainly doesn't make Linda a happy customer. This is an important part of the journey, both for Linda and for the selling company. The credit management process should therefore work well. We can obtain process performance indicators (PPIs) for this from the linked SAP Signavio Process Intelligence widgets in Figure 6.9.

Once the internal processes are complete, the goods are shipped and arrive at Linda's door. This is another moment of truth for Linda and crucial to whether she becomes a returning customer. If she is satisfied with the merchandise and the overall buying process, chances are she will order from the company again. If she is dissatisfied with the process or the goods, she will migrate to the competition.

In this use case, we would now like to put ourselves in the shoes of the company that wants to attract and retain Linda as a customer. Using SAP Signavio Process Manager, we can take a closer look at the end-to-end process of lead to cash and the associated credit management subprocess and provide insight into their operations. A key stakeholder in these processes is Linda, our customer. To understand how Linda experiences the lead-to-cash process and to create a corresponding outside-in view for our organization, we'll now use SAP Signavio Journey Modeler based on the data described in Table 6.2 (see Figure 6.8).

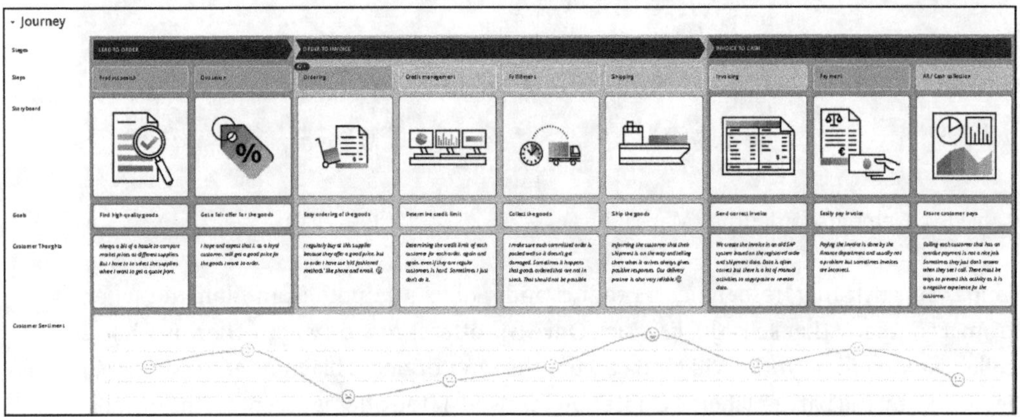

Figure 6.8 Customer Journey Model

This customer journey model represents the current situation and includes two personas: the customer (Linda) and the sales representative (Alan). It shows the main phases/stages and the main steps that need to be completed. Scrolling down, we see other elements of the model such as goals, customer thoughts, customer sentiment (the actual experience!), touchpoints, moments of truth, associated processes (these processes are provided through the link to SAP Signavio Process Manager), performance information as well as process intelligence (this information is provided through the link to SAP Signavio Process Intelligence), stakeholders, and systems. The customer journey model is fully configurable depending on your needs. For example, any type of section can be added as many times as needed next to another section or at the end of the journey. Common section types were described in Table 6.3.

Because our company has also implemented active process management and process performance reporting, the customer journey model also shows all the processes associated with the journey step and the performance of each process through the various widgets. You can see an illustration of this at the bottom of the screen in Figure 6.9. So, on a single page, you can see which steps make up the customer journey, which processes support this journey, and what impact these processes have on the actual CX.

How do you read the customer journey model more closely? First, let's look at some of the columns in the screen shown in Figure 6.8 in more detail by scrolling further down. As we do so, we'll take a closer look at the individual process steps of the customer journey.

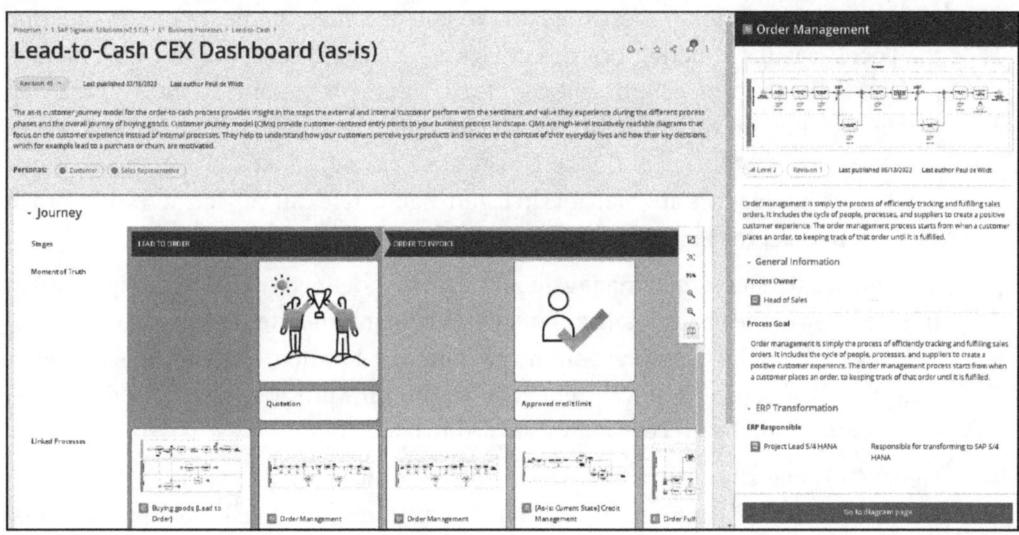

Figure 6.9 Process Model Linkage in the Customer Journey

In the first column, **Product search**, the process step of the product search is visualized. The customer is researching the goods to buy. The customer has some goals and thoughts about the process. The overall experience is expressed by the diagram in the **Customer Sentiment** row. As you can see, the customer's sentiment in this process step is rather mediocre, which is expressed by the neutral looking smiley. To complete the step, a website is used to collect information. The internal process associated with this is called purchasing goods. We can see process details for this from the linked process **Buying goods (Lead to Order)** under **Linked Processes** in Figure 6.9. The company doesn't have relevant performance information, so the space for process intelligence widgets is empty (see also Figure 6.10). At the end, we see the list of stakeholders and IT systems in use (see Figure 6.11).

In the second process step (**Quotation** column in Figure 6.8), the customer asks for a quote for the goods she wants to order. The quotation itself is a moment of truth because it decides whether the customer orders the goods or not. The quote must be created by the sales representative, as shown in the attached process order management. Scroll further down the screen, as shown in Figure 6.9, and click on the linked **Order Management** process. The page area with more process details will open. Optionally, the complete process can be opened with SAP Signavio Process Manager to show all tasks of the sales representative. In this example, however, we'll stay with the customer journey model.

If the offer doesn't meet the customer's expectations, the order won't be placed. Therefore, it must be ensured that the CX, as shown in Figure 6.8 and illustrated by a happy smiley icon, remains positive. Otherwise, the customer journey would end at this point.

However, if the customer is satisfied and accepts the offer, the next phase is ordering the goods (see the **Ordering** column in Figure 6.8). Based on the dissatisfied-looking smileys in the **Customer Sentiment** column in Figure 6.8, we can see that this is a step where customer sentiment is low, which lends itself to improvement in the ordering activity for the customer. For this and the next steps, some linked PPIs from SAP Signavio Process Intelligence are available in Figure 6.10. These can be seen at the bottom of the screen under the **Process Intelligence Widgets** section.

Credit management (**Credit management** column) is defined according to Figure 6.8 as the next step in the process, and although this is an internal step in our customer journey, the value in the **Customer Sentiment** column is low here as well. The sales representative has some skeptical thoughts about this process, which leads to the conclusions from the process analysis in Figure 6.10.

Figure 6.10 generally illustrates the moments of truth, linked process models, and process intelligence widgets per process step. For example, we can click on the individual process models or widgets to get the corresponding detailed information. To get to this screen, we simply scroll down a bit in the previous screen.

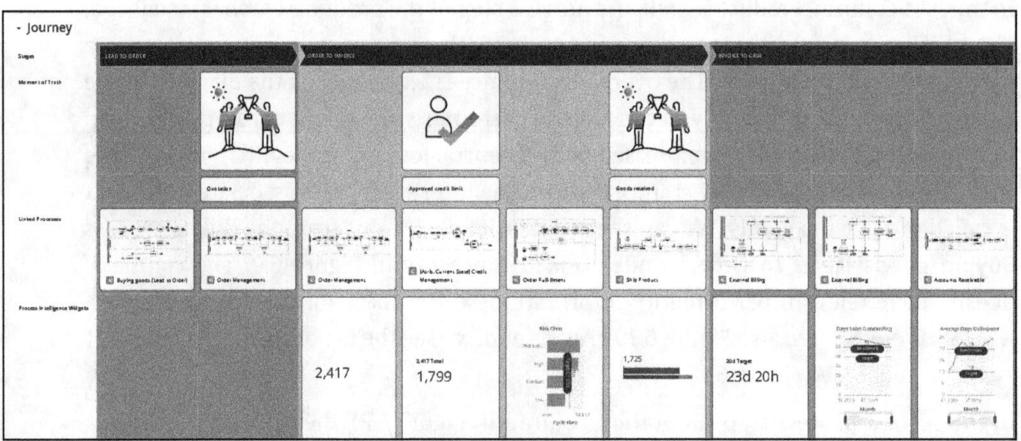

Figure 6.10 Excerpt of Customer Journey Model Elements

The other columns in Figure 6.8 represent various further steps of the end-to-end lead-to-cash process with the associated detailed subprocesses and linked performance information. This provides a clear overview of which activities contribute to the CX and how they are perceived.

In conclusion, the elements in Figure 6.11 provide an overview of the stakeholders involved and the IT systems used for the respective process steps. To get to this screen, simply scroll down further in the previous screen. Our journey contains linked processes, and we've added the **Organizational Units** and **IT Systems** areas, among others.

Thus, all organizational units and IT systems detected in the linked process models will automatically be displayed in this section. The first element lists all organizational units or IT systems that are recognized in a linked process. Using the IT systems example, in our case, **Salesforce**, **SAP FICO**, **SAP SD**, **SAP FSCM**, **SAP MM**, and **Digital Payment System** are involved in the processes. If a column contains a linked process, organizational units or IT systems in this process are indicated by a solid line. The automatically recognized organizational units or IT systems can neither be removed nor further processed.

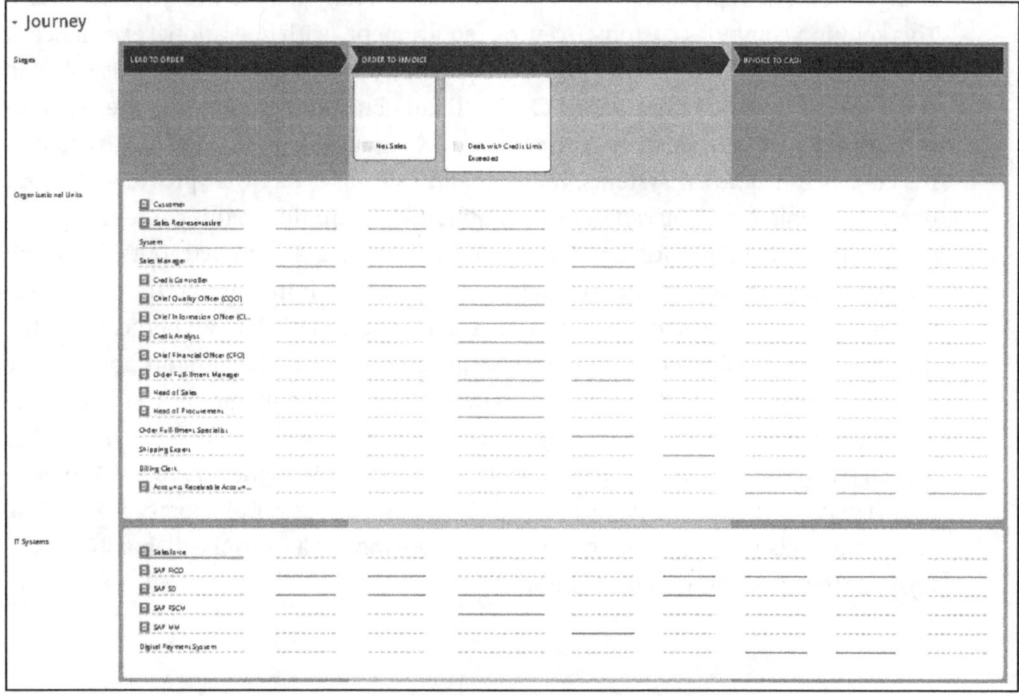

Figure 6.11 Organizational Units and IT Systems Involved in the Customer Journey

Adding IT Systems

In addition to the automatic recognition of IT systems by linked processes, the IT systems can also be added manually to your journey. To do this, you add a new row in the system and enter the name of the desired IT system. The IT system is listed in the first column, and all columns are marked with a dashed line. You activate manually added IT systems for a single column by clicking on the dashed line. Activated systems are indicated by a solid line. If you delete a manually added IT system from the first column, it's removed from the journey model. The same logic applies to manually adding organizational units.

6.4 Summary

SAP Signavio Journey Modeler is a helpful tool that supports you in achieving customer excellence. You can use SAP Signavio Journey Modeler to achieve customer excellence by developing customer journey models for touchpoint improvements, aligning business processes with CXs, and sharing customer-centric journeys for feedback and potential experience improvements.

SAP Signavio Journey Modeler is a software solution that allows you to visualize how your customers, representatives, or employees actually experience your company. This solution combines customer-centric requirements with operational excellence at scale, creating a solid connection between journeys, processes, and data. This solution supports the operationalization of CXs by linking business processes to the way customers really experience them. You can use SAP Signavio Journey Modeler to customize your organizational systems, metrics, and roles in a way that provides the best possible interactions. You can mine data from process mining in this context to gain a competitive advantage for CX. The solution promotes a complete outside-in and inside-out perspective and helps you better understand, improve, and transform customer interactions with your business by translating experiences and process analytics into operational reality. This includes generating more insights on how to give customers what they really want. This is always based on the integrated CXs in the customer journey. SAP Signavio Journey Modeler unfolds its full potential when linked to SAP Signavio Business Transformation Suite, especially SAP Signavio Process Manager, SAP Signavio Process Intelligence, and SAP Signavio Process Collaboration Hub. The latter is especially interesting for companies looking for a central collaboration platform for the entire process management.

Chapter 7
SAP Signavio Process Collaboration Hub

SAP Signavio Process Collaboration Hub is the core of SAP Signavio Transformation Suite and serves all users as a central entry point into your process landscape. The tool facilitates the holistic collaboration of all stakeholders not only during process design but also during process execution and continuous process improvement.

7

In Chapter 1, Section 1.3, we noted that complex situations are usually best solved through collaboration and by taking into account a wide variety of viewpoints and ideas. Via SAP Signavio Process Collaboration Hub (called "the Hub" for short), process models and additional information can now be made available to all employees (see Figure 7.1).

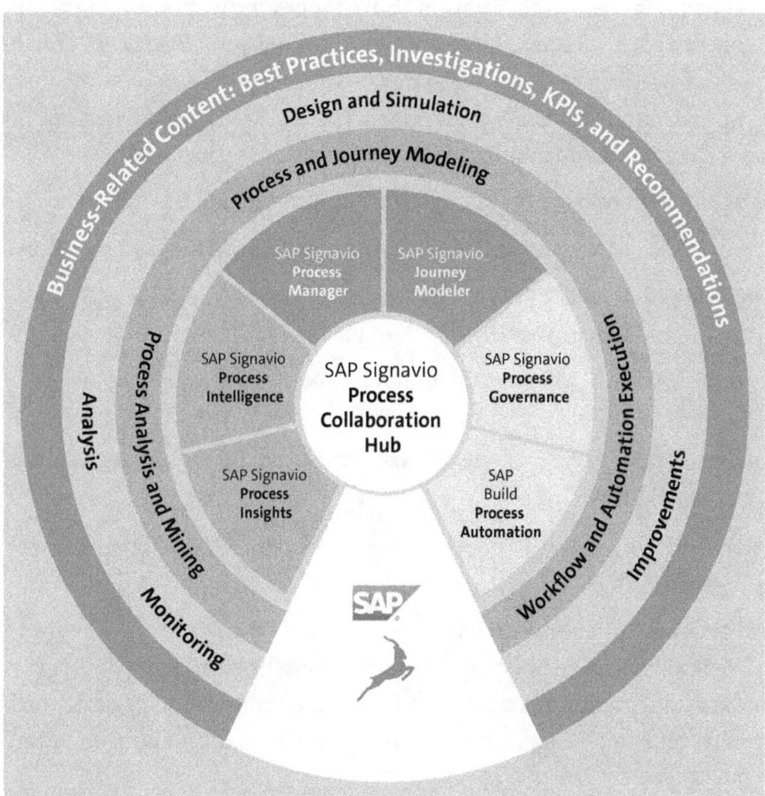

Figure 7.1 SAP Signavio Process Collaboration Hub in Focus

When it comes to collaborative process management, SAP Signavio Process Collaboration Hub is the core. This central hub enables effective communication within your organization and across organizational boundaries, while ensuring collective knowledge and collaborative work efforts. This tool supports real change for companies across all industries and geographies. Users can leverage commenting and sharing capabilities to foster a next-generation collaborative and transparent work environment by leveraging a single source of process truth across the enterprise. The Hub further drives your organization's digital and cloud strategy through end-to-end visibility, including SAP S/4HANA migrations. Let's take a closer look at that.

In Section 7.1, we first address the various features of SAP Signavio Process Collaboration Hub. Section 7.2 shows different potentials of the solution. Subsequently, we'll illustrate the use of SAP Signavio Process Collaboration Hub in practice using an example (Section 7.3).

7.1 Functions

SAP Signavio Process Collaboration Hub provides a holistic platform to leverage the knowledge of the masses and to continuously innovate and optimize processes. Accordingly, SAP Signavio Process Collaboration Hub changes the way employees work together in your company. This means, for example, that SAP Signavio Process Collaboration Hub acts as a single source of truth of your processes, providing you with live insights and giving your employees an enterprise-wide and transparent focal point where they can pool their work and expertise related to business processes.

The core features of SAP Signavio Process Collaboration Hub include the following:

- Receive updates on changes to relevant content, requested feedback, and mentions.
- Understand, track, and manage joint projects.
- Manage process content from the entire SAP Signavio Process Transformation Suite.
- Optimize presentations and inform stakeholders about updates in an easy-to-understand way.
- Obtain a comprehensive overview of the process landscape, including documentation, analysis, and key figures.
- Promote initiatives across the company with the help of a uniform level of information.

With SAP Signavio Process Collaboration Hub, you can stay up to date on changes to your process models in real time and display all process content clearly and easily. This enables you to understand, track, and manage collaborative projects and organize process content across the entire SAP Signavio Business Transformation Suite, including

SAP Signavio Process Intelligence, SAP Signavio Process Manager, SAP Signavio Journey Modeler, and SAP Signavio Process Governance. The Hub drives initiatives across the enterprise by ensuring all your stakeholders are on the same information level, streamlining presentations, and providing easy-to-understand updates to all employees.

SAP Signavio Process Collaboration Hub can also be personalized. You can assign people to specific groups. These group-based entry points mean no more time wasted searching for information relevant to the actual work. You can control exactly what the hub looks like for specific user groups and how it's laid out. You can also curate your own newsfeed by subscribing to the process models, conversations, or workspaces that are most important to you. By customizing the Hub view, you can better structure your tasks and improve user satisfaction and productivity. Access to the SAP Signavio Process Collaboration Hub features is managed using audiences, visibility, and licenses. Table 7.1 provides an overview of the numerous application features of SAP Signavio Process Collaboration Hub.

Feature	Description
Read access	View unpublished content with read-only access in SAP Signavio Process Collaboration Hub: ■ Navigation via process hierarchies ■ Display of process details ■ Full-text search ■ Bookmark capability
Comment access	Comment on process content: ■ Invite to comment ■ Intuitive commenting view ■ Status for comments (e.g., **New**, **Incorporated**, **Ignored**)
Notifications	Get notified about new comments and revisions to articles you've created, edited or saved as a favorite. Modelers are notified about all revisions; SAP Signavio Process Collaboration Hub users are notified only about actions on the published revision. You'll be notified about the following changes: ■ Someone adds a comment to one of your diagrams ■ Someone clarifies one of your comments ■ Someone reopens one of your comments ■ Someone mentions you in a comment ■ Someone replies to one of your comments

Table 7.1 Functions of SAP Signavio Process Collaboration Hub

Feature	Description
Launchpad	Set up by administrators, users benefit from a personalized entry point tailored to the needs of their user group. This way, SAP Signavio Process Collaboration Hub users can focus on the elements that are most relevant for their tasks and respective roles. Launchpad elements are displayed in the sidebar on the left side of the screen. The main elements of the launchpad include the following: ▪ **Home**: See the content you last viewed and the content you've saved as favorites. ▪ **Newsfeed**:See what new content has been published in your workspace. ▪ **Favorites**: Find saved items such as diagrams, dictionary items, dictionary categories, files, and folders that you've marked as favorites. ▪ **History**: Find content that you've recently opened. ▪ **Tasks**: You can open your inbox in SAP Signavio Process Governance with SAP Signavio Process Collaboration Hub by clicking **Tasks**. ▪ **Processes**: See all the diagrams published in your workspace. Access depends on the settings for your workspace. ▪ **Investigations**: See your SAP Signavio Process Intelligence investigations. ▪ **Dictionary**: View, edit, and create dictionary items for your workspace.
User menu	Access the user menu by clicking on your user name. The following options are available in the user menu: ▪ **View**: Switch between the **Preview** and **Published** views. ▪ **Products**: Access other SAP Signavio products. ▪ **Workspaces**: Switch between workspaces if you're a member of different workspaces. ▪ **Target groups**: See your current target group. If you're a member of multiple target groups, you can switch between them. ▪ **Content languages**: Select in which language the content should be displayed. The availability of a language for an element depends on the settings for the element. ▪ **Logout**: Log out in this area.

Table 7.1 Functions of SAP Signavio Process Collaboration Hub (Cont.)

Feature	Description
Actions	Find all the actions available for a page in the upper-right corner of the SAP Signavio Process Collaboration Hub screen. Depending on what is open at the moment, you can do the following: ■ Print an element. ■ Save an element as a favorite. ■ Copy a diagram link to the clipboard. ■ Open the comment area.
Create	Display all available content types. When you click a content type, the editor opens in a new tab in the browser. If you don't have write access to the folder you're viewing, a dialog prompts you to select a location before opening the editor. If available, you can also switch to SAP Signavio Process Governance or SAP Signavio Process Intelligence using the **Create** button.
Search	Search for all content from SAP Signavio Business Transformation Suite. To refine the search results, you can add search filters. Search filters are available for diagrams, dictionary items, files, and folders.
Views	Modelers can switch between two views: **Preview** and **Publish**. In **Publish**, only the published versions of elements are displayed. In **Preview**, labels in the diagram table and in the diagram page header indicate the current status of a diagram. For modelers, SAP Signavio Process Collaboration Hub opens in **Preview** by default. The user menu can be used to switch between views.
Publication mechanism	Publish process content: ■ Diagram publishing (integration with version management) ■ Batch publishing ■ Visualization of published diagrams
Diagrams	Click the **Processes** button to view all published diagrams in your workspace. When you click on a diagram element, the corresponding element attributes are displayed in the details pane. If a diagram is linked to other diagrams, you can also view the linked diagrams in SAP Signavio Process Collaboration Hub.
Manage folders and charts	Choose from the following options: ■ Publish charts and withdraw publications ■ Rename, move, and delete folders and diagrams

Table 7.1 Functions of SAP Signavio Process Collaboration Hub (Cont.)

Feature	Description
Embedding function	Embed process content in web pages: ■ Embed diagrams in wikis and blogs via generated HTML snippets ■ Embed PNG diagrams
Role-based publishing	Configure different read permissions: ■ Connection to the company's internal Active Directory (single sign-on [SSO] for SAP Signavio Process Collaboration Hub users) ■ SSO also possible for modeling users
Embedding in Microsoft SharePoint	Embed process content in Microsoft SharePoint: ■ Integration of the reading view within the Microsoft SharePoint interface ■ Customizability of the user interface (UI) according to the Microsoft SharePoint environment ■ Signavio WebPart for Microsoft SharePoint
SSO via Microsoft SharePoint	Log in via Microsoft SharePoint: ■ Secure access to the SAP Signavio Process Collaboration Hub from Microsoft SharePoint ■ Role-based publishing based on Active Directory or Microsoft SharePoint roles
SSO via SAML	Authenticate using the Security Assertion Markup Language (SAML) identity provider: ■ Role-specific publication of diagrams, based on SAML user identities ■ Supported providers: SAP ID Service, Google SSO, Okta, Microsoft Active Directory Federation Service (AD FS)

Table 7.1 Functions of SAP Signavio Process Collaboration Hub (Cont.)

We now take a deeper look at some of the features of the extensive launchpad menu bar on the left side of the SAP Signavio Process Collaboration Hub screen. Under the menu item **Newsfeed** in Figure 7.2, we can see all the news, for example, regarding process model changes as well as releases. This way, we're always up to date regarding our business processes.

Under **Favorites** in Figure 7.3, we keep track of our favorite process models. This allows an even faster entry into business processes that are important for us.

In Figure 7.4, we navigate further to the menu item **Recent**. Consequently, we can view the last activities such as the last edited or published process models.

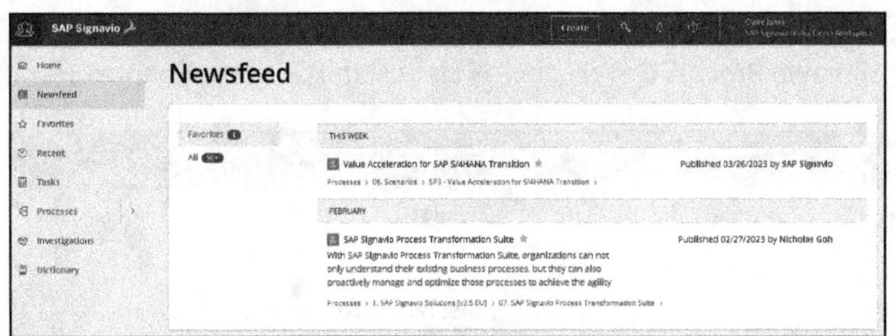

Figure 7.2 SAP Signavio Process Collaboration Hub: Newsfeed

Name	Type	Description	Last published
2EM - XX - Just-In-Time Supply to Customer	BPMN	—	12/17/2021
Business Operating Model (as-is)	Value Chain	A value chain is a set of activities that a firm operating in a specific industry performs in order to deliver a v...	09/13/2022
Company on a Page	Navigation Map	Visualize the main End-to-End Business Processes of the company and their performance levels.	07/18/2022
Customer Journey map	Picture	—	—
ERP Transformation Initiative (SAP Activate)	Navigation Map		10/11/2022
Hire-to-Retire	Folder	—	—
Lead-to-Cash - Business & Digital Transformation Cycle	Navigation Map		06/27/2022
Lead-to-Cash - Business & Digital Transformation Cycle (...	Navigation Map	—	06/27/2022
New Process - BPMN design	BPMN	—	06/16/2022
Order Fulfillment	Folder		
SAP Signavio Platform Enablement Framework	Value Chain	The Enablement Platform allows you to select and view various videos that will enable you to use the SAP Si...	07/08/2022
SAP Signavio Process Transformation Suite	Navigation Map	With SAP Signavio Process Transformation Suite, organizations can not only understand their existing busi...	02/27/2023
Value Acceleration for SAP S/4HANA Transition	Navigation Map	—	03/26/2023

Figure 7.3 SAP Signavio Process Collaboration Hub: Favorites

Name	Type	Description	Viewed
Company on a Page	Navigation Map	Visualize the main End-to-End Business Processes of the company and their performance levels.	2 minutes ago
[As-Is] Fashion retailer sourcing process	BPMN	Company: global fashion retailer Situation: the business has been continuously in the press, being ac...	5 minutes ago
Sustainability	Navigation Map	—	5 minutes ago
Enterprise Transformations	Navigation Map		6 minutes ago
Process Excellence	Navigation Map	—	7 minutes ago
Lead-to-Cash - Business & Digital Transformation Cyc...	Navigation Map	—	7 minutes ago
Source to Pay	Value Chain	The generic process "Source to Pay" (S2P) includes all activities associated with managing the compre...	12 hours ago
End-to-End Processes	Value Chain	A Process represents a structured set of sub processes, process steps or activities designed to accom...	12 hours ago
Plan to Optimize Fulfillment (generic)	Value Chain	The Business Value Flow Diagram is a structured collection of value generating Business Activities tha...	12 hours ago
SAP Signavio Process Explorer	Value Chain	SAP Signavio Process Explorer provides access to accumulated process knowledge in a central location...	13 hours ago
L2C CM - Transformation Business Objectives	Navigation Map	This diagram lists the context, key issues, key goals, and key measurement indicators for the transfor...	21 hours ago
Home	Navigation Map		21 hours ago
Lead-to-Cash CEX Dashboard (as-is)	Journey	—	Yesterday
Request to Resolution (generic)	Value Chain	The Business Value Flow Diagram is a structured collection of value generating Business Activities tha...	Yesterday
Lead to Cash	Value Chain	The generic process "Lead to Cash" covers all activities related to marketing and selling products and ...	Yesterday
Order to Fulfill (generic)	Value Chain	The Business Value Flow Diagram is a structured collection of value generating Business Activities tha...	Yesterday

Figure 7.4 SAP Signavio Process Collaboration Hub: Last Activities

Under **Tasks** in Figure 7.5, we get an overview of various tasks assigned to us. At this point, a link to SAP Signavio Process Governance is also established (see Chapter 8).

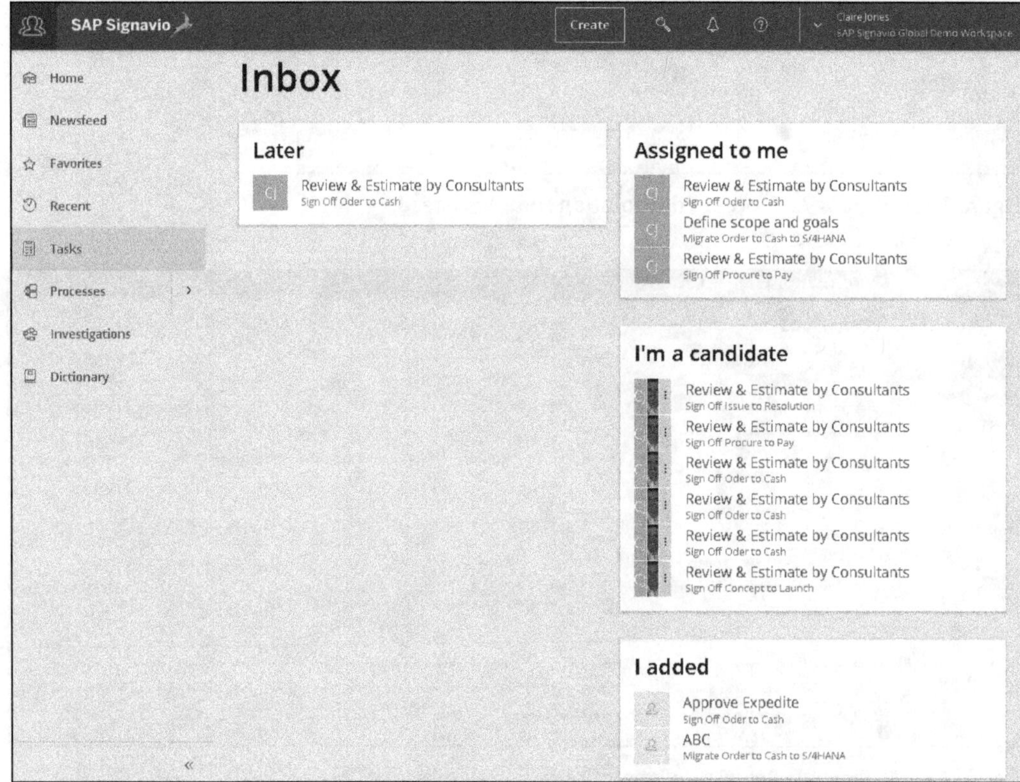

Figure 7.5 SAP Signavio Process Collaboration Hub: Tasks View

The **Processes** menu item in Figure 7.6 is entirely dedicated to our business processes. We can display all process and journey models directly within SAP Signavio Process Collaboration Hub. In connection with this, we also get various process information, for example, regarding quality and risk management, scope and goals, or number of revisions. We can comment on the processes and edit, simulate, and share them with stakeholders via the SAP Signavio Process Manager modeling tool.

If we want to get an overview of our process analyses, we click on the **Investigations** tab. In Figure 7.7, you can see all sorts of process investigations that link directly to SAP Signavio Process Intelligence (see Chapter 4).

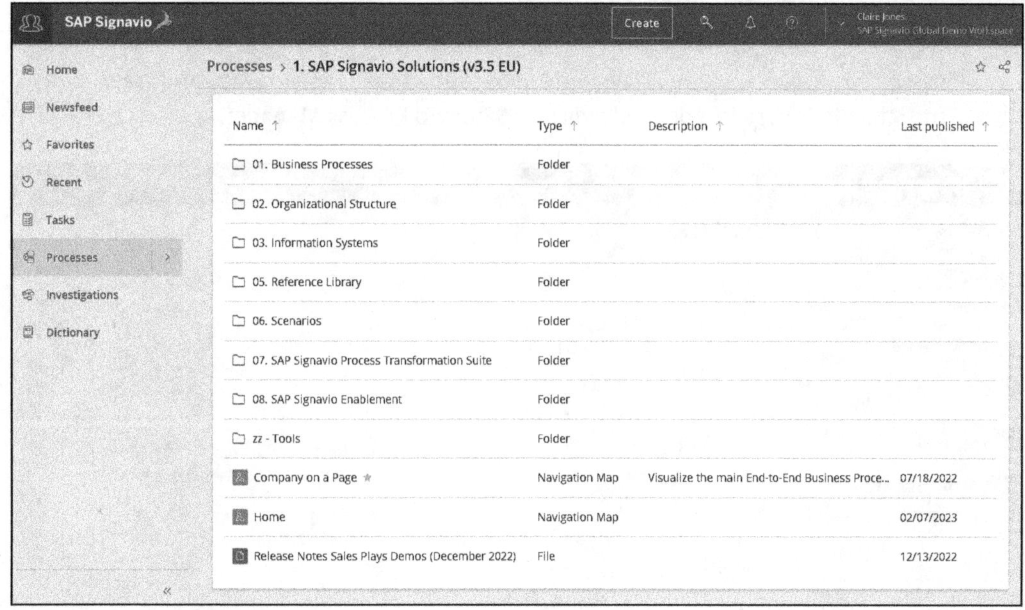

Figure 7.6 SAP Signavio Process Collaboration Hub: Business Processes

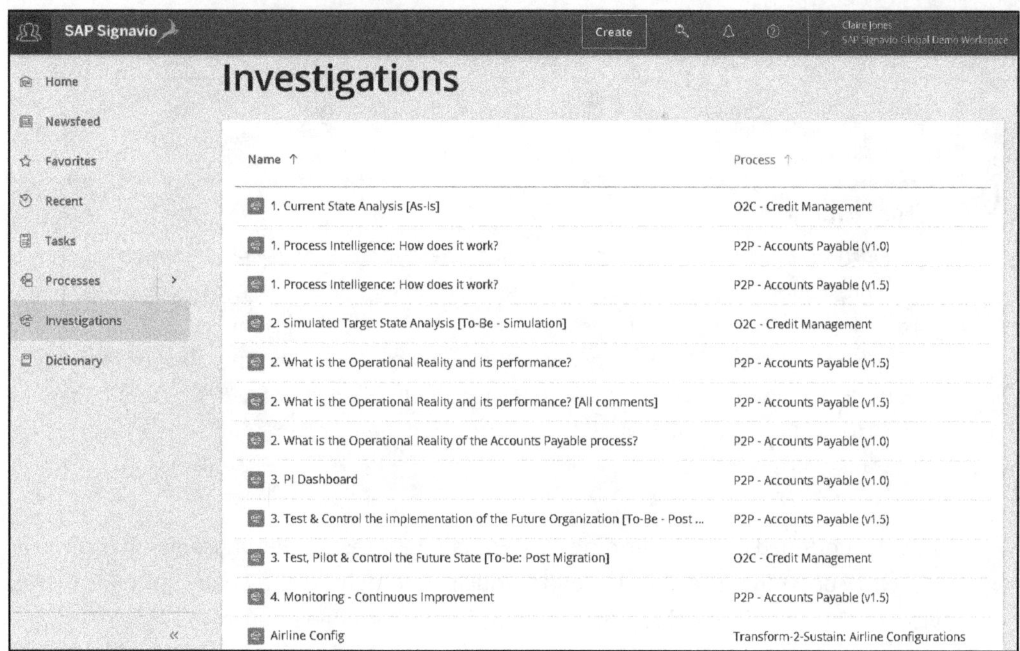

Figure 7.7 SAP Signavio Process Collaboration Hub: Process Analyses

Finally, the menu item **Dictionary** in Figure 7.8 provides exciting insights into the dictionary, in which a wide variety of terms are defined in relation to the business processes. Examples include IT systems used, documents, risks, goals, and roles. Entries into the dictionary can be made via SAP Signavio Process Manager.

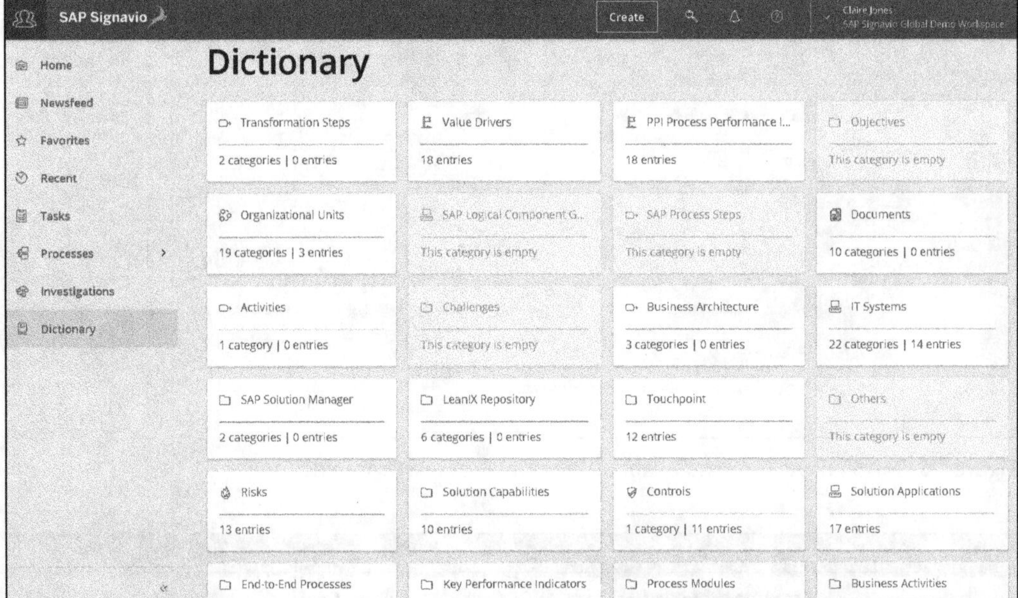

Figure 7.8 SAP Signavio Process Collaboration Hub: Dictionary

The Dictionary

The dictionary is the central object management repository in SAP Signavio. A dictionary item represents an object that is relevant to one or more of your processes. You can use the dictionary to manage and reuse specific modeling elements. The dictionary also helps you ensure that all modelers use the same terms and the same elements in your organization-specific modeling environment. The dictionary is a critical component for achieving consistent and well-structured management of business objects in your diagrams.

As we mentioned at the beginning, SAP Signavio Process Collaboration Hub is the central entry point into the SAP Signavio world. Accordingly, as shown in Figure 7.9, we can also access the individual SAP Signavio solutions from there.

Figure 7.9 Calling SAP Signavio Solutions

7.2 Benefits

The core benefits of SAP Signavio Process Collaboration Hub include the following:

- Strengthen teams and improve collaboration.
- Find out central process data source.
- Cooperate through all-round vision.
- Present complex processes with ease.
- Keep track of changes and feedback.
- Dismantle silos.
- Promote knowledge sharing and engagement.

Primarily, SAP Signavio Process Collaboration Hub enables effortless collaboration to inform users about the latest results of your business process activities and thus keep all stakeholders up to date. The tool has modern and particularly user-friendly features for holistic collaboration regarding your business processes. The intuitive UI makes it easy to use for employees of all knowledge levels. As they are individually notified of relevant updates and conversations, managers can close the gap between accountability and action. As a result, accelerated information sharing gives you the opportunity to generate more ideas, optimize processes, and lower inhibitions to change.

> **Employees in Focus**
> The key to sustainable success lies with your people. With SAP Signavio Process Collaboration Hub, you can collaborate in a unified way by creating a single source of process truth for all teams, breaking down business silos and creating a better understanding of your key performance indicators (KPIs), tasks, and projects.

Furthermore, SAP Signavio Process Collaboration Hub creates a single source of truth about your processes and thus ensures a common understanding throughout the entire company. Everything you and your employees need is in one central place. Data, process models, and analyses from SAP Signavio Process Intelligence, SAP Signavio Process Manager, and SAP Signavio Process Governance can be accessed directly via the Hub. The search function across the entire SAP Signavio Business Transformation Suite ensures that you can find all process information relevant to your roles and see what your colleagues are currently working on. You can filter user-specific results across the suite to clean up large data logs with a simple click. The Hub's simple version control means you can keep track of the status of work in progress and easily see the end result of your efforts. Figure 7.10 shows, for example, a purchase-to-pay process model in SAP Signavio Process Collaboration Hub, including all revisions. The person responsible can publish the latest revision of the model directly through the solution so that the revised version is visible to the entire audience.

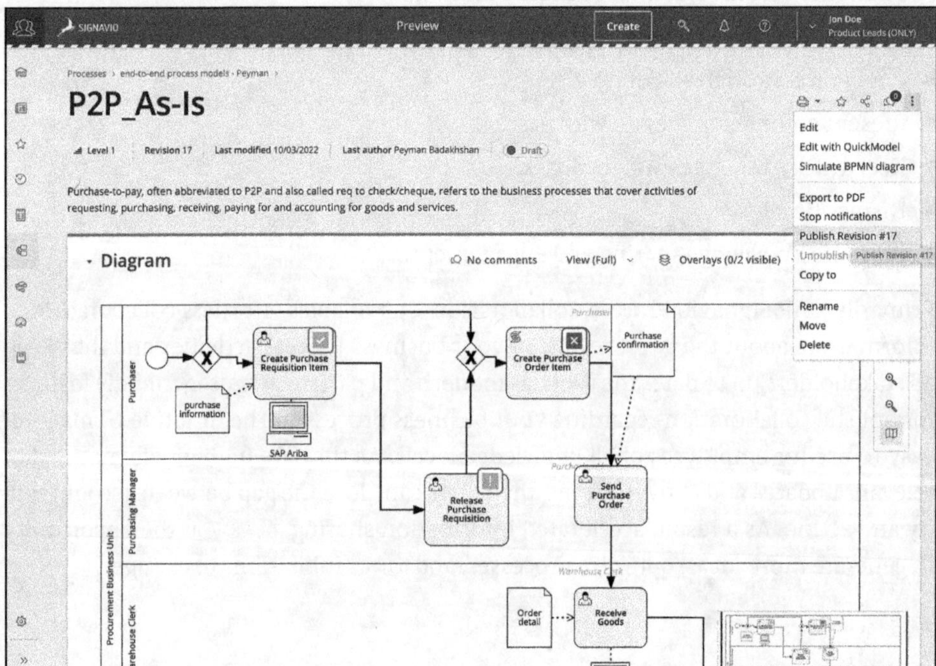

Figure 7.10 SAP Signavio Process Collaboration Hub: Publish Process Models

SAP Signavio Process Collaboration Hub also allows you to make all of your company's processes effectively and, most importantly, easily accessible via the intuitive presentation mode. Process specialists and business users can collaborate to increase the adoption of new processes in your organization. Simple step-by-step instructions of even complex business processes enable your employees to understand and improve process excellence. This tool not only gives you an overall view of your process landscape but also lets you effortlessly navigate your entire process repository. You provide detailed full-screen presentations in the integrated overview mode. You can also show the most important process interactions as well as switch smoothly between a general overview and detailed content information.

Figure 7.11, for example, illustrates a holistic overview of all business processes within a company, from order-to-fulfill processes to operate-to-maintain processes and from sales to asset management. Widgets from SAP Signavio Process Intelligence analyses are linked to keep an eye on the KPIs for each business process at a glance. This enables even more efficient and targeted management of your business processes.

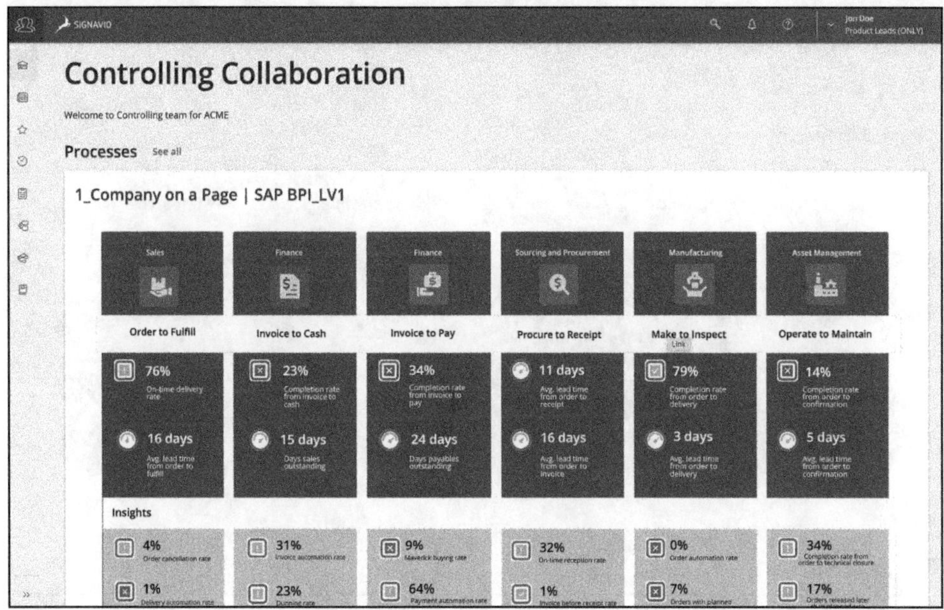

Figure 7.11 SAP Signavio Process Collaboration Hub: Company on a Page (Example)

SAP Signavio Process Collaboration Hub places a special emphasis on increasing both the efficiency and effectiveness of your business. Everyone in your organization is up to date and informed about the what, why, and how of applicable work. As alluded to in Part I, organizational silos, with associated problems such as conflicting priorities and goals, diluted effort, and employee resistance to decisions based on incomplete or unclear data, are among the challenges of successful business transformation. SAP Signavio Process Collaboration Hub provides a corresponding solution to each of these problems, resulting in improved productivity across the enterprise. In this way, you can set new milestones in operational excellence and improve the way teams work with clearly defined goals and structured responsibilities.

SAP Signavio Process Collaboration Hub provides you with a 360-degree view of your process landscape from documentation to analysis to metrics. You can publish diagrams to inform your organization about key processes, improving alignment between process owners and the rest of the organization. Navigation maps that you can create and share make it easy for everyone to understand important process-related information and metrics. By creating and sharing these navigation maps, you can guide employees through important process-related information and metrics accordingly. In short, with SAP Signavio Process Collaboration Hub, you can provide process experts and professionals with access to all SAP Signavio solutions from a single location, ensuring a full all-round view regarding your business processes. The initial screen shown in Figure 7.12 is freely configurable.

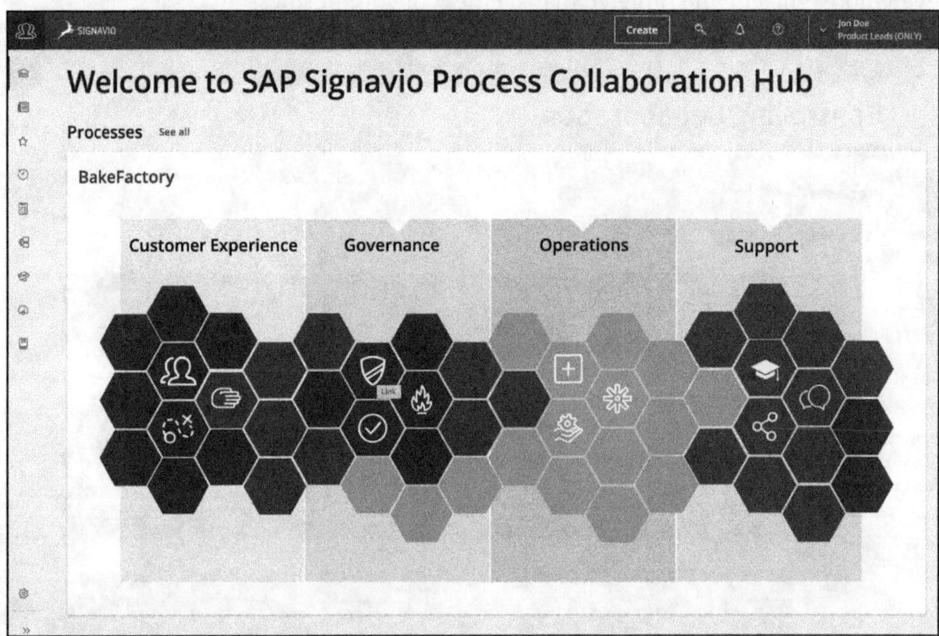

Figure 7.12 SAP Signavio Process Collaboration Hub: Initial Screen (Example)

With SAP Signavio Process Collaboration Hub, you can curate and customize content relevant to specific audiences. This simplifies the use of the tool and subsequently improves the collaboration of your teams. For example, capture charts can differentiate themselves with respect to your finance and HR departments. You can define both the groups and the content available to them. In addition, you can configure target groups that relate to each group in SAP Signavio Process Collaboration Hub. To do this, you can configure the appearance of SAP Signavio Process Collaboration Hub under

Settings in the sidebar. You can add your own logo, customize the SAP Signavio Process Collaboration Hub color scheme, set up personalized launchpads for different target groups, and manage attribute groups and attribute visibility for each target group. To add additional target groups, you first need to add user groups to your workspace. You can access the SAP Signavio Process Manager user management via **Settings** because access rights to content are set directly in SAP Signavio Process Manager. To use this feature, you need an administrator account. With a target group-specific start page, you can subsequently customize specific metrics and entry diagrams that are relevant for the users' work. In short, you're in full control over how specific audiences in your organization should browse relevant content (see Figure 7.13).

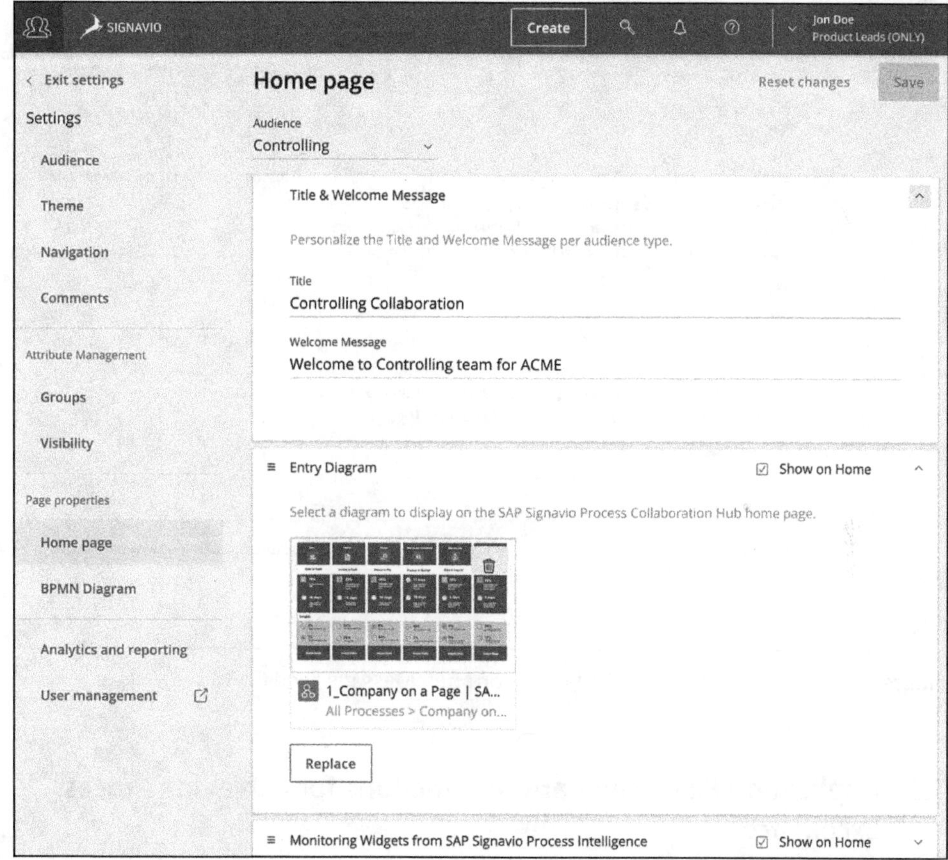

Figure 7.13 SAP Signavio Process Collaboration Hub: Configuration of the Target Group

With SAP Signavio Process Collaboration Hub, you can keep track of changes in content, feedback, and mentions regarding your business processes. For example, you can view process-related content that has been recently changed in your organization. Once you've been mentioned in a comment or content, you follow has been updated, you'll receive notifications directly in SAP Signavio Process Collaboration Hub. You'll receive additional email notifications when you've been invited to provide feedback on a diagram (see Figure 7.14). This is how teamwork works!

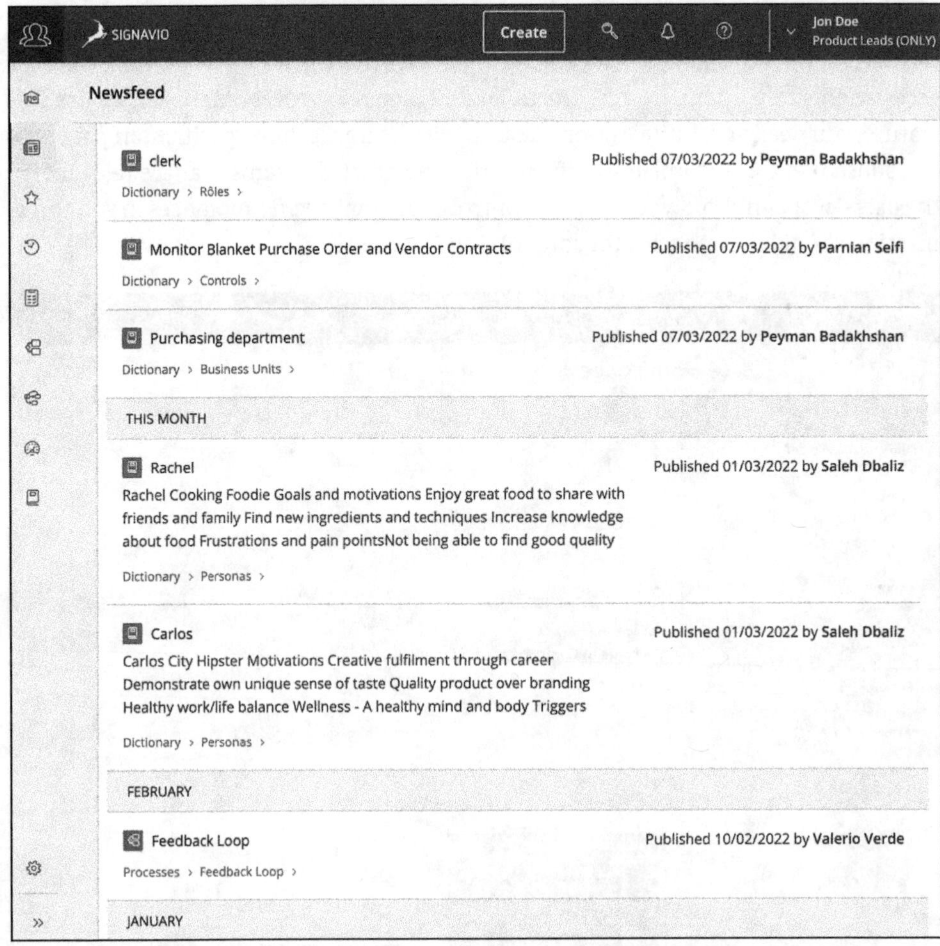

Figure 7.14 SAP Signavio Process Collaboration Hub: Message Feed

7.3 Application Example: Various Functions for Achieving Process Excellence

We'll now show you how to use SAP Signavio Process Collaboration Hub by presenting a practical application example. This example is structured very broadly to get as many impressions of the solution as possible. The goal is to let you experience the various possibilities of SAP Signavio Process Collaboration Hub in action, independent of industry and across roles. We assume that we've already modeled the first business processes using SAP Signavio Process Manager (see Chapter 5) and analyzed them with SAP Signavio Process Intelligence (see Chapter 4).

In Figure 7.15, you can see the initial screen of SAP Signavio Process Collaboration Hub. From here, you have the option to create new process documents via the **Create** button.

Figure 7.15 SAP Signavio Process Collaboration Hub Initial Screen

A window opens where we can choose from various document types (see Figure 7.16). We can choose between a BPMN process model (**BPMN** button), a value chain (**Value Chain** button), or an event-driven process chain (**Event-driven process chain (EPC)** button).

In our example, we don't want to create any further process documents and consequently close the tab. Instead, we would like to focus on the usage and the multiple features of SAP Signavio Process Collaboration Hub. We're first of all interested in ensuring process excellence in our company. To do this, within the initial screen in Figure 7.17, click on the **Process Excellence** button. As you've already learned, this initial screen can be fully personalized.

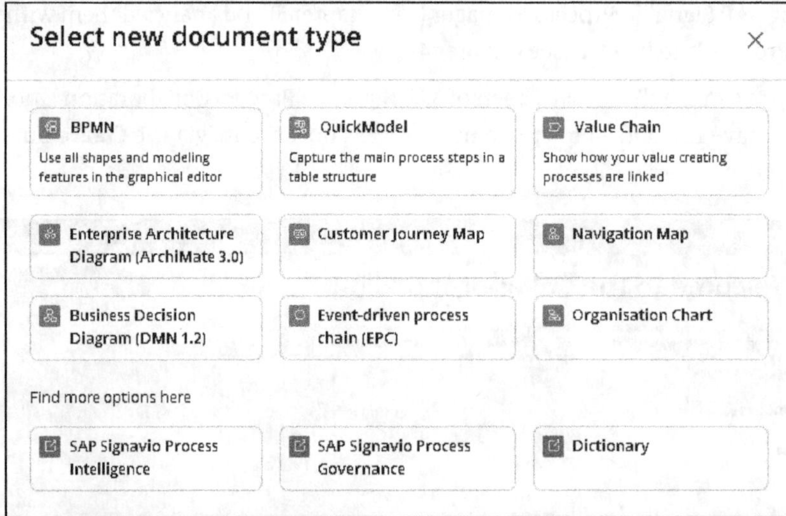

Figure 7.16 Create New Documents

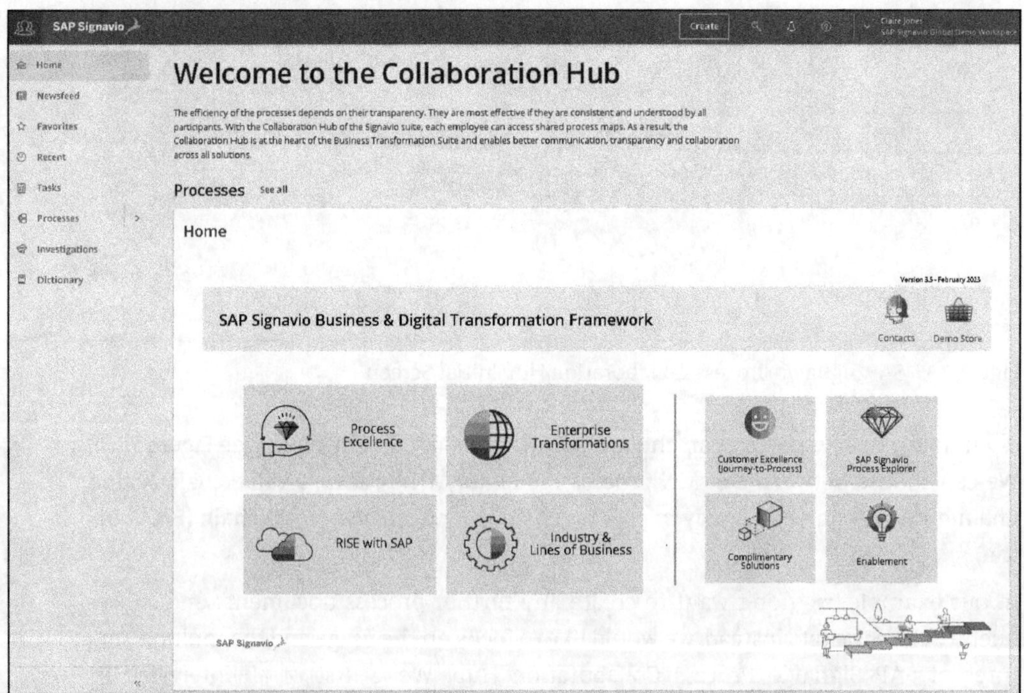

Figure 7.17 Focus on Process Excellence

From SAP Signavio Process Collaboration Hub, you can view all the company's business processes that have been published through the platform. At this point, we assume that we're responsible for the end-to-end lead-to-cash process, and want to transform it. To do this, first open the corresponding process by clicking on the **Lead-to-Cash** tile (see Figure 7.18).

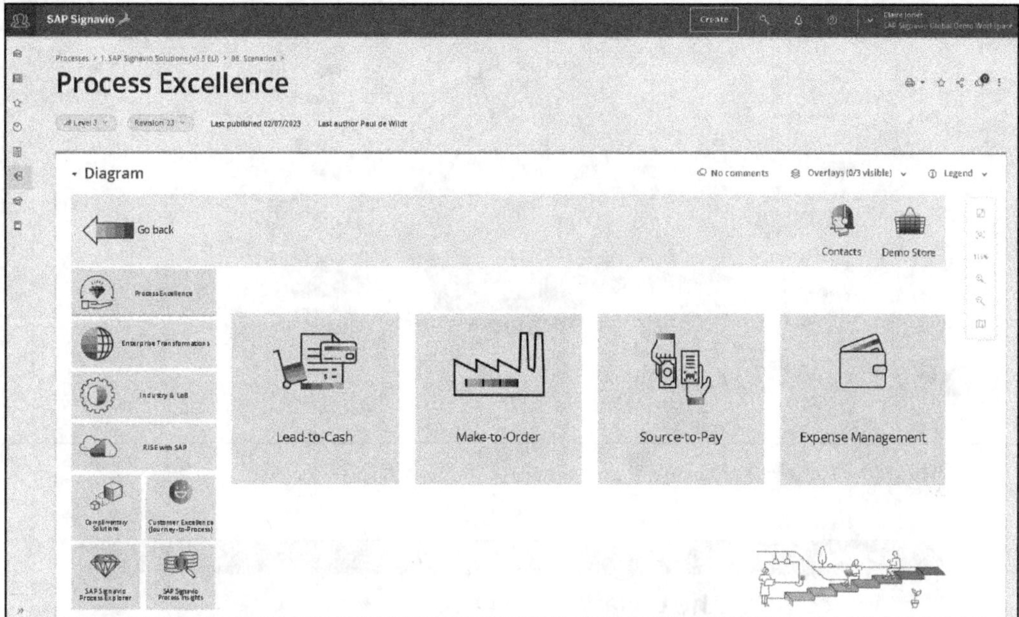

Figure 7.18 Select Desired Business Process

The UI in Figure 7.19 is configured so that the process transformation from our lead-to-cash process is divided into SAP Activate project phases. For example, if you click on the second project phase **Prepare**, you get all process-relevant information on the right side of the screen. This includes linked process models and analyses as well as customer journey models. From here, you can easily get an overview of the transformation project and manage the lead-to-cash process holistically.

> **What Is SAP Activate?**
>
> SAP Activate offers your company a transparent process with structured and solution-specific procedures. You decide how to introduce and extend new, differentiating functions in your organization. In Part III of this book, we go into more detail about the individual SAP Activate project phases and show how SAP Signavio actively supports them.

In our example, assume that you're the contact person for business transformation within the company. In this course, as shown in Figure 7.20, click on the **Enterprise Transformations** button.

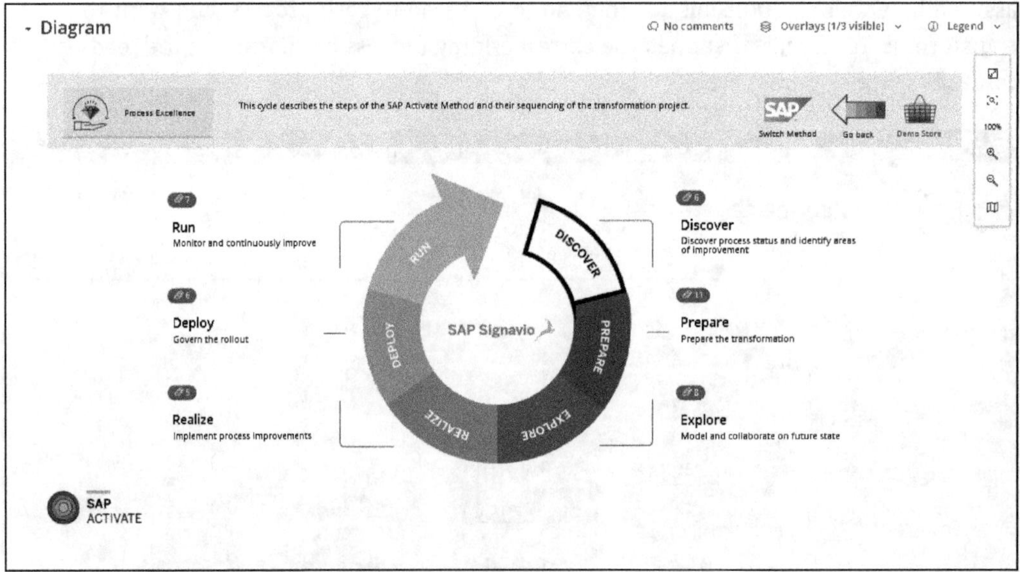

Figure 7.19 Overview of Linked Process Files

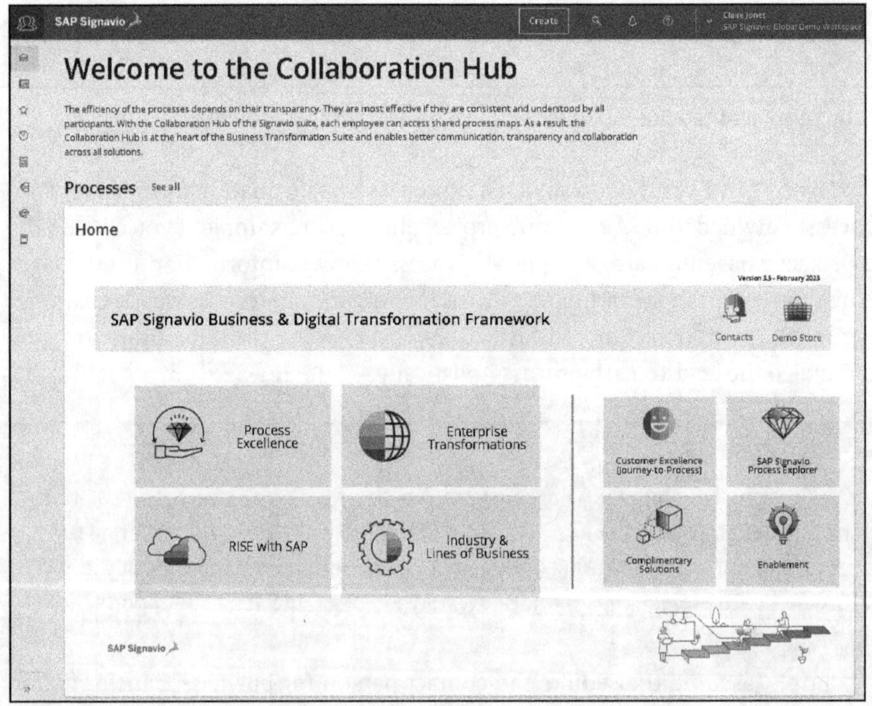

Figure 7.20 Focus on Business Transformation

Now you want to make your company fit for the future and make all business activities as sustainable as possible. To do this, click on the **Sustainability** button (see Figure 7.21).

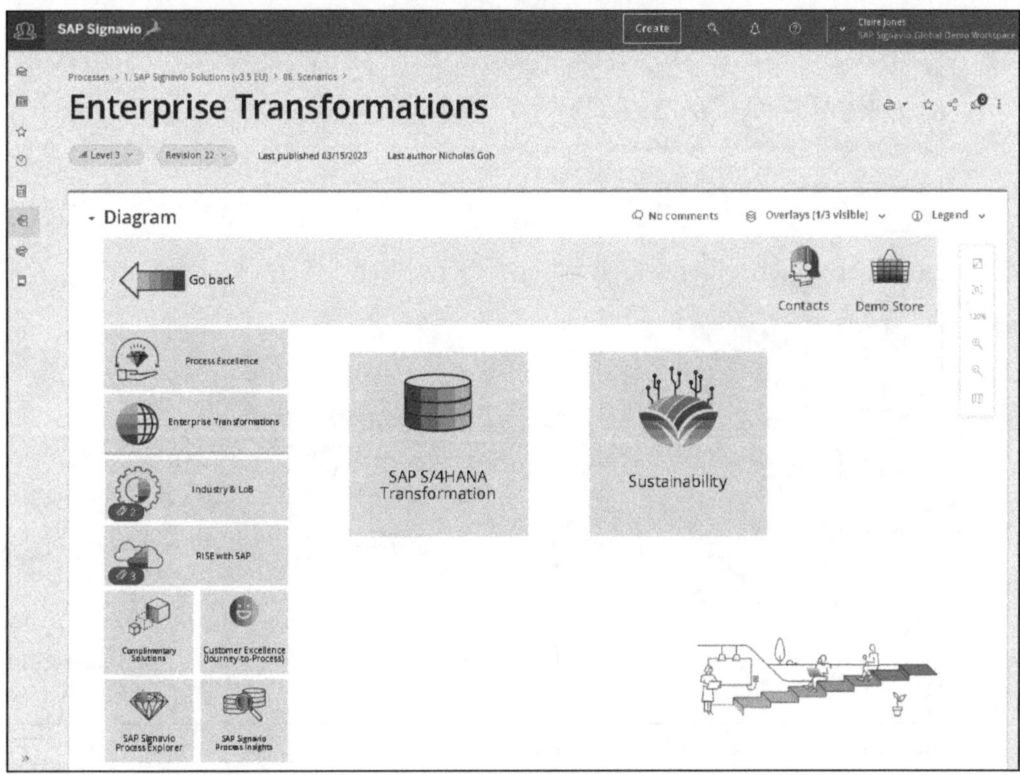

Figure 7.21 Select Desired Scenario

[+]

Keeping an Eye on Your Migration to SAP S/4HANA

With SAP Signavio Process Collaboration Hub, you also have the option to manage your complete SAP S/4HANA transformation. You can read more about how to do this in Chapter 12.

Next, you select a concrete use case that we'll address in more detail (see Figure 7.22). Decide on the transformation to a sustainable purchasing process, and click on **The Universal Process Action Case**. The goal is to view the current situation based on the actual data and the target state, including the target process analysis to operate more sustainably within the purchasing process.

You arrive at the UI in Figure 7.23. Here, all actions of the actual purchasing process of your company that are relevant from a sustainability perspective are reflected. In this context, you can also directly call up a target process analysis linked to SAP Signavio Process Intelligence and/or filter or view individual activities according to roles and responsibilities. To open the target process analysis for review, click on the link under **Navigation**. To filter individual process activities by roles or responsibilities, set the respective filters in the **Activities** section.

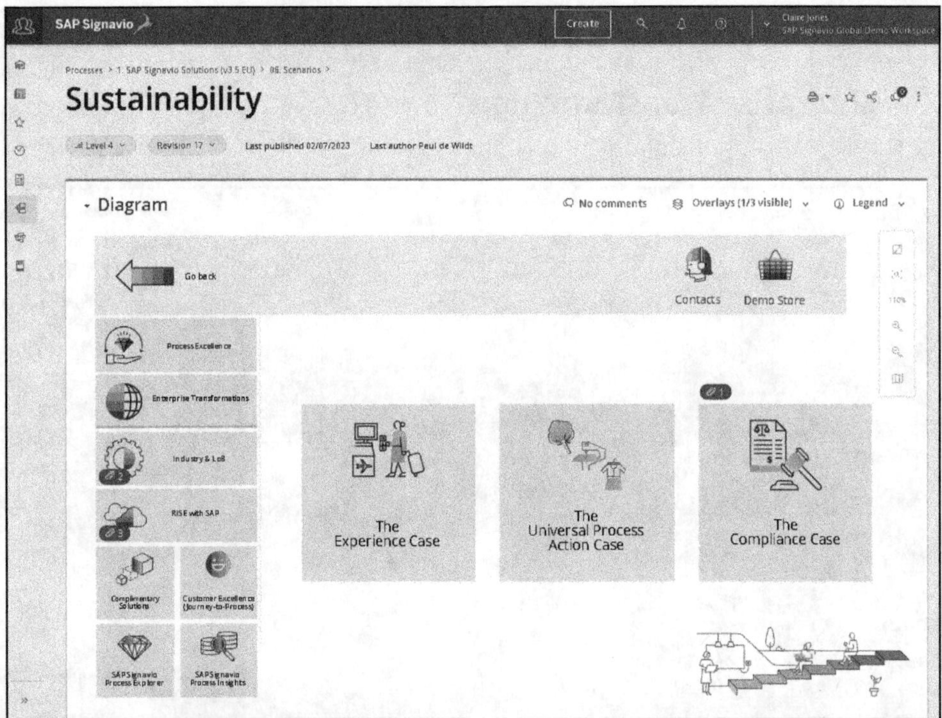

Figure 7.22 Selecting the Desired Use Case

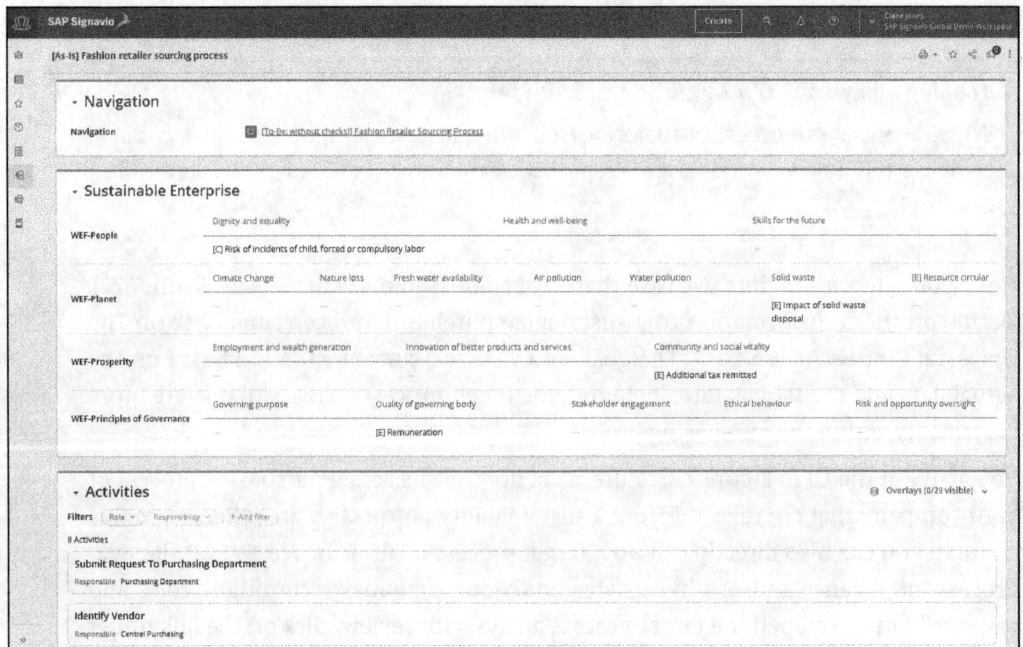

Figure 7.23 Insights into Linked Process Files

You can see what the published actual process model for this looks like in SAP Signavio Process Collaboration Hub in Figure 7.24. You can get to this process model by simply scrolling within the preceding UI.

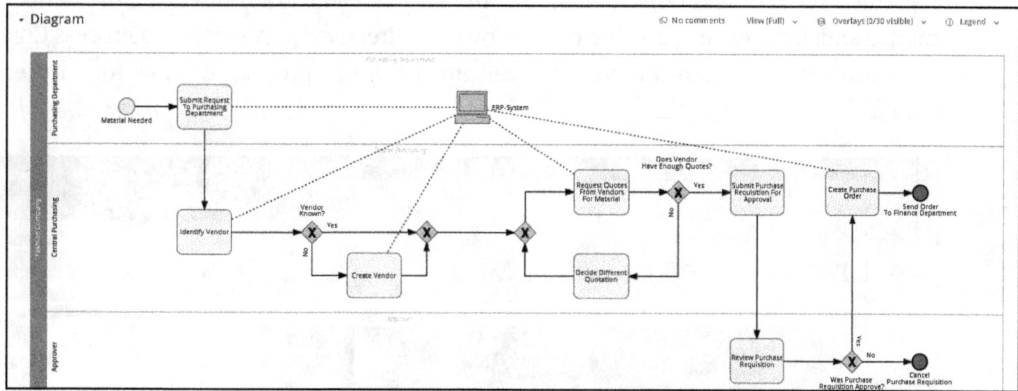

Figure 7.24 Published Actual Process Model in SAP Signavio Process Collaboration Hub

If you click on the link to the SAP Signavio Process Intelligence investigation, as described, the complete target process investigation opens, as shown in Figure 7.25. As a result, you can get insights about the target state of the purchasing process directly from SAP Signavio Process Collaboration Hub.

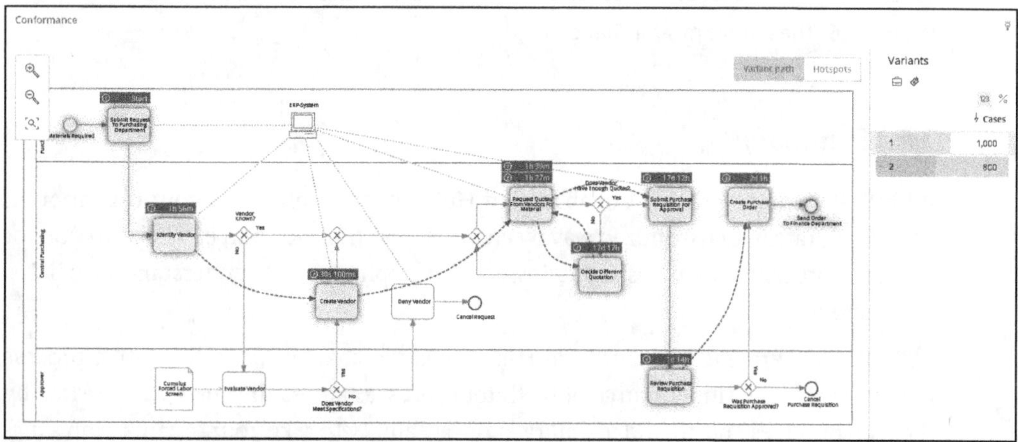

Figure 7.25 Published Target Process Investigation in SAP Signavio Process Collaboration Hub

Figure 7.26 illustrates another example of SAP Signavio Process Collaboration Hub: **Company on a Page**. The goal of this example is to visualize the essential business processes as well as the process performance of the company in a compact way. The interface is structured by business process area, role, and IT system used. Additionally, widgets from SAP Signavio Process Intelligence are integrated (see Chapter 4). The result is a sound overview of the most essential business metrics for each end-to-end process. You can see at a glance how the business operations are performing. As a CFO, for example, you would be interested in the global profitability of the company. The associated metric is shown in orange. When you click on it, you get rough information regarding the metric on the right side of the screen. You can comment on the performance and, if necessary, involve process owners directly via SAP Signavio Process Collaboration Hub. In addition, you can call up the entire process analysis for further details.

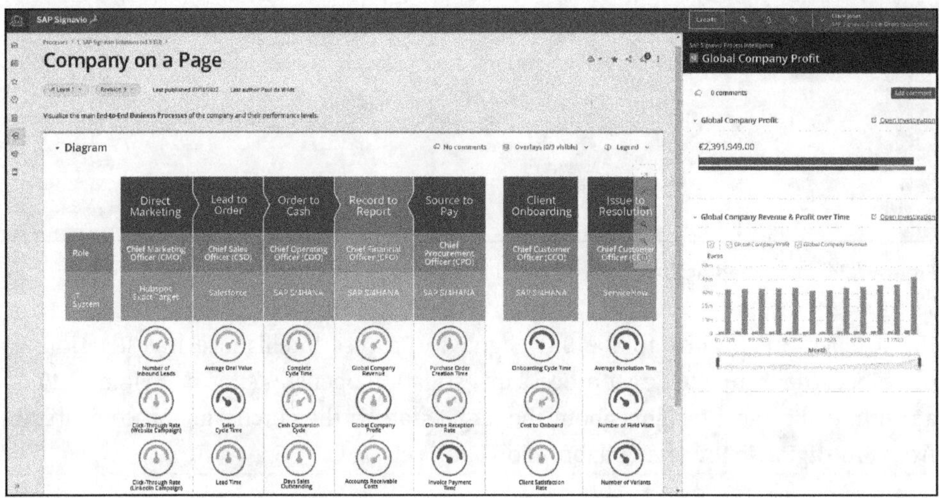

Figure 7.26 The Company at a Glance

7.4 Summary

With SAP Signavio Process Collaboration Hub, all stakeholders in your organization can collaborate in a customized way by creating a centralized source of information for all teams, breaking down business silos, and developing a better understanding of KPIs, tasks, and projects.

SAP Signavio Process Collaboration Hub is a game changer in collaborative process management. The single source of truth for processes across the enterprise created by the Hub provides end-to-end visibility and traceability to take your organization's digital strategy to the next level. The tool allows you to read and comment on published content and process models, as well as to preview new models before they are published. In this context, it avoids cumbersome email chains or searching folders for the latest version of documents. In addition, you can see responsibilities, roles, process hierarchies, and process steps at a glance and find all relevant process information using the integrated search function. The Hub enables seamless content creation, consumption, and collaboration across the entire SAP Signavio Process Transformation Suite, including SAP Signavio Process Intelligence, SAP Signavio Process Manager, SAP Signavio Journey Modeler, and SAP Signavio Process Governance.

Chapter 8
SAP Signavio Process Governance

SAP Signavio Process Governance is a workflow tool that allows you to manage the lifecycles of your business processes and maintain process compliance and governance throughout your organization. With the help of this solution, you can reduce the cost of your workflow management and, if necessary, modify it to meet regulatory requirements.

SAP Signavio Process Governance supports process consistency and efficiency across your entire organization (see Figure 8.1).

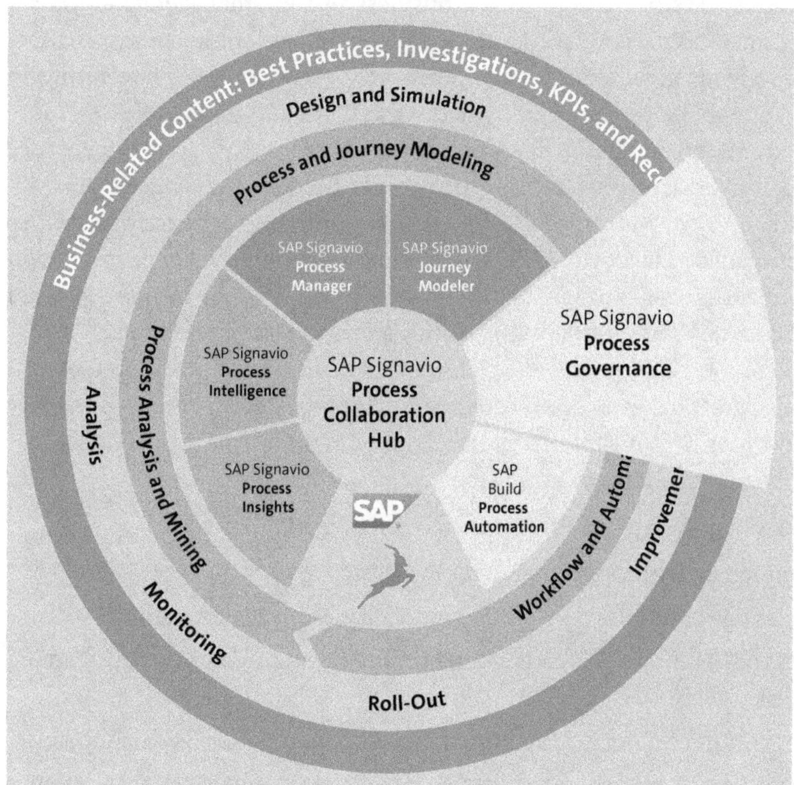

Figure 8.1 SAP Signavio Process Governance in Focus

It's an easy-to-use and accessible workflow engine that leverages what drives your business productivity: fewer manual tasks, fewer roles, lower costs, and higher quality and speed. Business process models can be transformed into standardized process control workflows that can then be rolled out across the enterprise. Organizations can prototype and complete workflows for rapid adoption of their process models and integrate task management capabilities to capture, connect, control, and communicate how work gets done. Creating workflows requires no programming skills. You'll benefit from shared task lists, seamless integration with other cloud systems, and form creation for targeted task assignment.

In Section 8.1, we'll first look at the various functions of SAP Signavio Process Governance. Section 8.2 shows various potentials of the solution. Subsequently, we'll illustrate the use of SAP Signavio Process Governance in practice by means of an example (Section 8.3).

8.1 Functions

SAP Signavio Process Governance is a web-based platform for modeling and executing workflows. Despite its kinship with classic business process management systems (BPMSs), SAP Signavio Process Governance greatly simplifies workflow automation. You'll find SAP Signavio Process Governance useful for describing and collaborating on routine tasks. The tool helps you coordinate tasks and handoffs, issue releases, route documents, and handle complex business processes. SAP Signavio Process Governance offers ready-to-use workflow capabilities for process modeling and accelerated approval. In this context, the term *workflow* means a repeatable sequence of activities that are part of a larger task. With SAP Signavio Process Governance, for example, you can quickly and easily create automated workflows based on your business process models. Managing tasks and initiating maturity and regulatory assessments across process documentation and implementation teams becomes easier with this tool. You can also track disparate business operations in a single place and ensure accountability for assigned activities. That means you can save time and money and be confident that your business processes always run the same way.

The core features of SAP Signavio Process Governance are as follows:

- Enforce comprehensive process governance with approvals.
- Intuitively configure and trigger workflows for collaborative optimization.
- Flexibly expand your workflows as your business grows, with no programming required at all.
- Schedule process maturity assessments for process owners and ensure compliance.
- Use simple and timesaving drag-and-drop form editors to enhance user experience (UX).
- Reduce variability in work execution, and ensure consistent and high-quality results.

[«]

8

Control Individual Process Cases with SAP Signavio Process Governance

Within an organization, processes are used to manage activities and specify the tasks and actions that must be completed to achieve a specific goal. For example, when hiring new employees, the tasks of scheduling an interview, interviewing candidates, and sending a job offer must be completed. After you've published a process model, you can subsequently start many individual cases. SAP Signavio Process Governance now tracks which tasks and actions you need to perform for each case. With cases, you can gather relevant information for colleagues that provides the necessary context to execute tasks. For more information about cases, see Table 8.1.

SAP Signavio Process Governance integrates seamlessly with other SAP Signavio business process transformation solutions to help you manage your process lifecycle more effectively. You can export workflow data from SAP Signavio Process Governance directly into SAP Signavio Process Intelligence and live insights. This transfer enables internal exploration of enterprise data using powerful process intelligence analytics that provide real-time insight into how to improve your workflows and business processes. Table 8.1 provides an overview of the numerous application features of SAP Signavio Process Governance.

Feature	Description
Tasks	A task is a job that must be done by a person. Employees can perform tasks within individual cases. Cases usually contain multiple tasks, typically the tasks defined in the process. You can also add ad hoc tasks to a case. In SAP Signavio Process Governance, you assign a task to a specific user, set a due date, and add subtasks.
Cases	A case is a shared scope of work to achieve a specific goal (e.g., hiring staff or signing a contract). Cases are typically larger than a single task for one person, but smaller than an entire project. In a case, the goal is broken down into specific action items (or tasks) so you can work on it with other people. A case includes a set of tasks, discussions, and documents, and it allows stakeholders to share all relevant contextual information for the tasks.
Processes	Recurring work can be described using processes. Think of a process as a recipe for performing routine activities. For example, imagine a process for recruiting. Each time this process is started, three tasks must be performed: review résumé, schedule interview, and conduct interview. Each time a person starts the process, a new case is created in SAP Signavio Process Governance. You can create and edit executable processes using the process editor. You can think of an executable process as a type of software.

Table 8.1 SAP Signavio Process Governance Functions

Feature	Description
Processes (Cont.)	Processes facilitate automation, and automating processes with SAP Signavio Process Governance enables business users to create meaningful processes without technical knowledge and without access to an IT development department.
Labels	To get a better overview of all processes in your organization, you can use labels to categorize the processes, e.g., by department or status.
Inbox	Display your tasks in your personal inbox. There, you'll find the tasks that have been assigned to you or for which you're responsible. The inbox also contains tasks that you've created ad hoc or all subtasks created for these tasks.
Collaboration (collaboration space)	Invited users will receive an email with a link to the login page where they can create an SAP Signavio Process Governance user who will become a member of the organization. This enables collaborative work on a job.
Access rights	With access rights in SAP Signavio Process Governance, you can restrict who can access a process, edit cases, or perform certain tasks within a process. By default, all processes and tasks are public, so all users in your organization can access them. By configuring access rights, you can restrict access to specific users or groups.
Analytics (reporting)	Create and share reports with aggregated overviews. Each report is launched on demand and summarizes the cases of a process in tabular and graphical form in the web user interface (UI). You can also export your case data to analyze it in SAP Signavio Process Intelligence.
Search	The search function considers entries in the **Name** field of categories, process descriptions, and comments. To find tasks, cases, processes, reports, or comments, use the search function in SAP Signavio Process Governance.
Action types	The following action types are available: ■ **User task**: Indicates that a person is performing a task. ■ **Multiuser task**: Specifies that a group of users perform the same user task. ■ **Send email**: Sends an email to a specified user. ■ **Script task**: Allows developers to add JavaScript code during process execution. ■ **Subprocess**: If you create a general overview of the main process, you can model subprocesses as separate workflows and link these subprocesses to the main process, which is also called the *parent process*.

Table 8.1 SAP Signavio Process Governance Functions (Cont.)

Feature	Description
Action types (Cont.)	■ **Create document**: Creates files containing case information. ■ **Document template**: The user is assigned a task where he is prompted to enter the required information into a custom task form. The information from the task form is applied to the uploaded model. ■ **Assign variables**: Copies the value of a variable to another variable. ■ **Box - Upload File**: Saves one or more files to a Box account you select. ■ **Google Drive - Upload File**: Sends one or more files to an account of your choice. ■ **Google Drive - Add rows to spreadsheets**: Adds a row to a Google Sheets spreadsheet. ■ **Google Drive - Add Calendar Event**: Adds an event to a Google calendar. ■ **DMN tasks**: Allows you to execute decision logic from a Decision Model and Notation (DMN) diagram in SAP Signavio Process Governance. ■ **Model state set**: Automatically updates the diagram state in SAP Signavio Process Manager and SAP Signavio Process Collaboration Hub, e.g., to mark the diagram as **Approved** or **In Progress**. Accordingly, you need access to SAP Signavio Process Manager for this action.
Forms	Use forms to enter information during the execution of a process in SAP Signavio Process Governance. Forms can be part of form triggers as well as part of user tasks.
Trigger	Use triggers to describe the way your processes start. With manual triggers, processes start manually when you select a process and start a new case. Use form triggers to start a process using a form with public or private form triggers or trigger a new case with an email trigger.
Control flow	To specify the order in which actions are executed in a process, use transitions, gateways, and events. Transitions specify that the workflow engine executes the target element only after the source element has been completed. In Business Process Model and Notation (BPMN), a transition is called a *sequence flow* and is represented as an arrow from the source element to the target element. Exclusive gateways allow selection between different execution paths. Parallel gateways are used to model tasks that can be executed in parallel or in any order. Events can be process start events, process end events, and process intermediate events.

Table 8.1 SAP Signavio Process Governance Functions (Cont.)

Feature	Description
User menu	The user menu consists of several sections. In the first area, you can access your personal settings for SAP Signavio Process Governance by clicking **My Profile**. If you have administrator rights, you'll find organization settings as well as services and connectors in this area. If you have access to additional SAP Signavio Business Transformation Suite products, you can find them in the **Products** section. The currently selected product is highlighted. Clicking the product name opens the selected product in a new browser window. The **Organizations** area lists all organizations you belong to. When you click on an organization, it opens in the same browser window. You won't be logged out of the currently selected organization. The **Help** section provides links to the user guide and a collection of sample workflows. You can also provide feedback in this area. Click **Logout** to log out from SAP Signavio Process Governance.
Dictionary integration	For working with SAP Signavio Process Governance, you can integrate the SAP Signavio Process Manager dictionary. This allows you to extract data from dictionary items and use it in your workflows. To use this feature, your organization in SAP Signavio Process Governance must be linked to a workspace in SAP Signavio Process Manager.
Notifications	Receive notifications about task or case updates. SAP Signavio Process Governance sends a series of email notifications to inform process participants about the cases you're handling and to avoid delays in task handover when someone assigns a task. The notifications are related to a case, task, comment, user, or account.
Connectors	Connect to other SAP Signavio products or third-party offerings: ▪ Email connector: Send an email to the specified user. Specify sender name, recipient and reply, subject, attachments, and message text as you wish. ▪ Google connector: Use a Google connector to upload files, print files, manage calendar events, and more. ▪ SAP Signavio Process Manager connector: The integration of SAP Signavio Process Manager with SAP Signavio Process Governance enables users of SAP Signavio Process Manager to execute approval workflows for their process models with SAP Signavio Process Governance. ▪ Salesforce connector: SAP Signavio Process Governance integrates with Salesforce workflows. You can configure files so that changes in Salesforce automatically trigger processes in SAP Signavio Process Governance.

Table 8.1 SAP Signavio Process Governance Functions (Cont.)

Feature	Description
Connectors (Cont.)	■ Custom connectors: Integrate dynamically structured data from other IT systems into your workflows. The workflow system fetches data from a third-party system via a connector that you implement and host yourself.

Table 8.1 SAP Signavio Process Governance Functions (Cont.)

8.2 Benefits

SAP Signavio Process Governance provides a better workflow when it comes to process management. The tool gives you the ability to automate repetitive process governance workflows so your employees can focus on value-added tasks. By creating forms, checklists, and decision points; setting up email notifications for deadlines; and collaborating on shared task lists, you can identify and eliminate unnecessary work. In this way, you can limit email traffic and time-consuming meetings. As a result, you can complete the steps that really matter to your company's success more quickly and efficiently.

The following is an overview of the core benefits of SAP Signavio Process Governance:

- Quickly configurable workflows without expert knowledge
- Easy management of process release cycles
- Maturity assessments
- Process-integrated risk and control management
- Process standardization based on workflows
- Automated tracking of tasks
- Fewer delays (due to automatic triggers and actions)
- Control in the places where it's needed
- Clear communication of handovers
- Traceability: Who did what?
- Clarity: Who has to do what?
- Saving resources

[«]

Do You Need Process Modeling Skills to Use the Tool?

SAP Signavio Process Governance is extremely user-friendly and can be used by anyone who needs to describe and collaborate on routine tasks, regardless of whether they are familiar with process modeling or not. Even if you have no process modeling experience, you'll quickly become familiar with the solution.

SAP Signavio Process Governance also helps you keep pace with constant change. The tool is the perfect starting point for the most complex tasks and activities. The intuitive UI makes it easy to turn process models into workflows, so you only need the essential features of your business processes, such as tasks and decisions, to get started. You can add more details later (e.g., custom notifications, access control, or automatic reminders), and workflows scale as your business grows. Accordingly, 360-degree process governance is established with approvals and maturity assessments while eliminating spreadsheets by tracking tasks directly. All workflows can be created easily, quickly, and without expert knowledge. Figure 8.2 illustrates how configurable workflows can be created with SAP Signavio Process Governance via **Processes • Actions**.

SAP Signavio Process Governance automates and standardizes manual decision-making by providing an automated approval solution. This not only reduces the risk of a wrong decision but also allows the processes based on it to be carried out more efficiently. The automated approval process includes rapid response to changes. This ensures that business decisions are recorded, which in turn is particularly important to ensure compliance with internal as well as external regulations and policies.

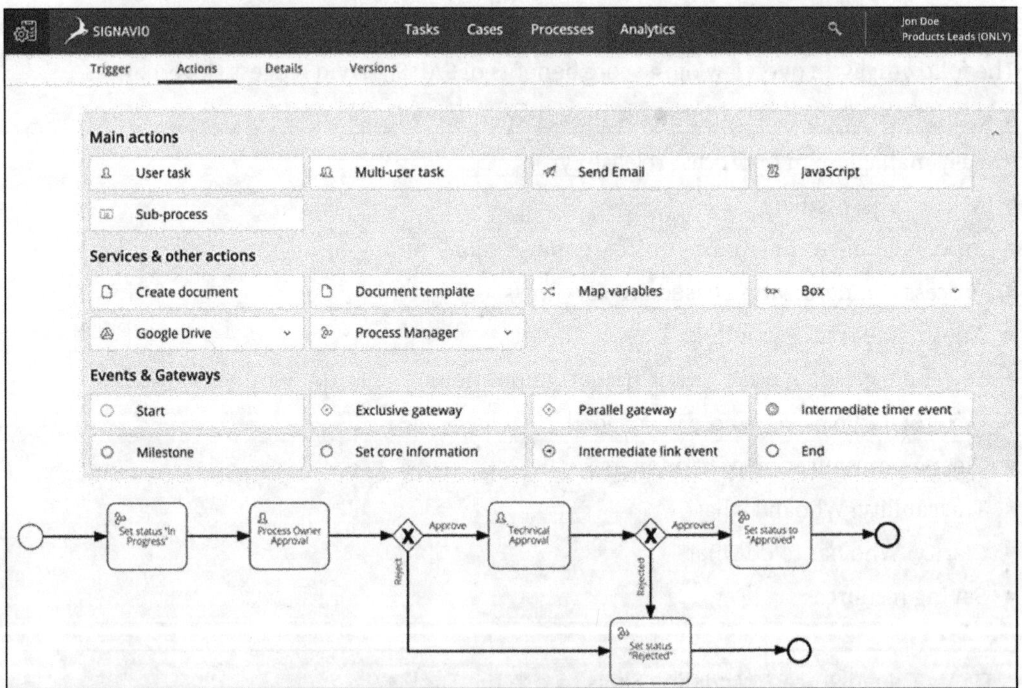

Figure 8.2 Create Configurable Workflows

SAP Signavio Process Governance supports better communication and coordination of various business tasks. The solution gives you a clear indication of how far a particular process has progressed and what the next step will be. This results in fewer delays due to neglected or forgotten tasks, more accountability in decision-making, and more flexibility to consistently implement process improvements. In addition, onboarding new team members becomes much easier as you can see at a glance exactly what needs to be done and where to find all relevant resources.

> **The Added Value of SAP Signavio Process Governance**
>
> SAP Signavio Process Governance helps you gain complete visibility and control over your workflows, prototype for rapid deployment, and scale workflows without coding, and it reduces business process variation and rework.

SAP Signavio Process Governance helps you automate your process approvals and publishing processes. You can set up process owners, risk owners, and quality managers and their respective tasks by selecting *process attributes*. Attributes describe additional information and properties of a process. You can then assign and notify these users using attributes for process approvals, risk and control assessments, and maturity ratings. Process compliance is easily increased with this tool by taking process approvals and maturity assessments directly from the SAP Signavio Process Manager solution. Once an approval has been granted, you can automatically publish processes. At the same time, all approvals are documented.

You can also open the tasks assigned to you with SAP Signavio Process Collaboration Hub by clicking on **Tasks**. As shown in Figure 8.3, your tasks are displayed in your personal inbox.

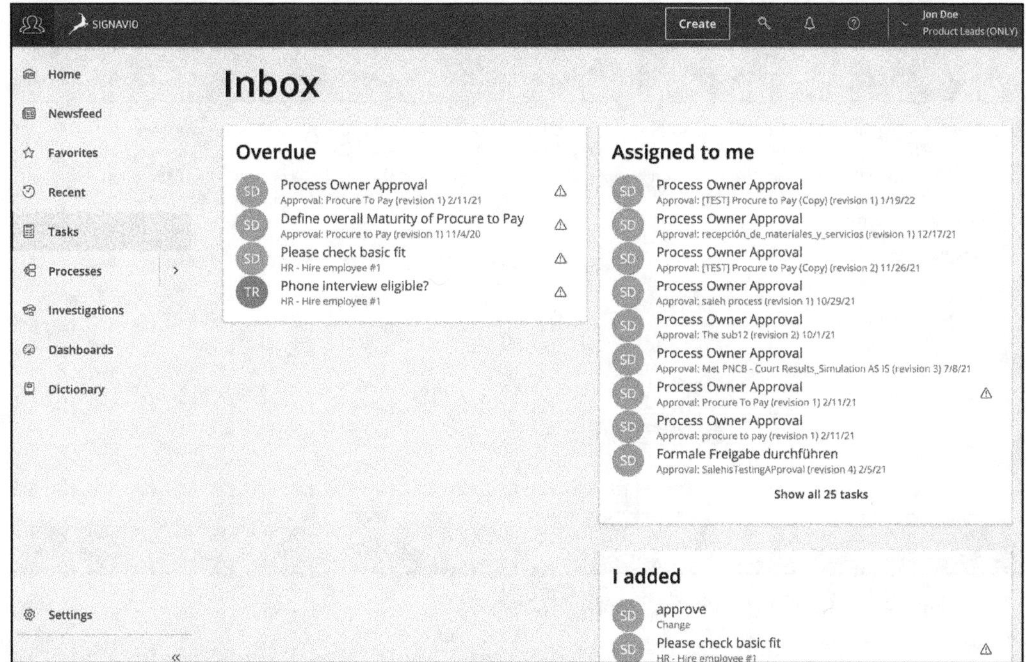

Figure 8.3 Inbox for Personal Tasks within SAP Signavio Process Collaboration Hub

You can find tasks from SAP Signavio Process Governance that have been assigned to you or tasks for which you're a candidate. The inbox also contains tasks that you've created ad hoc and any created subtasks related to them. With all these capabilities, you can manage your process release cycles effectively.

Likewise, SAP Signavio Process Governance helps you enforce your business process maturity assessment on multiple levels. With workflows configurable via the BPMN standard, you can reliably track your tasks without the need for spreadsheet programs. In combination with SAP Signavio Process Intelligence, you can analyze the data extensively and derive actions directly. With SAP Signavio Process Governance, you automate and standardize manual decisions based on workflows. Thus, you reduce the risk of wrong decisions and promote even more efficient operations. In addition, this tool allows you to integrate your workflows and specific tasks with commonly used cloud-based software, such as Salesforce or Google Drive, as well as your own IT systems.

SAP Signavio Process Governance gives you the ability to centrally manage approval workflows. Accordingly, during the approval workflow, different states can be assigned to the revision of a diagram. As shown in Figure 8.4, an icon next to the diagram indicates its state. You can either use the diagram states defined by SAP Signavio (**Approved, Rejected, Published, in progress**) or create your own list of diagram states. The **Publish in Hub** option specifies that a revision is automatically published to SAP Signavio Process Collaboration Hub when this state is assigned to the revision. The **Reset expiration** option specifies that the expiration date of the approval is recalculated from the assignment of this status.

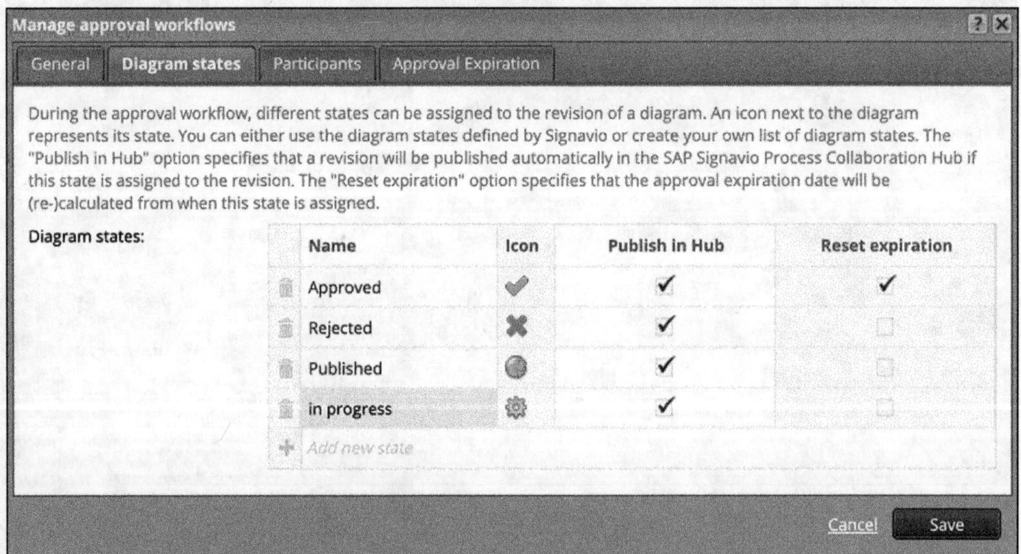

Figure 8.4 Configuration of Diagram States

Furthermore, SAP Signavio Process Governance allows you to set up process-integrated risk and control management. In this context, you can manage risks and controls centrally in one place using the dictionary in SAP Signavio Process Governance. As a result, you can more easily organize risks and controls into structured folders of your choice. Process owners in your organization can review where each risk or control element is used within your business process. In addition, any item you create can be linked to any process or task. For improved collaboration, you can view and use all risk and control elements within SAP Signavio Process Collaboration Hub (see Figure 8.5).

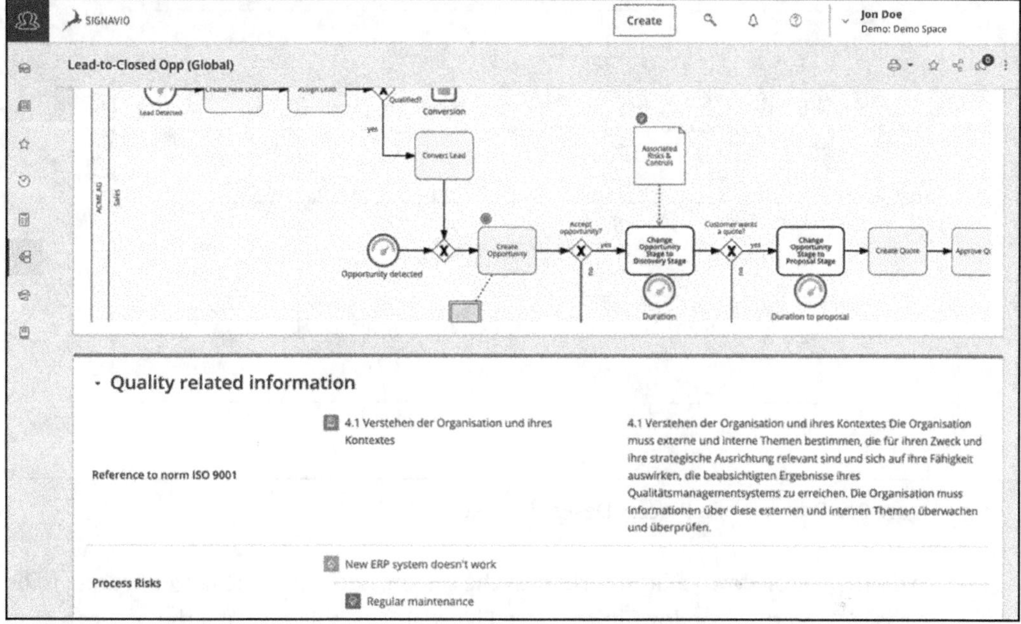

Figure 8.5 Risk Control

8.3 Application Example: Workflow Setup for Approval Process

In this section, we'll present the application of SAP Signavio Process Governance using a practical example. Our example is about a transformation project regarding SAP S/4HANA for manufacturing company ACME Inc. In this project, our company wants to use SAP Signavio Process Governance to integrate approval workflows into the process models. The goal is for SAP Signavio Process Governance to strengthen process compliance and reduce workflow administration costs, eliminating the need for heavy email traffic and manual intervention. In addition, the tool is expected to help with risk control and alignment between business and IT, both internally and externally.

In Figure 8.6, you can see a workflow for approving a business process design for order to cash. This can be used to inform appropriate stakeholders as well as to obtain approval for the created target process. We've opened the workflow itself directly via the navigation in SAP Signavio Process Collaboration Hub by clicking on the corresponding link.

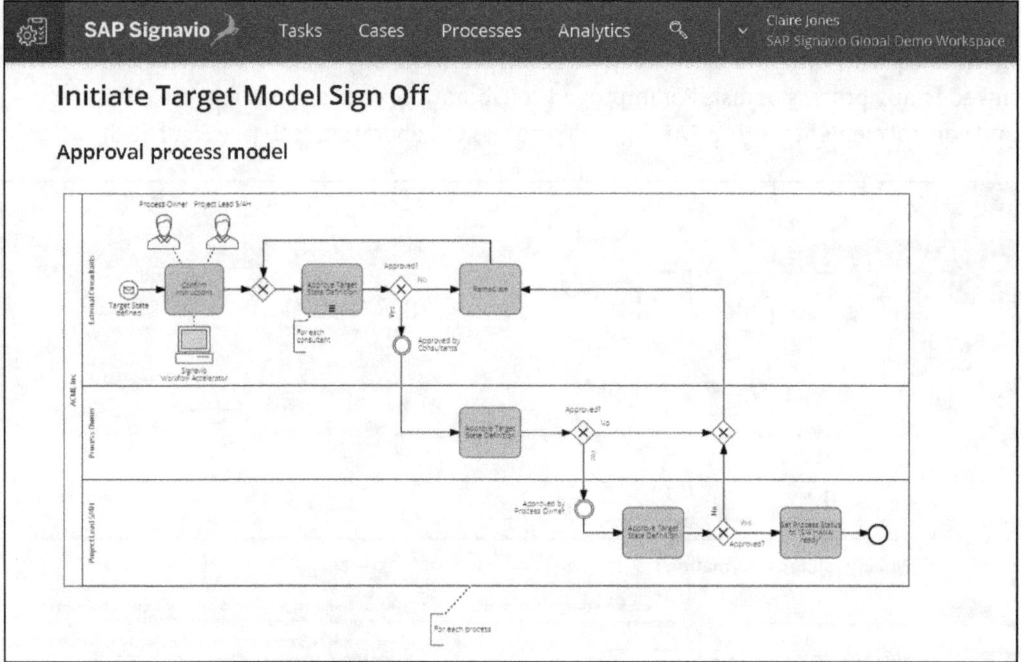

Figure 8.6 Example of a Process Design Release

To do this, scroll down a little further in the screen, and select the target process to be released from the dropdown menu (see Figure 8.7). In this case, it's the order-to-cash process.

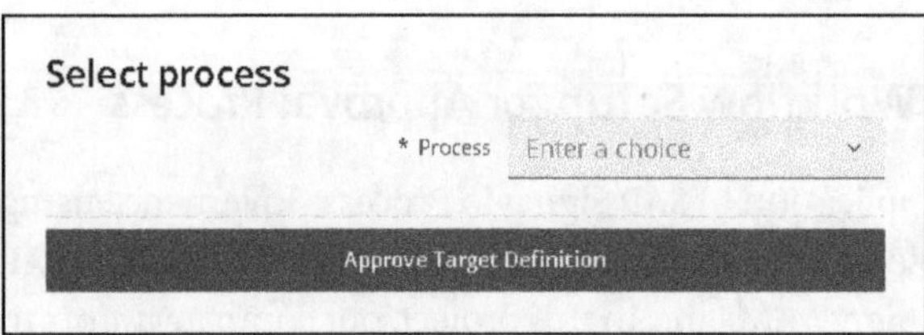

Figure 8.7 Select Target Process to Be Released

After you've selected the **Order to Cash** target process, the illustrated view shown in Figure 8.8 appears automatically. Add some basic information such as the budget (**Budget** field), the process goals (**Goals** field), and the process owners (**To be Approved by** area). For the latter, ERP owners, consultants, and process owners can be included.

As shown in Figure 8.9, you can subsequently enter some more detailed information, such as the due date for the target process to be released, and store the process model for viewing. According to the project management information, enter March 28, 2024, as the due date (**Define ERP Implementation Project** area, **Estimated Due Date** field). Finally, the target process can be approved (**Approve** button) or rejected (**Reject** button).

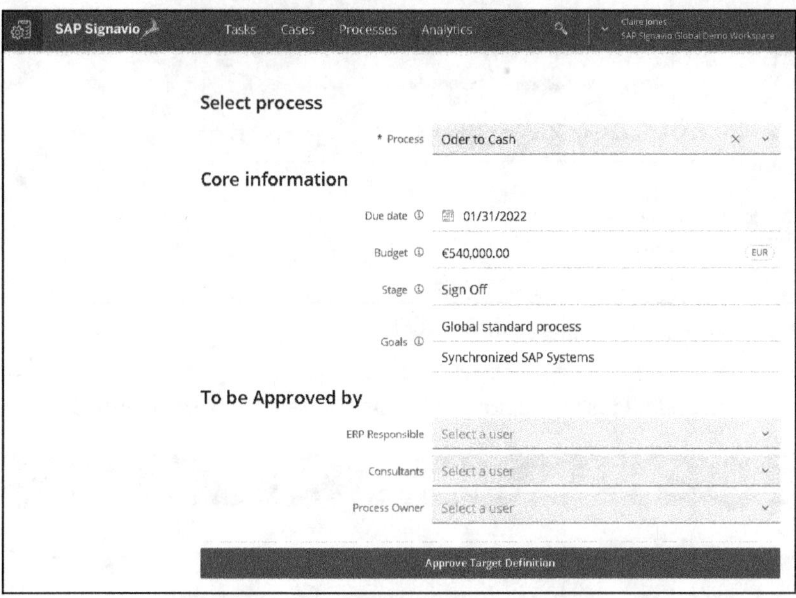

Figure 8.8 Core Information for the Target Process to Be Released

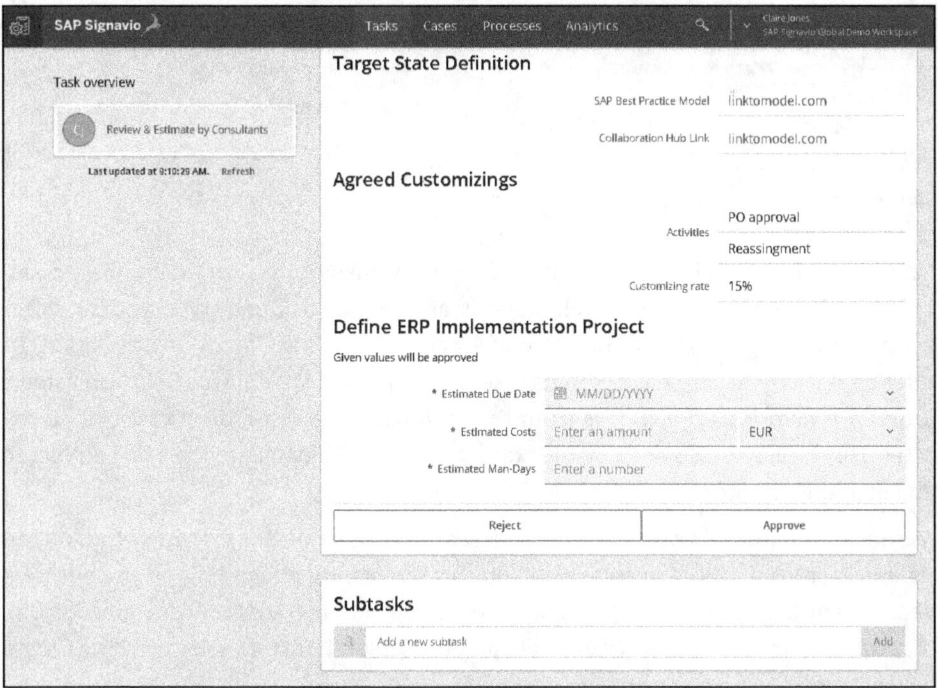

Figure 8.9 Detailed Information for the Target Process to Be Released

In this example, Claire is responsible for the release of the order-to-cash target process model. As described, March 28, 2024, has been defined as the due date with regard to the process release (**Task due date** field). This is shown in Figure 8.10.

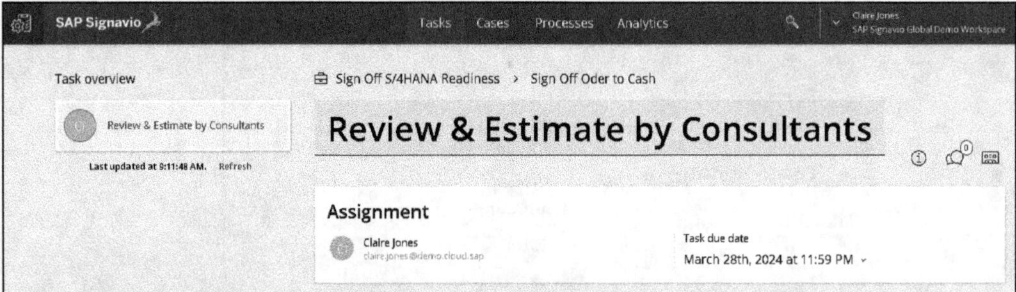

Figure 8.10 Process Design Release at a Glance

Claire can view and access all tasks assigned to her on the **Tasks** tab (see Figure 8.11).

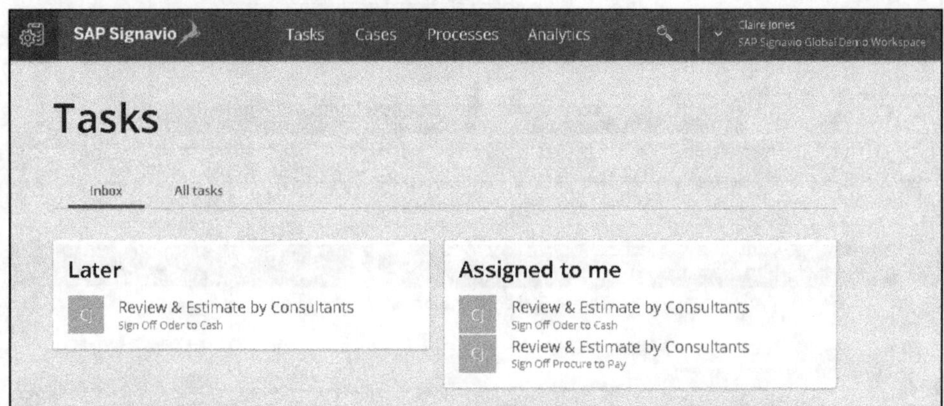

Figure 8.11 Tasks at a Glance

On the **Cases** tab in Figure 8.12, Claire also gets an insight into various process cases. These can be filtered by type as well as status, as desired. To filter by type, you can select various cases of processes in the **Cases of Process** field using a dropdown selection. To filter by status, you can select various case statuses in the **Case Status** field also using a dropdown help. You can choose from the options for all cases, open cases, or closed cases. This filter can also be used, for example, to create separate reports for completed and pending activities.

As shown in Figure 8.13, you get an overview of various processes for which you can initiate new cases on the **Processes** tab. You can filter these depending on the business department (e.g., finance, HR, IT, etc.), process owner, and triggers (e.g., email, form, manual, etc.). In addition, you can create new processes. To test this feature, click **Create new process**.

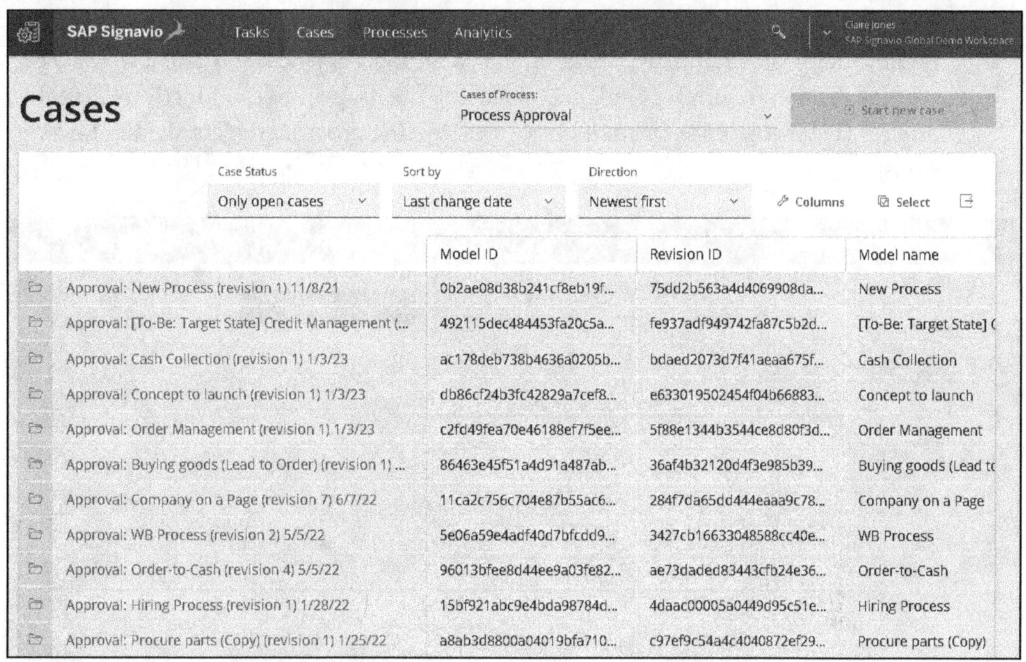

Figure 8.12 Process Cases at a Glance

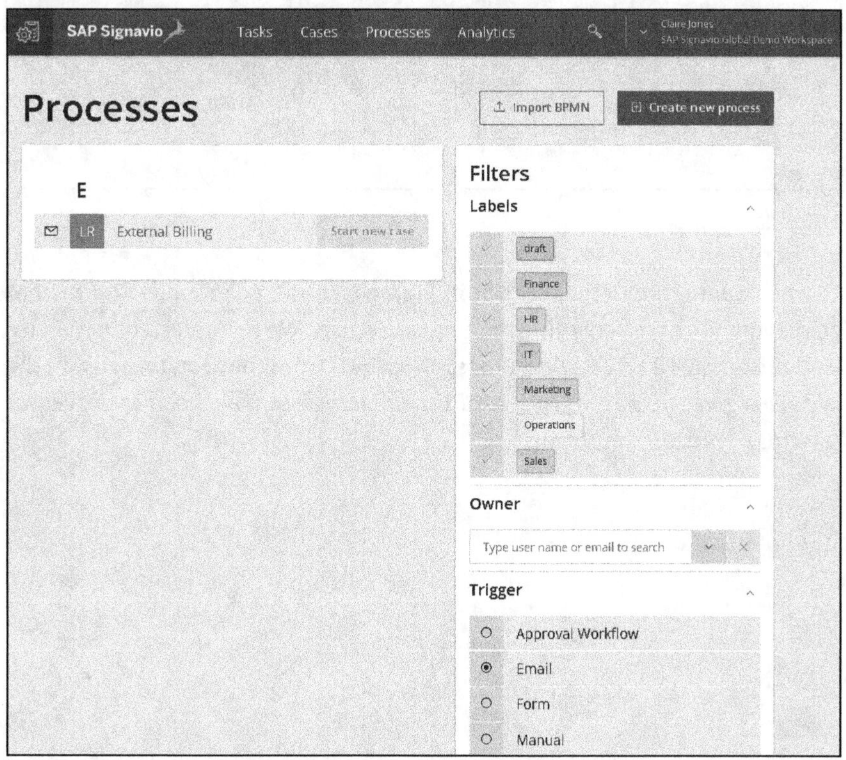

Figure 8.13 Processes at a Glance

First, you define the name of the process and assign appropriate labels. Our test process is relevant for the finance department, so use the label **Finance**. You can now easily create a process from different action types using the drag-and-drop function to create a process (see Figure 8.14). In addition, you can define process triggers, details, and versions.

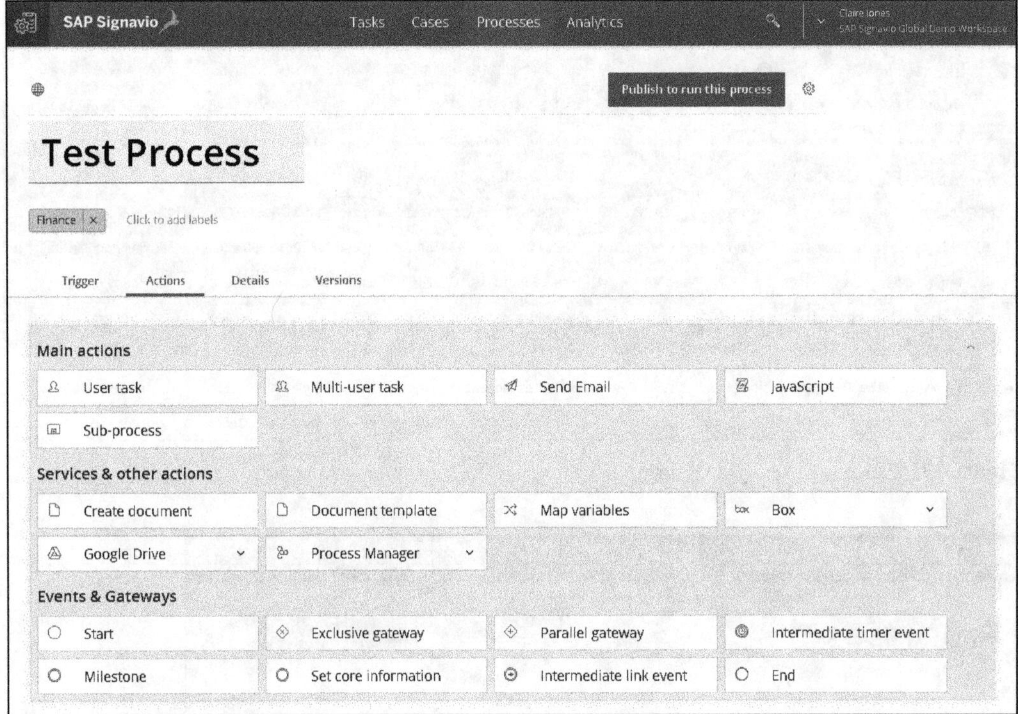

Figure 8.14 Create Processes

Finally, the individual processes, as shown in Figure 8.15, can be grouped and further analyzed in the **Analytics** tab according to their case status. We're interested in the process approval process. To do this, select **Process Approval** from the dropdown list in the **Process** field. This gives you an overview of how many cases have been approved or rejected in terms of process approval.

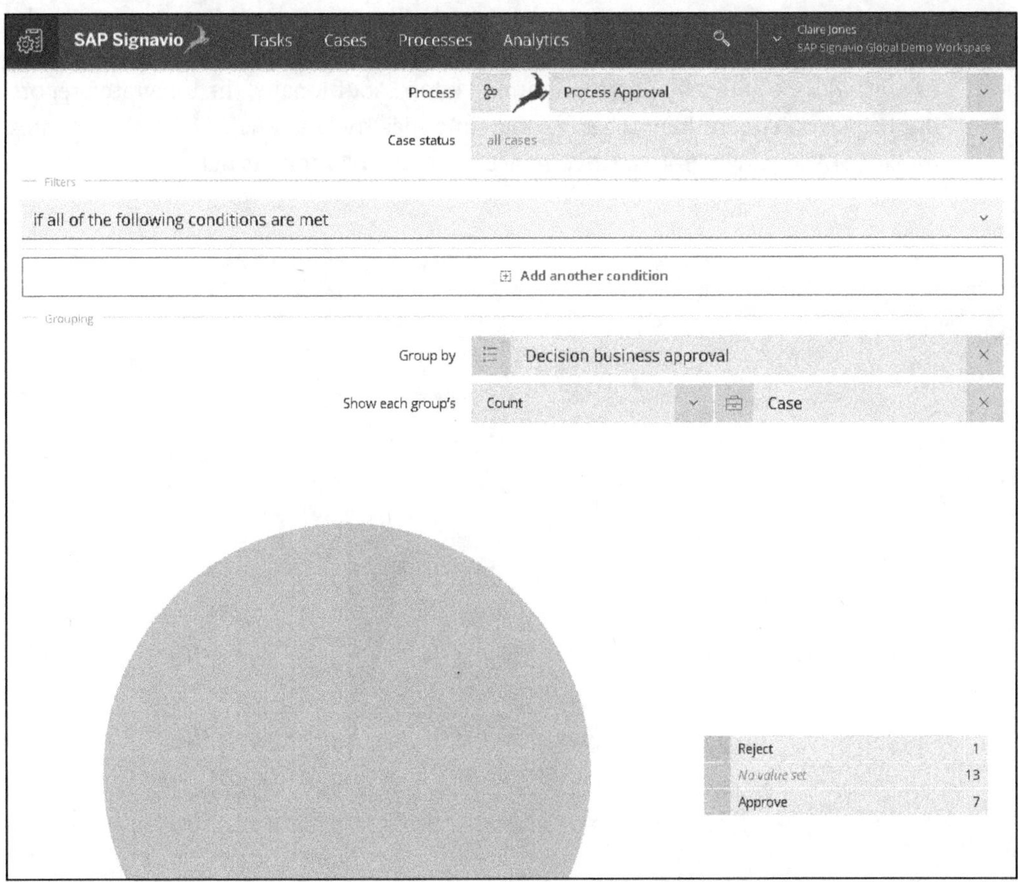

Figure 8.15 Analyze Processes

8.4 Summary

To ensure the success of your business transformation, you can turn to SAP Signavio Process Governance. That's because SAP Signavio Process Governance helps you gain complete visibility and control over your organization's workflows, prototype workflows for rapid deployment and scaling without coding, and reduce business process variation and rework.

SAP Signavio Process Governance transforms your business process models into standardized workflows that can be deployed across the enterprise. You can coordinate tasks, keep track of work, and ensure consistent accountability and collaborative process governance. The tool ensures reliable, higher-quality process outcomes by reducing variability in work execution, providing better service to internal and external stakeholders by completing tasks faster, and creating an appropriate and accessible audit trail of actual work by tracking decision points and completed tasks. The easy-to-use and accessible workflow engine enables 360-degree process governance with approvals, maturity assessments, and more. SAP Signavio Process Governance enables rapid deployment of workflow models with an intuitive UX. The solution is easy to configure and scale without any programming effort. Additionally, the integrated reporting helps you streamline processes and simplifies the analysis of critical mandates, making SAP Signavio Process Governance a true trailblazer in its field.

Chapter 9
SAP Build Process Automation

SAP Build Process Automation enables you to create practical automations and workflows using process automation capabilities without programming. As a result, the solution helps you meet changing business requirements in an even more agile way.

9

Volatile economic conditions continuously change your business requirements. To build resilience and support future growth, your organization must evolve by improving and transforming processes and practices to better meet current demands. However, with limited developer resources and increasingly complex IT landscapes, this can be difficult. *SAP Build Process Automation* can help you stay innovative while getting the most out of your IT investments (see Figure 9.1).

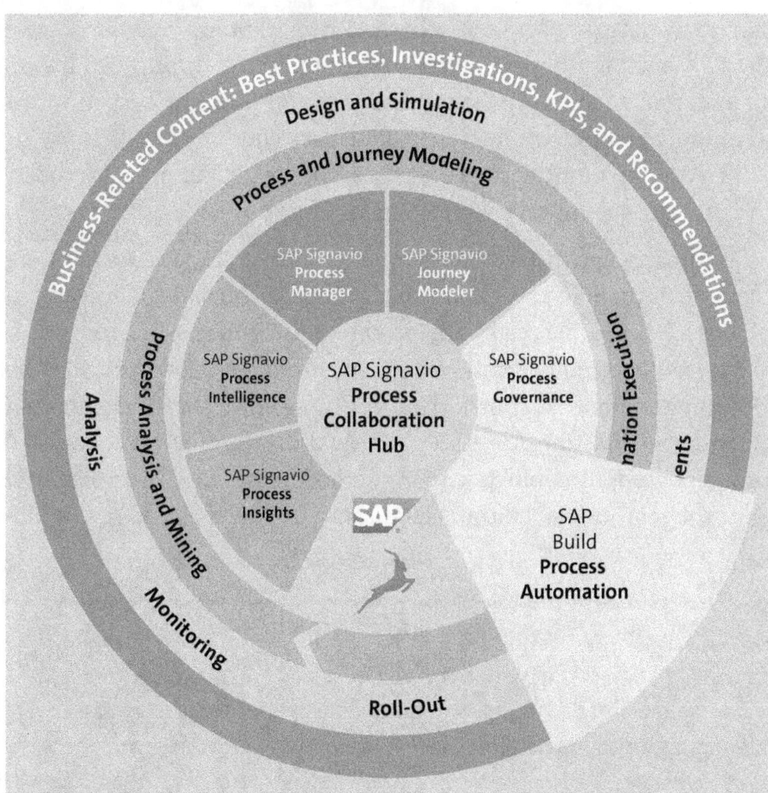

Figure 9.1 SAP Build Process Automation in Focus

SAP Build Process Automation belongs to the SAP Build product family but is functionally integrated into the SAP Signavio portfolio and provides a complement to SAP Signavio Process Governance, for example. Therefore, SAP Build Process Automation is also covered in this book.

In Section 9.1, we'll first look at the various functions of SAP Build Process Automation. Section 9.2 shows the various benefits of the solution. We then use an example to illustrate how SAP Build Process Automation is used in practice (Section 9.3).

9.1 Functions

Many business users and technologists have extensive process knowledge and technical acumen but lack programming skills. With the SAP Build Process Automation solution, your experts can become *citizen developers*, or end users who act as developers. They are transformation drivers who can adapt, improve, and innovate your business processes. By combining functions from SAP Workflow Management and SAP Intelligent Robotic Process Automation (SAP Intelligent RPA) services with an intuitive citizen developer experience, the tool enables employees to manage workflow and process automation with visual functionality that requires no programming skills.

The solution offers native integration with other SAP applications; prebuilt process content, including RPA bots; and process intelligence supported by artificial intelligence (AI). Building on proven enterprise-class solutions, SAP Build Process Automation can help you simplify process automation and increase your innovation capacity. It also helps you increase the number of people who can create and modify process enhancements to continuously improve your SAP solutions.

SAP Build Process Automation provides an intuitive no-code user experience (UX) and enables your professionals to build process automations based on SAP applications in your IT landscape. Citizen developers can build processes using forms, business rules, and automation. They can also embed actions and advanced workflows provided by IT—all in a unified environment, of course. Your business experts can use prebuilt functions and content, reuse existing processes and scenarios, and modify processes to meet changing business needs. Combined with best practices from SAP, partners, and the SAP community, the solution can help accelerate and simplify the process of building and changing business processes.

SAP Build

SAP Build Process Automation (along with *SAP Build Apps* and *SAP Build Work Zone*) is one of the core components of SAP Build. SAP Build enables business users to easily develop and extend enterprise applications, automate manual processes, and create engaging enterprise websites.

SAP Build users can share reusable artifacts such as UX components, workflows, automations, data models, and business logic across teams and projects. To access business data from other systems, developers can encapsulate application programming interfaces (APIs) as simplified service interfaces, called *action projects*, and make them available in the SAP Build library. These action projects reflect individual operations or methods of an API but hide the complexity from users. An example of a simple action project is a REST API that returns a list of to-dos. With the help of action projects, users can use them without any technical background and without having to deal with the underlying complexity of APIs, while always ensuring compliance with IT security requirements. Similarly, citizen developers can access artifacts such as data models created in the SAP Business Application Studio integrated development environment (IDE) via SAP Build solutions. All of this simplifies the development of modular and mature applications using the skills, expertise, and resources of your fusion teams. *Fusion teams* are multidisciplinary teams of developers, executives, and end users who leverage their specific areas of expertise and domain knowledge to develop solutions that address common goals.

In addition to action projects, you can also create business process projects. Think of a business process project as a way to group tasks and processes for a business scenario. It could consist of several steps that can be included as skills. For example, the business process could include *automations*, *approval forms*, *conditions*, *business rules*, and *process visibility dashboards*. Along the way, the action projects just introduced encapsulate external APIs required for your business scenarios. These APIs are typically created by the IT team and made available to citizen developers for use as part of the business process projects. When a new business process project is created, you can create a process in this context and begin to assemble the sequence of activities or capabilities required to automate the business scenario.

SAP Build Process Automation provides automation of tasks and processes using bots for robotic process automation (RPA), workflows, forms, integrated AI, business rules, and decisions. The solution supports the following capabilities:

- **Digitize processes**
  Create or customize end-to-end processes with an intuitive user interface (UI).
- **Create interactive forms**
  Create form-based workflows with drag-and-drop functionality and connected data sources.
- **Create and run automations**
  Manage process automations to maximize efficiency in your business.
- **Manage decisions**
  Develop and manage decision logic using decision tables.

- **Achieve end-to-end process visibility**
  Support event-driven, real-time visibility into comprehensive process instances using process visibility dashboards.
- **Discover and manage predefined content**
  Make even smarter decisions by proactively identifying process bottlenecks and issues.

All these capabilities of SAP Build Process Automation are especially enhanced with the features listed in Table 9.1.

Feature	Description
No-code process builder	With simple drag-and-drop capabilities, digitize your workflows by creating forms, managing decision logic, and creating, customizing, and organizing process flows.
Robotic process automation (RPA)	Automate repetitive manual tasks with no-code and low-code functions or the integrated recorder for automations. Common examples of tasks that lend themselves to automation: ■ Manual tasks: Tedious, time-consuming, and error-prone routine tasks such as copy-and-paste operations; tasks regarding data extraction, data entry, and data creation; and consolidation and processing of data from multiple data sources. ■ Extensive tasks: Recurring steps that must be performed in sequence, e.g., as part of data migrations. ■ Cross-system tasks: Tasks that require access to multiple applications that don't provide suitable APIs, such as web applications, legacy enterprise solutions, and software as a service (SaaS) offerings.
Bots	Develop supervised bots for tasks that must be actively initiated by people, as well as the development of unsupervised bots for tasks that can be performed without human intervention.
Integrated AI	Access built-in AI capabilities for intelligent document processing without involving data scientists. Common examples include the following: ■ Data extraction: Extract data from a large number of digital documents or scanned images and transfer them to your enterprise systems for processing. ■ Structured documents: Extract data from structured data sources, such as Microsoft Excel files. ■ Unstructured documents: Extract data from unstructured data sources, such as PDF, PNG, TIFF, JPEG, etc. ■ Data enrichment: Enrich extracted data, based on your own master data, and link incoming document-based extractions with a unique ID that your system can process further.

Table 9.1 SAP Build Process Automation Functions

In addition, to accelerate business transformation from process analysis to process automation, more than 135 tailored process optimization recommendations from SAP Signavio Process Intelligence can be used directly in SAP Build Process Automation (see Chapter 4). SAP Signavio Process Intelligence can also trigger automated actions via APIs to execute process automation with workflows and bots. Furthermore, SAP Build Process Automation complements SAP Signavio Process Governance (see Chapter 8), allowing users to manage approvals for process model changes in SAP Signavio Process Manager (see Chapter 5). Thanks to this integration, SAP offers a comprehensive set of capabilities for operational transformation and automation (see Figure 9.2).

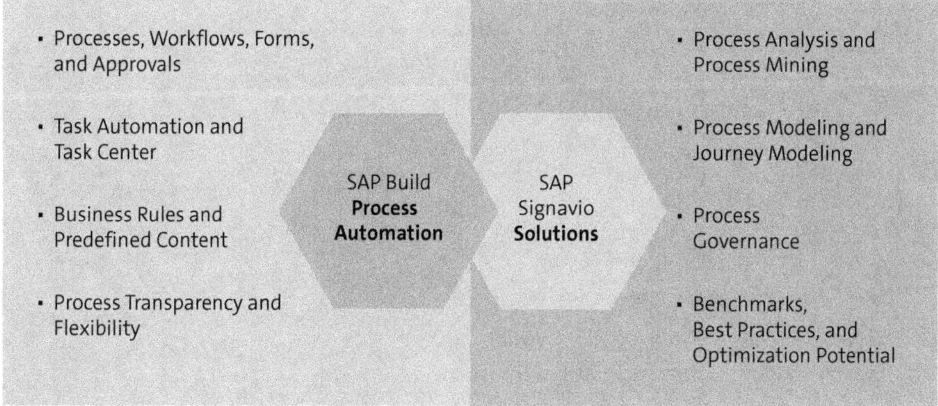

Figure 9.2 Integration of SAP Build Process Automation in SAP Signavio Solutions

In terms of the UI, SAP Build Process Automation provides four main tabs, which are explained in Table 9.2.

Tab	Description
Lobby	The **Lobby** tab is the starting point for all SAP Build products. The lobby provides standard functionality such as lifecycle management and monitoring. Created artifacts such as UX components, workflows, data models, and business logic can be shared across products and projects, enabling component reuse and the creation of end-to-end solutions. All products offer a wide range of prebuilt templates and modules that can be dragged and dropped into and out of the development environment. The lobby can be accessed via the SAP BTP cockpit.
Store	The tab allows you to refer to ready-to-use templates for quick implementation. You can choose from different project types, products, disciplines, and/or industries.
Monitor	The tab provides information about your processes, including the started processes and their status.
Settings	The tab shows basic settings of your SAP Build Process Automation solution.

Table 9.2 Important Tabs in SAP Build Process Automation

As an example, the **Store** tab has been selected in Figure 9.3. Here, we can use predefined content for further use, searching by project name, sorting search results as desired, or filtering by project type, for example.

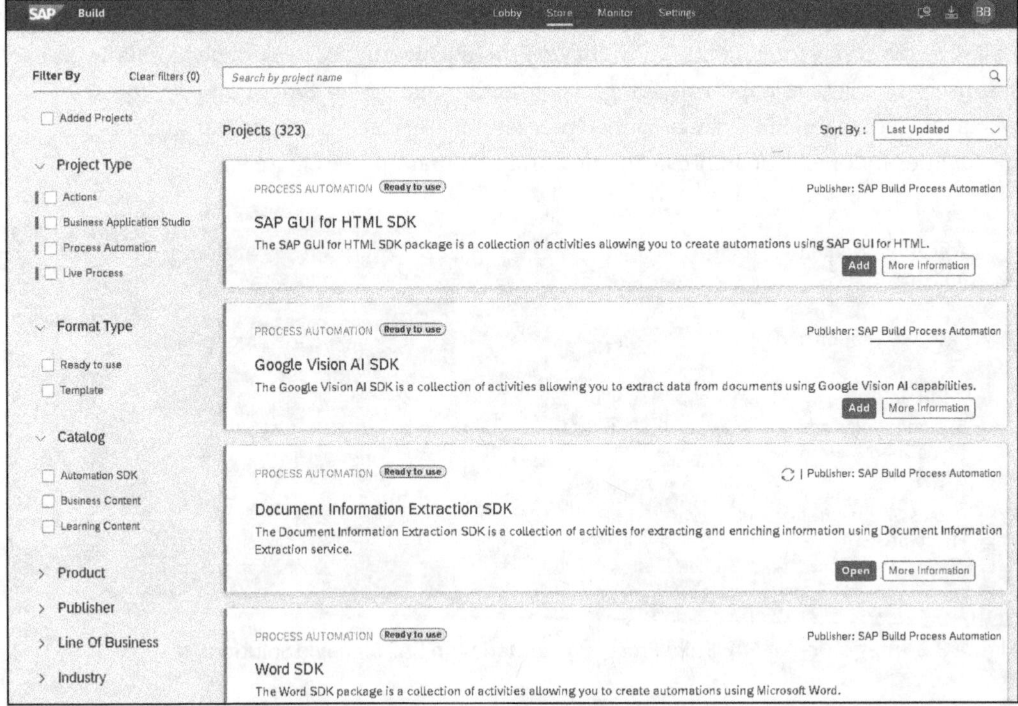

Figure 9.3 Store Tab in SAP Build Process Automation

9.2 Benefits

Let's begin with an overview of the core benefits of SAP Build Process Automation:

- Carry out rapid adaptation, improvement, and innovation of processes.
- Automate processes with minimal support from IT.
- Promote active collaboration between business and IT.
- Promote knowledge sharing within the organization.
- Increase the level of process automation.
- Increase process efficiency.
- Increase business flexibility.

SAP Build Process Automation draws on integration and automation content from SAP Business Technology Platform (SAP BTP) and provides native integration with solutions in the SAP application stack. You can also benefit from content developed by the broad ecosystem of SAP partners and members of the SAP community for SAP and third-party solutions. The process automation solution offers prebuilt functions, templates, and content in a library specifically designed to work with SAP solutions. In Figure 9.4, for example, you can see a template for a release process in the sales order area. You can get to this template by using the functions described in Section 9.1 introduced in the **Store** tab.

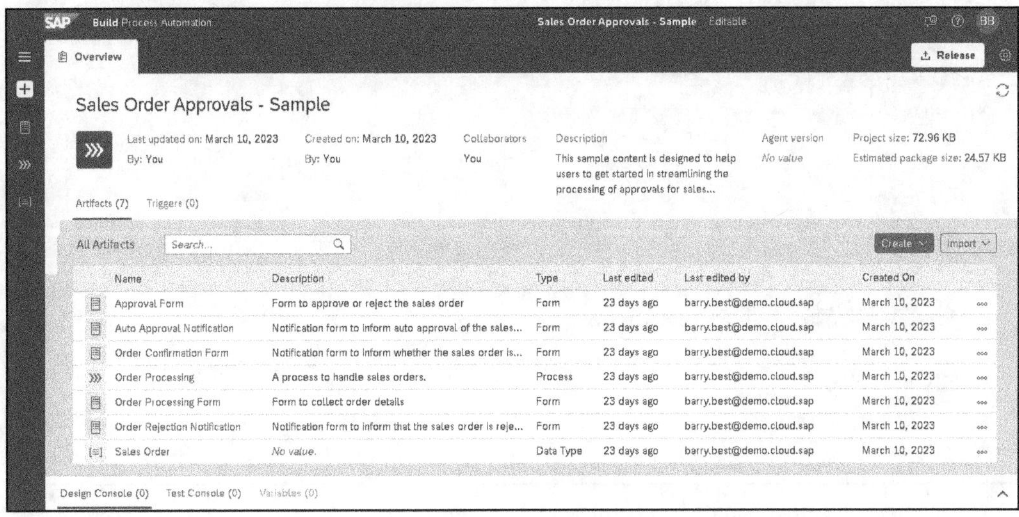

Figure 9.4 Template for a Sales Order Approval Process

With just a few clicks, citizen developers can combine and reuse a range of functions, including bots, workflow components, process steps, and actions. Features such as *Process Builder* and *Form Builder* provide drag-and-drop functionality that simplifies the effort required to create new automated processes or revise existing processes in SAP applications. With SAP Build Process Automation, your organization can build on SAP expertise to achieve new efficiencies and business value without the complexity of traditional process automation.

Workflow management capabilities help citizen developers digitize workflows and orchestrate or extend structured processes tailored to specific business needs. In addition, business rules are used to automate and flexibly adapt decision logic. Workflow management capabilities also increase process flexibility by helping industry professionals configure and manage process variants, decision logic, and process transparency dashboards with a low-code or no-code approach. Users can choose from a variety of content packages. These content packages can be accessed from the lobby and further through the store (see Figure 9.5) to support your projects. These packages include process variants, workflows, business rules, dashboards, and UIs. The solution supports extensions to standard processes from SAP S/4HANA, SAP Ariba solutions, and various SAP applications—including the SAP ERP application—as well as common usage patterns for cross-business unit workflows. A unified launchpad and task center acts as a central entry point for assigned workflow items and approvals.

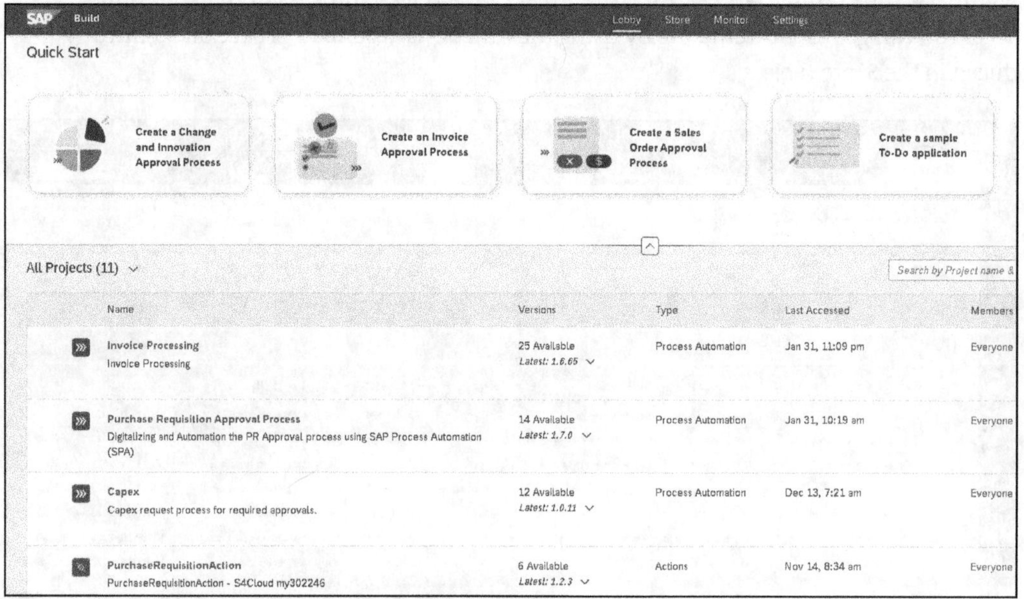

Figure 9.5 Predefined Contents for Process Automation

SAP Build Process Automation also includes embedded *RPA capabilities* that help citizen developers improve process flows through task automation using RPA bots. The RPA capabilities help professionals automate repetitive manual tasks by mimicking user interactions with any system. Users can automate data transfers between legacy and web-based systems that lack sufficient integration capabilities. These capabilities help you accelerate task processing, flexibly scale your systems to meet changing needs, and reduce error rates. Users can choose from a wide range of prebuilt best-practice bots in the **Store**—both unattended and attended—for different business areas. Ready-to-use bots help you get fast results from RPA capabilities.

SAP Build Process Automation also provides unified, enterprise-grade *AI capabilities* that help citizen developers create intelligent business processes. With this solution, you can leverage machine learning capabilities to create intelligent actions and recommendations based on the correlation of workflow data and attribute importance. Business users receive data-driven recommendations with confidence levels, including information on influencing factors, to help them make the best decision for the business. By eliminating parallel reviews, this feature helps streamline your processes. Citizen developers can also configure and train machine learning explainability to ensure the system provides the best recommendations for each end user decision. They can also configure the solution to periodically retrain the scenario based on completed workflows, which helps produce recommendations with high confidence. SAP

Build Process Automation also supports document extraction using bot automation. You can classify business documents and route them to the right process, extract data from various sources, and create appropriate records. Together, these capabilities help you solve business problems by combining industry expertise with intuitive technology.

As described earlier in Section 9.1, SAP Build Process Automation and SAP Build are important components of SAP BTP. SAP BTP provides low-code and no-code development tools that enable all developers to quickly build automations and enterprise applications to meet changing business needs. As a result, SAP Build Process Automation also leverages these capabilities. From a single point of entry, citizen developers and experienced developers can access the tools that best meet their needs. For example, citizen developers can use their knowledge of business processes to create simpler automations using SAP Build Process Automation. Figure 9.6 illustrates a flow model of an automated invoice approval process that we opened via the lobby on the **Invoice Approval Process** tab.

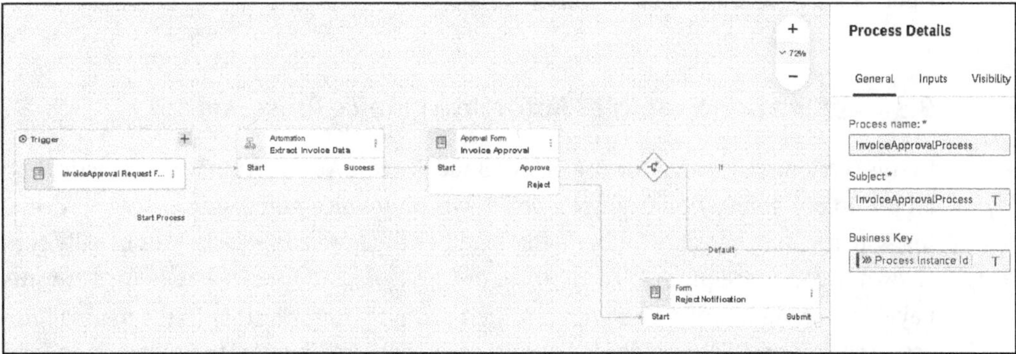

Figure 9.6 Sequence of an Automated Release Process

As automation requirements become more complex, these aspects can be easily handed over to professional developers within your organization. Professional developers can leverage the low-code development capabilities of SAP Business Application Studio, a modern cloud-based IDE tailored for efficient business application development. In this context, they can access advanced workflow features to create more complex process flows, forms, or dashboards for process transparency (see Figure 9.7). With SAP Build Process Automation, professional developers can in turn make these complex functions available to citizen developers. The integrated, persona-optimized collection of developer tools makes it possible to reduce any backlog in process development by accelerating projects. Because the tools are all part of the same development framework, they help break down silos between citizen developers and professional developers.

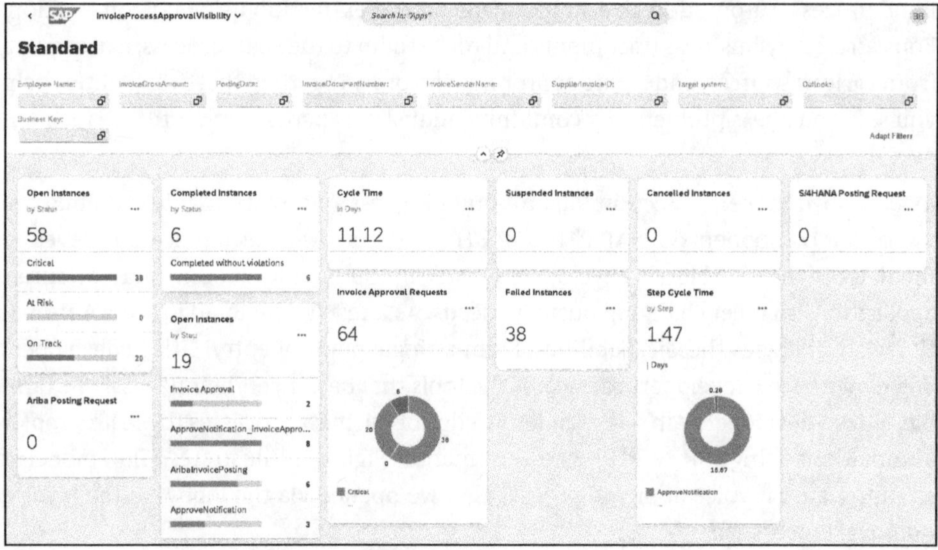

Figure 9.7 Default Dashboard for Citizen Developers

9.3 Application Example: Automate Invoice Processing

In this section, we present the use of SAP Build Process Automation with a practical example. In our application example, we'll focus on invoice processing. Invoice processing is an accounts payable function that handles the invoice lifecycle. This usually consists of invoice receipt, invoice approval or rejection, and invoice payment. Accounts payable receives many emails with attached invoices from vendors. In the manual process, the responsible employees have to scan each email, open the attachment, and post the invoice to the target system. In addition, several reviews as well as approvals may be required before posting to the target system. The entire process can be very cumbersome due to many repetitive activities, lack of visibility, unstructured information capture, and very manual processing of data. But this entire process can be automated and digitized with SAP Build Process Automation to improve team productivity.

This application example shows an easy-to-understand invoice processing process created with SAP Build Process Automation. The process starts with a simple request form where our employee Barry Best has various options to select the location of the invoice document (local folder location). Depending on the location of the document, SAP Build Process Automation extracts the invoice PDF using the *Document Information Extraction* service and then forwards the invoice to the person responsible for approval. Once the request is fully approved, SAP Build Process Automation triggers the automatic posting of the invoice to SAP S/4HANA Cloud. At the end of the process, Barry Best receives a confirmation with the posting document ID. All forms used throughout the process were created using SAP Build Process Automation's integrated Process Builder.

To represent our scenario visually, in the initial screen in Figure 9.8, we first select the **Process Trigger** tile to fill out an invoice approval form or send it to the person responsible for approval.

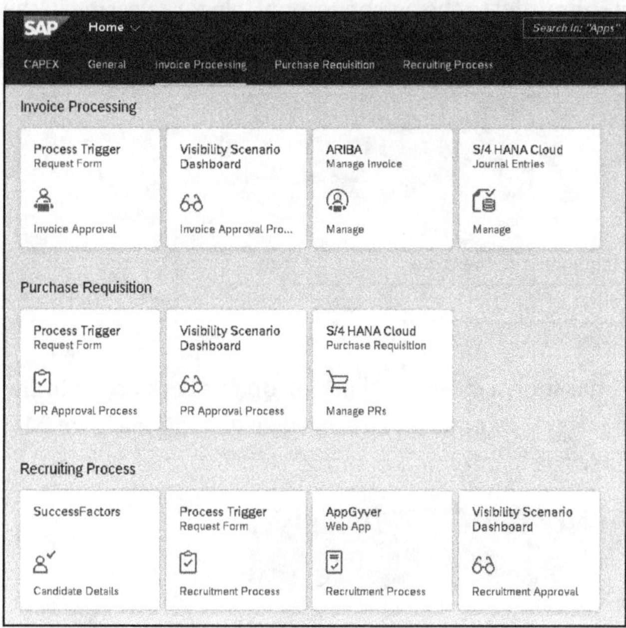

Figure 9.8 Initial Screen: Process Trigger

To start the release process, we add the parameters shown in in Figure 9.9 such as the name of the employee, the date, and a short comment about the issue.

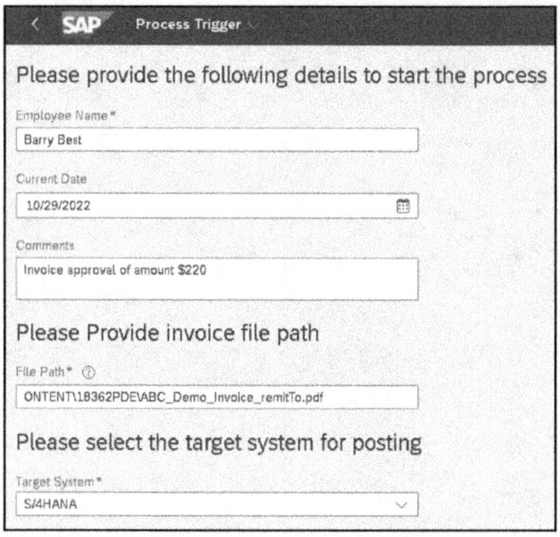

SAP Process Trigger

Please provide the following details to start the process

Employee Name *
Barry Best

Current Date
10/29/2022

Comments
Invoice approval of amount $220

Please Provide invoice file path

File Path *
ONTENT\18362PDE\ABC_Demo_Invoice_remitTo.pdf

Please select the target system for posting

Target System *
S/4HANA

Figure 9.9 Request Form for Invoice Approval

We also specify the path to the invoice to be released and the target system. In our case, the target system is SAPS/4HANA Cloud, where the invoice will eventually be posted. Once all the information is filled in, we can submit the form via the **Submit** button.

When the form is successfully submitted to the person responsible for releasing it, the message **Your Form is submitted successfully** is displayed (see Figure 9.10).

Figure 9.10 Successfully Submitted Application Form

Now we switch back to the initial screen by simply clicking on the back arrow in the upper-left corner of the taskbar. Here, we now notice a message inbox (see Figure 9.11). Clicking on the **My Inbox** tile takes us to our inbox.

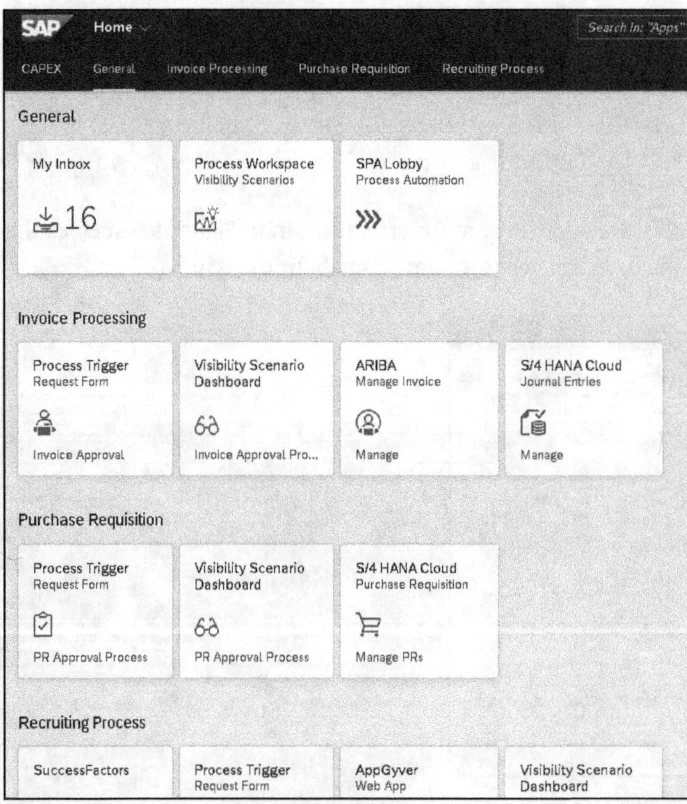

Figure 9.11 Initial Screen: Inbox

In our example scenario, we are the applicant and the person responsible for release at the same time. Accordingly, we received a notification with the request to approve the invoice. We comply with this request by clicking on the corresponding task on the left and giving our approval in the **Approver Comment** field. We enter "Approved." and confirm our entry by clicking on the **Approve** button (see Figure 9.12).

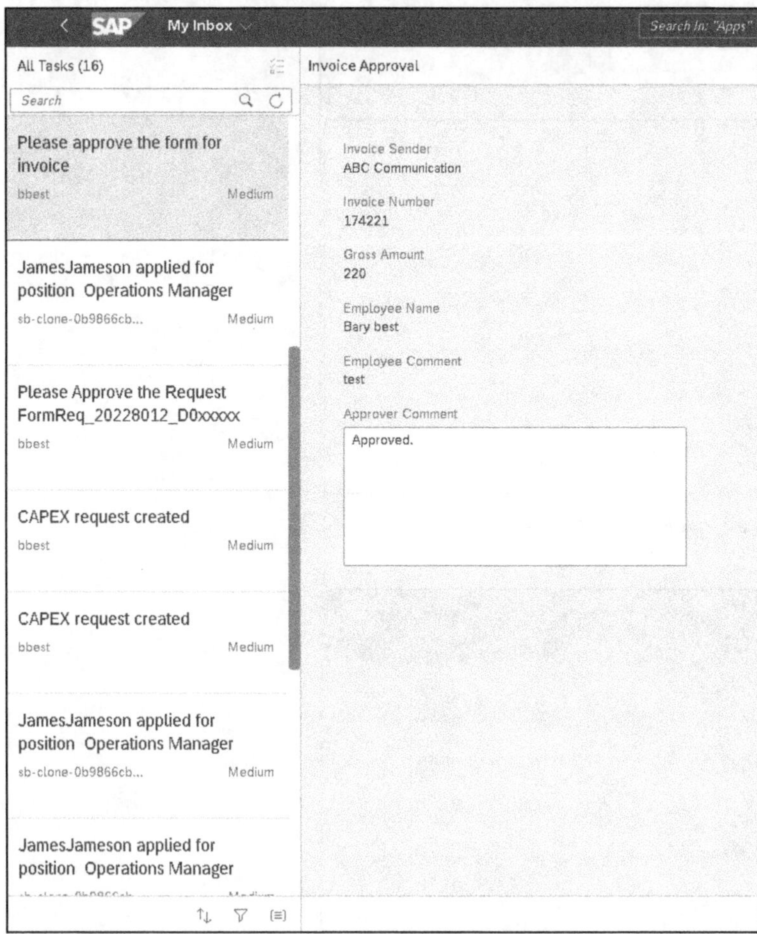

Figure 9.12 Approve Invoice Release

Subsequently, as shown in Figure 9.13, we receive a confirmation that the invoice has been released and already posted to the target SAP S/4HANA Cloud system. For this purpose, the number of the posting document is also provided (**Posting Document ID**), which we note for now.

We exit the inbox by simply clicking the back arrow in the top left of the taskbar and go directly to the target SAP S/4HANA Cloud system to verify that the invoice was successfully posted (see Figure 9.14).

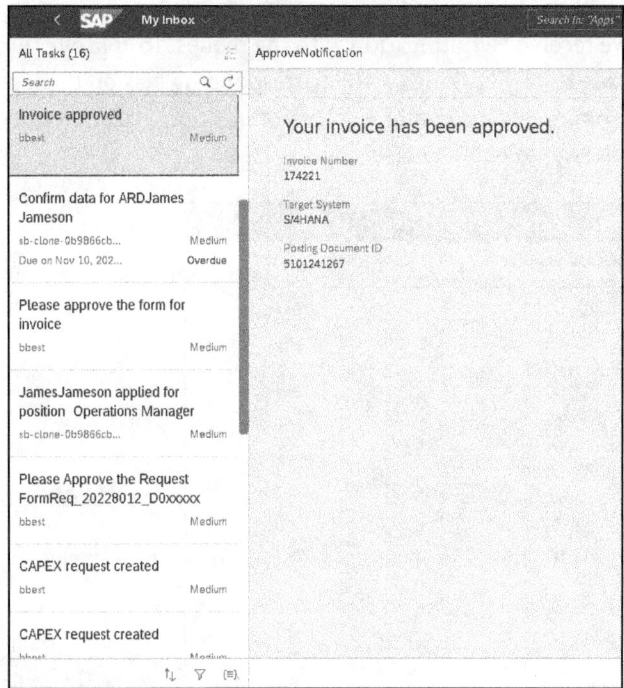

Figure 9.13 Approved Invoice Release

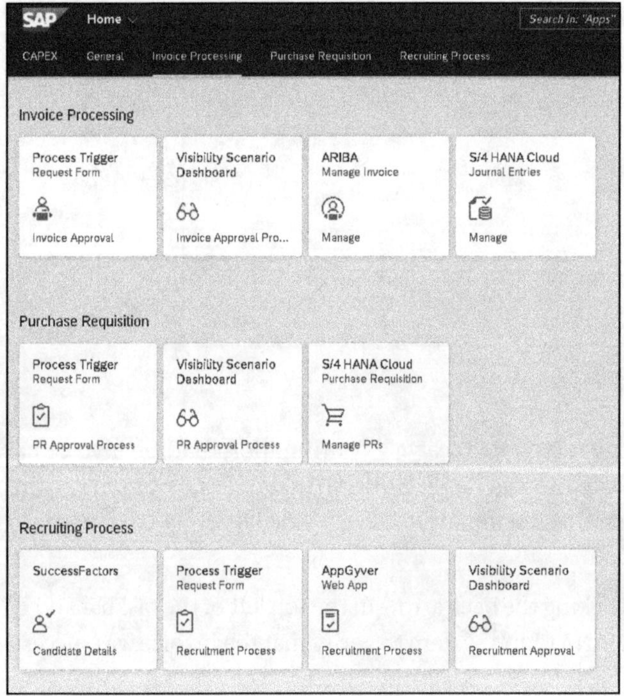

Figure 9.14 Initial Screen: Target System

We click the **S/4 HANA Cloud Journal Entries** tile. In the **Journal Entries** area, we search specifically for the booking entry using the booking number that was communicated to us in the invoice release confirmation (see Figure 9.15). The entry is visible in the system, and we click on the entry line to get more details about the entry.

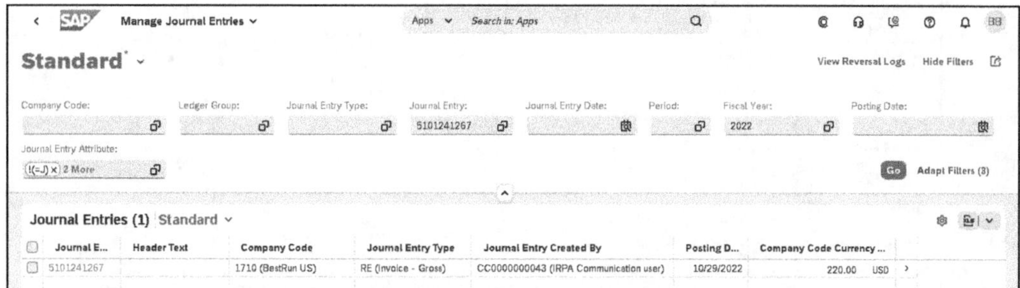

Figure 9.15 Search for Journal Entry in SAP S/4HANA Cloud

In Figure 9.16, we now see the complete booking entry, including amount and booking date. Invoice-relevant details were taken directly from the PDF invoice attached to us.

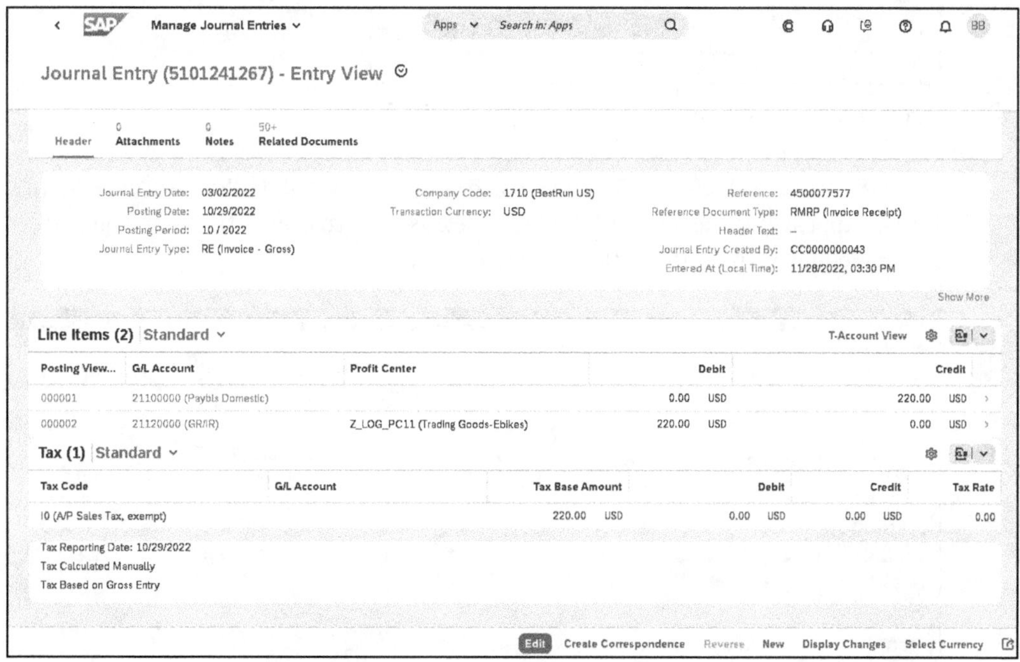

Figure 9.16 Display of the Journal Entry in SAP S/4HANA Cloud

This example shows how quickly we can post an invoice using SAP Build Process Automation. As soon as the invoice was released, it was automatically posted to the target system. This saves employees valuable time to focus on other activities.

To get an overview of all invoice release processes, on the initial screen in Figure 9.17, click on the **Visibility Scenario Dashboard** tile.

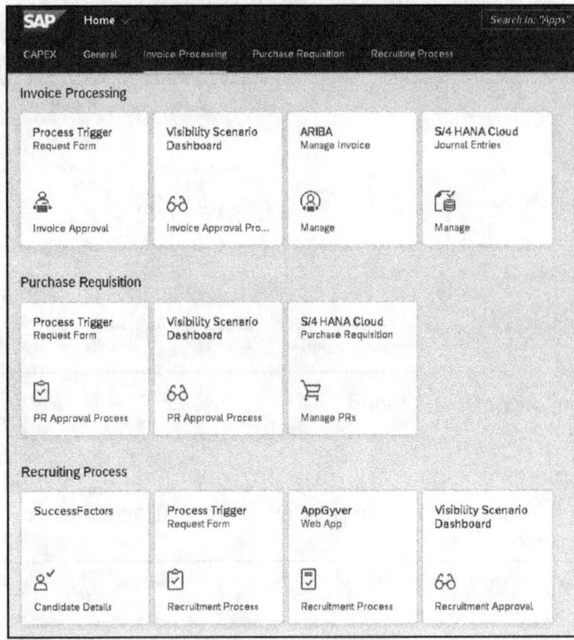

Figure 9.17 Initial Screen: Dashboard

The dashboard in Figure 9.18 shows us the most important information regarding invoice approval at a glance. This includes details on turnaround times, the number of release requests, and open or closed instances.

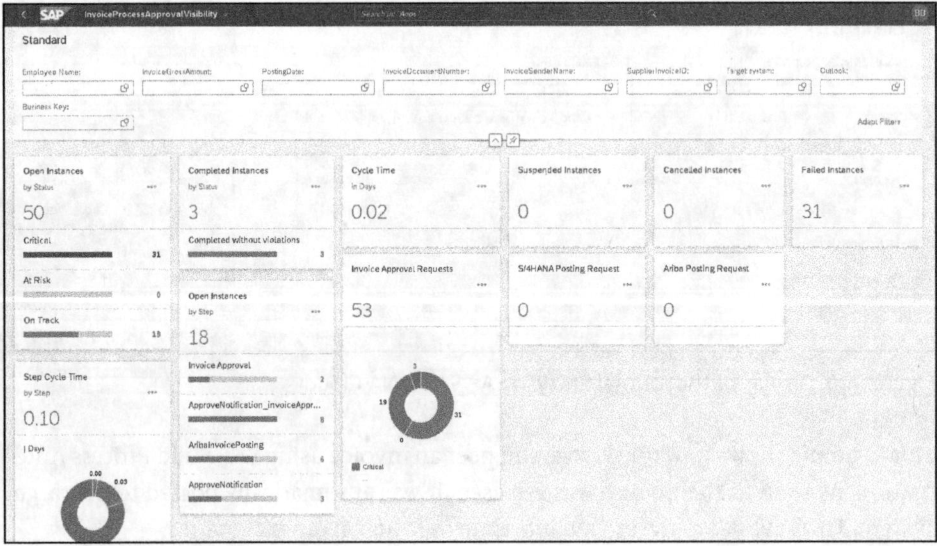

Figure 9.18 Invoice Approval Dashboard

9.4 Summary

SAP Build Process Automation helps you simplify and accelerate the deployment of process automation in your organization. For example, your professionals can build process automation on SAP applications without programming skills and without depending on IT support.

SAP Build Process Automation provides workflow management and task automation capabilities in a single no-code process builder. Prebuilt functions, templates, predeveloped content, and bots from an extensive SAP partner ecosystem and the SAP community provide rapid value.

The solution runs in the SAP BTP cloud environment and integrates natively with other SAP applications without requiring an additional layer of integration. SAP Build Process Automation can also help improve the return on investment (ROI) of your SAP applications, as the tool increases both the usage of these business systems and the level of process automation without adding new layers of complexity.

PART III

How to Successfully Navigate the Transition to SAP S/4HANA with Business Process Transformation

PART III

How to Successfully Navigate the Transition to SAP S/4HANA with Business Process Transformation

Chapter 10
Transitioning to SAP S/4HANA as a Gateway to Business Process Transformation Management

In this chapter, you'll get an overview of the basic principles of SAP S/4HANA, possible transition scenarios, opportunities and challenges, and why the move to SAP S/4HANA is the ideal time to get started with business process transformation management.

10

Business processes are the heart of your company. *Business process transformation management* is a prerequisite so that you can improve them at any time. It allows you to translate your corporate strategy into the necessary adjustments to your processes, which in turn are implemented in IT systems such as an ERP system. This is done continuously in conjunction with application lifecycle management (ALM). Employees from the respective departments and IT also gain a common view of the business processes in their company. Figure 10.1 illustrates this interaction.

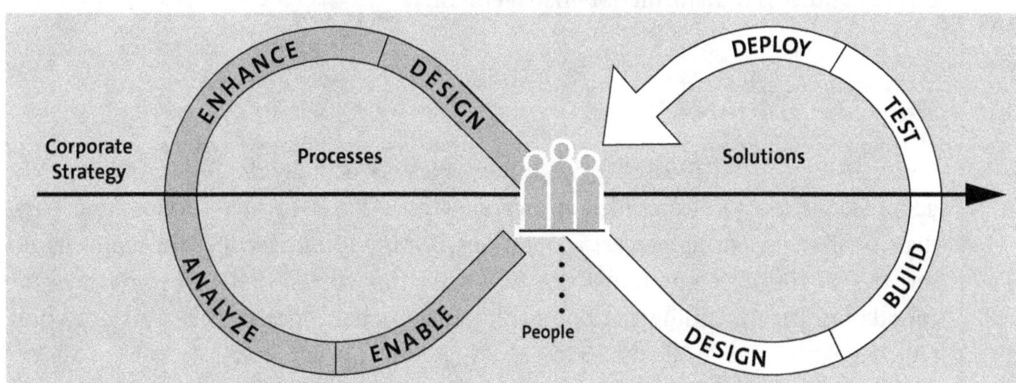

Figure 10.1 Interaction of Corporate Strategy, Processes, and Systems

Business process transformation management not only enables you to improve business processes at any time, but it's also a prerequisite for a successful transition from SAP ERP to SAP S/4HANA. If you're already using SAP ERP, you can fundamentally redesign your existing processes while transitioning to SAP S/4HANA. The redesign of existing processes generates the real benefit of the transition to SAP S/4HANA for the company and each individual employee. Business process transformation management supports you in analyzing your as-is processes, defining the target processes, and carrying out change management activities. However, it not only serves the value-creating transition to SAP S/4HANA but also the continuous process improvement. This becomes necessary due to new requirements from business units or the need to improve existing process flows. Adjustments can be made not only on the basis of the new SAP S/4HANA functions but also using SAP Business Technology Platform (SAP BTP) and the rest of the SAP portfolio.

For new SAP customers who opt for SAP S/4HANA as their ERP solution, business process transformation management also supports all project phases. In the case of a new implementation, however, the focus is less on analyzing the as-is processes and more on an as close-to-standard design based on the SAP standard processes. After a successful go-live, the process analysis ensures that the implemented processes lead to efficient business processes. Combining the implementation of SAP S/4HANA with the establishment of business process transformation management is therefore also ideal in this case.

In this chapter, you'll first get an overview of the key differences of SAP S/4HANA compared to SAP ERP for the topic of business process transformation management, the possible transition scenarios, and the opportunities that arise when transitioning to SAP S/4HANA (Section 10.1). Section 10.2 is devoted to the definition of business process transformation management and the fundamental benefits. Finally, in Section 10.3, we look at why the transition to SAP S/4HANA is the right time to get started with business process transformation management.

10.1 SAP S/4HANA

The transition to SAP S/4HANA not only enables but actually requires processes to be adapted. Section 10.1.1 provides an overview of SAP S/4HANA's development paradigms and the resulting process adaptations. Section 10.1.2 describes the main characteristics of the different transformation scenarios to SAP S/4HANA and lays the foundation for describing the opportunities and challenges in the transformation, which are discussed in Section 10.1.3.

10.1.1 The Basic Principles of SAP S/4HANA

In 2011, SAP launched the *SAP HANA* in-memory database platform to the market. SAP HANA is based on the principle of holding data in the main working memory and thus performing analyses at a speed that wasn't possible with classic disk-based databases.

The capabilities of SAP HANA also made it possible to accelerate complex ERP processes. This was implemented in 2012 in the so-called *sidecar approach*, in which SAP

ERP continued to run on a non-SAP HANA database. In this approach, data was replicated to the SAP HANA database, and read accesses were redirected to the SAP HANA database via a secondary database connection. This significantly improved read response times, speeding up transactions. Whereas profitability analyses could previously only be performed with restrictive selection criteria, and, for example, material requirements planning (MRP) could be carried out in a job processing overnight, it was now possible to work in dialog mode without restrictions. The speed advantages thus formed the basis for rethinking processes.

To avoid redundant data storage as well as data replication and to improve the response times of not just selected transactions, SAP HANA was released for Enhancement Package 7 of SAP ERP in August 2013. This accelerated a large number of transactions as well as customer-specific developments.

However, optimizing the processes solely by accelerating the response times of the transactions was only possible to a limited extent. A prerequisite for a fundamental redesign of the processes was to question the table structures that had grown over the years. SAP HANA offered the opportunity to eliminate summary and index tables that were no longer necessary. SAP started the redesign in the area of accounting and released the ERP add-on *SAP S/4HANA Finance 1503* in March 2015. By reducing the number of tables and simultaneously combining tables from the classic Financial Accounting (FI) and Controlling (CO) components, the reconciliation effort was reduced. The redesign of processes was thus no longer only based on faster response times but also on revised table structures and the elimination of transactions.

Nevertheless, the SAP ERP development strategy had a fundamentally nondisruptive character. Revising a database design created over many years, eliminating redundant existing transactions, redefining fundamental development paradigms, and delivering these changes to customers in the form of further add-ons would have resulted in too much complexity for SAP and its customers alike.

To ensure both investment protection for existing customers and to carry out a fundamental redesign of processes, the decision was made to develop a new independent ERP product: *SAP S/4HANA*.

The following five basic development principles have been defined for the development of SAP S/4HANA:

1. **Deployment in the cloud and on-premise**
   Operation of the solution is offered in the public cloud, in the private cloud, and on-premise.
2. **Principle of one**
   A business process is supported by a transaction.
3. **SAP HANA as database**
   SAP HANA is supported as the only database.

4. **Role-based, process-oriented design**
   New developments are role-based and not functional (transactional) in the form of SAP Fiori apps.
5. **Upgrade-like transformation scenarios**
   Change for existing customers isn't exclusively in the form of a new implementation.

The principle of SAP HANA as a database enabled a simplified database design, making SAP HANA the only supported database for SAP S/4HANA. This made it possible not only to accelerate transactions, as already described, but also to perform complex evaluations, which in the past had been outsourced to a data warehouse such as SAP Business Warehouse (SAP BW), directly in the SAP ERP system again. Users no longer need to switch between two systems for transactional activities such as processing a sales order and analytical tasks such as revenue analysis. Working methods and the underlying processes can thus be fundamentally rethought. Working in one system also enables the principle of *insights-to-action*. Users are shown the most important key figures in SAP Fiori in their specific role context, for example, as the person responsible for MRP, without first having to start transactions with sometimes complex selection variants in SAP ERP or queries in SAP BW. Trends and exceptions are visualized immediately. In the next step, users can not only analyze the key figure but also display the material or the source of requirements, for example, via a transaction or app.

In SAP ERP, module-specific reporting was supported by various tools such as the Report Writer, the Report Painter, or SAP List Viewer (ALV). What all these solutions had in common was that after specifying data selection criteria, some of which were complex, the user generally had to analyze the data in tabular form, which in turn often resulted in data being exported to office applications. The time required for selecting, analyzing, and exporting data is significantly reduced by SAP S/4HANA customers in the area of reporting, which offers the opportunity to significantly simplify not only processes but also system landscapes. As a rule, these analysis options are still available, but can be harmonized and simplified by the aforementioned SAP Fiori-based analysis options.

Accelerated response times from SAP ERP or from SAP GUI-based transactions and SAP Fiori apps made it possible to rethink processes, especially in the area of reporting. However, an even greater benefit was achieved by replacing the transactional SAP GUI user interface with role-based SAP Fiori apps. The combination of a web-based user interface that can be customized by users or administrators and can also be used on mobile devices offered the opportunity for a revolutionary, rather than evolutionary, redesign of processes. SAP ERP was based on a transactional user design. For many employees, operating a large number of transactions—some of which were complex—was a major challenge. Companies therefore reduced the number of transactions by creating new transactions in the form of custom developments and combining multiple transactions in them to increase usability for users. Because SAP ERP, and therefore

SAP transactions, support differently complex business processes in a variety of industries, the design of transactions is often very complex. A multitude of fields, divided into tabs, made the operation of the transactions time-consuming and error prone. Therefore, SAP standard transactions were often copied, and irrelevant fields were hidden with a high development effort. SAP GUI offered no possibility for personalization.

SAP Fiori

The use of SAP Fiori is a core principle in the development or enhancement of SAP S/4HANA. In the SAP Fiori apps reference library, customers receive tailored recommendations on which SAP Fiori apps are relevant for the use of their current SAP GUI transactions (see *http://s-prs.de/v917401*).

For more information on SAP Fiori, see these resources:

- Comparison of the sales order fulfillment process in SAP ERP and SAP S/4HANA: *http://s-prs.de/v917402*
- Comparison of the end-to-end inventory management process in SAP ERP and SAP S/4HANA: *http://s-prs.de/v917403*
- Recommendations for switching from SAP GUI to SAP Fiori: *http://s-prs.de/v917404*
- An overview of the insight-to-action approach: *http://s-prs.de/v917405*

Based on these principles, the first on-premise release—SAP S/4HANA 1511—was in November 2015 and the first cloud edition, SAP S/4HANA Cloud Public Edition 1908, was released in August 2019. In the following years, there were annual on-premise releases and initially quarterly cloud edition releases.

SAP S/4HANA also supports a combination of the two operating models in the form of a *two-tier architecture*. In this architecture, SAP S/4HANA is used at the group level, and SAP S/4HANA Cloud Public Edition is used in subsidiaries and foreign subsidiaries, such as sales offices. Several SAP S/4HANA instances can also be combined and used in the form of centralized finance or procurement.

In January 2021, SAP then launched *RISE with SAP*, which includes SAP S/4HANA in the form of both a public cloud variant and a private cloud variant. In March 2023, SAP also introduced the *GROW with SAP* offering, which is aimed at new customers in the midmarket segment. In addition to SAP S/4HANA Cloud Public Edition and activation services, the offering also includes SAP BTP. This allows the extension of standard processes with the help of the low-code/no-code solutions from *SAP Build*.

SAP S/4HANA Release Strategy

Information on SAP's release strategy can be found, for example, in the document *SAP S/4HANA Maintenance Strategy* (see *http://s-prs.de/v917406*) or in the *Product Availability Matrix* (see *https://userapps.support.sap.com/sap/support/pam*).

The basis for the development of SAP S/4HANA is EHP 7 of SAP ERP 6.0, which was necessary to enable existing customers to transition to SAP S/4HANA without the need for a new implementation and thus ensure investment protection.

Changes to SAP ERP during the development of SAP S/4HANA were documented in the form of simplification items and summarized in a simplification list for the individual components and industries. The simplification items describe functions that are no longer available in SAP S/4HANA. The following statuses exist for the alternatives provided to customers:

- Alternative is available
- Alternative is planned—part of the roadmap
- Alternative is planned—not yet part of the roadmap
- Alternative isn't planned

In addition to the simplification elements that must be taken into account immediately when transitioning to SAP S/4HANA, SAP defined a compatibility scope with the first SAP S/4HANA release in November 2015. This describes functions that can still be used by customers in a transition phase until the end of 2025, but which are also no longer part of the SAP S/4HANA target architecture in the long term. A phased adaptation is possible for customers, provided they don't fully implement the target processes during the initial transition. SAP plans to provide solutions based on SAP S/4HANA for the compatibility scope.

[»]

Simplification Elements and Compatibility Scope

For more information on the following topics, see, for example, SAP Help Portal at *https://help.sap.com/docs/SAP_S4HANA_ON-PREMISE*.

More information on the simplification elements and the scope of compatibility can also be found in SAP Notes and in various blog posts:

- Explanation of simplification element categories: *https://launchpad.support.sap.com/#/notes/2931193*
- Compatibility scope matrix for SAP S/4HANA: *https://launchpad.support.sap.com/#/notes/2269324*
- The most important questions and answers about the compatibility scope: *http://s-prs.de/v917407*
- Compatibility scope overview: *http://s-prs.de/v917408*

SAP Readiness Check for SAP S/4HANA provides an analysis of which simplification elements and which compatibility scope items are relevant for you. You can find more information in the documentation: *https://help.sap.com/docs/SAP_READINESS_CHECK*.

By far, the most important driver for the redesign of processes during the conversion to SAP S/4HANA is, in addition to the need to redesign processes, the large number of innovations. While the focus of development activities in the first few years was on simplification, the number of new functions is continuously increasing.

[«]

SAP S/4HANA Innovations

For more information on the latest SAP S/4HANA innovations, see the following blog posts:

- *http://s-prs.de/v917409*
- *http://s-prs.de/v917410*
- *http://s-prs.de/v917411*

10.1.2 The Change to SAP S/4HANA

The transition to SAP S/4HANA can be made in three different ways:

- New implementation
- System conversion
- Selective data transition, a hybrid of the first two options

For new customers or customers who also want to transition to the public cloud with the change to SAP S/4HANA, a new implementation is the only option. For existing customers who don't want to combine the transition to SAP S/4HANA with a change in operating model to the public cloud, the question is which transition option is right for them. Table 10.1 provides an overview of the transition paths.

Transition Path	Description
New implementation	■ New installation of an on-premise SAP S/4HANA system or decision for one of the two operating models: SAP S/4HANA Cloud Private Edition or SAP S/4HANA Cloud Public Edition ■ Implementation based on SAP best practices ■ Transfer of master data and open items
System conversion	■ Transition to on-premise SAP S/4HANA version or SAP S/4HANA Cloud Private Edition ■ Conversion of the existing SAP ERP system, in which the configurations, extensions, authorizations, interfaces, master data, and transaction data are retained
Selective data transition	■ Conversion of an existing SAP ERP system without master and transaction data analogous to system conversion or implementation of a new SAP S/4HANA system or use of a private cloud and transfer of configuration and enhancements based on transports ■ Selective migration of master and transaction data, including a chronological history

Table 10.1 Transition Paths to SAP S/4HANA

SAP supports its customers in the change to SAP S/4HANA with different tools that are used depending on the transition scenario. Table 10.2 provides an overview of the tools.

Change Path	Tool
New implementation	Data migration cockpit
System conversion	■ SAP Readiness Check for SAP S/4HANA conversion ■ ABAP Test Cockpit (ATC) ■ Software Update Manager (SUM) with the Database Migration Option (DMO)
Selective data transition	SAP Business Transformation Center

Table 10.2 Tools for the Change to SAP S/4HANA

If you're already operating one or more SAP ERP systems, as opposed to new customers, as already described, the question arises as to which of the transition scenarios is right for you. An initial, basic evaluation of the options can be made by answering the following questions:

- **Do the current business processes support the long-term corporate strategy?**
  If a fundamental redesign of the business processes is required to reflect the corporate strategy, a new implementation is favored.
- **Are companies able to map their business processes through the SAP standard process content?**
  If the company is able to adapt to the SAP standard, a new implementation is favored. Using the SAP standard makes it possible to reduce costs by eliminating the need to stockpile expertise for enhancements, among other things, and to maximize the benefit from the SAP S/4HANA investment because the reduction in testing efforts enables more frequent release upgrades.
- **Is the change to SAP S/4HANA an IT project or a business initiative?**
  If the transition to SAP S/4HANA is a project that is being driven significantly by IT, a system conversion should be considered. If established processes are to be analyzed and redesigned in the course of the transition to SAP S/4HANA, a new implementation is favored.
- **Is the transition to SAP S/4HANA possible in one step?**
  If the SAP ERP source release allows a transition in one step, a system conversion is favored.
- **Is there a need to keep all data?**
  If all master data and, above all, transaction data is required, a system conversion is preferable. If data cleansing may even be necessary, and if companies can start with open items and master data at the beginning of the transition, a new implementation is favored.

- **Is landscape consolidation and process harmonization required as part of the transition to SAP S/4HANA?**
  If both aspects must be taken into account when transitioning to SAP S/4HANA, a new implementation is favored.
- **How large is the number of interfaces (to SAP and non-SAP systems)?**
  If the number of interfaces is very large, a system conversion is favored.
- **Is the company capable of implementing a multiyear transformation plan?**
  System conversion without a comprehensive process design results in severely limited added value. Therefore, to maximize the benefits, a multiyear downstream transformation is necessary.

Answering the questions may already lead to a result as to which change path is preferred. For a more detailed evaluation of whether a new implementation or a system conversion should be chosen, and for an analysis of whether the selective data transition scenario is suitable as an alternative to a system conversion or a new implementation, you should nevertheless carry out a detailed examination. This involves defining various target architectures in addition to the basic assessment of transition paths that has already been performed, analyzing the initial situation, and comparing the resulting possible transition paths. For the decision-making process, you should also consider benefits, costs, and the required personnel resources.

With regard to the target architecture, two questions should first be considered that are also relevant for new customers:

- Which operating model is the right one?
- What is the right number of SAP S/4HANA instances?

With regard to the operating model, both customers already operating an SAP ERP system and new customers should first ask themselves whether they want to combine the transformation to SAP S/4HANA with a transition to a public cloud operating model. The prerequisite for this change is that the available process scope of SAP S/4HANA Cloud Public Edition basically supports the customers' requirements.

SAP supports customers in the analysis through the digital discovery assessment. Within the digital discovery assessment, customers define the required process scope on the basis of scope items, which, in addition to a textual description, include the supported business process in the form of a Business Process Model and Notation (BPMN) diagram. Adaptations of the standard processes are supported by an enhancement concept. In the decision-making process for SAP S/4HANA Cloud Public Edition, it's not only sensible to follow a fit-to-standard approach, as is the case with the SAP S/4HANA Cloud Private Edition or an on-premise system, but rather a necessary prerequisite for the success of the project. This means that the standard process scope is decisive for the process design and not initially a consideration of customer-specific requirements in the form of a fit-to-gap approach.

[»]

SAP S/4HANA Cloud Public Edition

For a blog entry that gives you an overview of the digital discovery assessment, visit *http://s-prs.de/v917412*. The different enhancement concepts of SAP S/4HANA Cloud Public Edition are described in a blog entry at *http://s-prs.de/v917413*.

In addition to the question of the operating model, you should analyze which is the right number of SAP S/4HANA instances for your company in preparation for the evaluation and decision regarding the transition path. This is particularly necessary if multiple SAP ERP systems are currently being operated. The following instance strategies are conceivable:

- **Single instance**
  A central, company-wide system.
- **Divisional breakdown/approach**
  One system per division or business model with a loose (nonsystem) coupling or a partial template.
- **Regional breakdown/approach**
  One system per region with a loose coupling, partial template, or common development template.
- **Functional division/approach**
  Setup of systems per function, for example, sales and finance.

Following are some decision criteria regarding which instance strategy is the right one:

- Process harmonization and synergies
- Transparency about processes and data
- Costs for operating the system landscape
- Interface complexity
- The ability to quickly implement process adjustments

You can see an exemplary evaluation of the process-related criteria in Table 10.3.

Instance Strategy	Process Harmonization and Synergies	Transparency about Processes and Data	Ability to Make Process Adjustments Quickly
Single instance	Very high	Very high	Low
Divisional distribution	High	Medium	Medium high
Regional distribution	Medium	Low	Medium
Functional division	High	Medium	Medium high

Table 10.3 Evaluation of Instance Strategies Based on Process-Related Criteria

The choice of operating model and instance strategy are the basis for the SAP S/4HANA target architecture. Possible target architectures include SAP S/4HANA Cloud Public Edition, SAP S/4HANA Cloud Private Edition, or on-premise SAP S/4HANA, as well as three other options:

- **SAP S/4HANA for central finance**
  Financial processes are operated in a centralized manner and in conjunction with the integration of SAP ERP, SAP S/4HANA, and non-SAP systems.
- **SAP S/4HANA for central procurement**
  Procurement processes are centralized in the same way as SAP S/4HANA for central finance.
- **Two-tier ERP approach**
  SAP S/4HANA Cloud Private Edition and SAP S/4HANA Cloud Public Edition systems run side by side.

Even if a company's core business processes are supported by SAP S/4HANA, for many customers, the transition to SAP S/4HANA is associated with a more extensive cloud or system transformation of the existing SAP landscape. Accordingly, the target architectures often include further solutions from the SAP product portfolio.

When analyzing the initial situation, existing SAP ERP systems are analyzed from both a process and technical perspective. Typical questions are as follows:

- Which simplification elements are relevant?
- What is the scope of compatibility to consider?
- How large is the share of the customer's own developments? Are they still used, and, if so, how extensive and how large is the adaptation effort?
- What enhancements have been made outside of SAP ERP to eliminate usability and response time vulnerabilities?
- What is the process performance of the SAP ERP system in terms of end-to-end processes and individual metrics?
- How many document types and process variants are there, how often are they used, and are they useful at all?
- How good is the quality of the master data?

Once you've defined possible target architectures (taking into account the operating models and the SAP S/4HANA instance strategy) and analyzed the initial situation, you can define company-specific transition options. These are to be evaluated on the basis of defined decision criteria. In practice, this often involves weighting the decision criteria.

[»]

More Information on Transition Options

If you need more information on SAP S/4HANA transition options, project success factors, and tools to help customers make the transition, the practical guide "Mapping Your Journey to SAP S/4HANA" provides a comprehensive overview (*http://s-prs.de/v917414*).

10.1.3 Opportunities and Challenges When Transitioning to SAP S/4HANA

The transition to SAP S/4HANA is a complex project for any company. Company-specific opportunities and challenges in the change to SAP S/4HANA result from, among other things, the target architecture, the initial situation, and the change path that the company decides on. A large number of questions need to be answered in these three dimensions as well as in others. In this chapter, as well as in Chapter 12, we look at opportunities and challenges that result from the need to adapt business processes. This is fundamentally necessary for two reasons:

- There are differences between SAP ERP and SAP S/4HANA due to the SAP S/4HANA design principles, which are described in terms of simplification elements and the scope of compatibility, resulting in mandatory adjustments.
- The aim is to achieve added value through enhancements, that is, adjustments that aren't mandatory, but without which the benefits are severely limited, and the possibilities offered by SAP S/4HANA aren't exhausted. Therefore, although these adjustments are optional in the narrower sense, they are actually quite important, as a transformation project without added value doesn't represent a sensible investment for companies.

Challenges that don't arise from the adjustments to the business processes aren't discussed in this and the following chapters, or only to a lesser extent, because business process transformation management doesn't provide any significant answers here. Examples include changing the operating model or reducing technical downtime, that is, the time window during which SAP ERP isn't available. These aspects aren't covered in this book or are only discussed marginally. The focus is on the opportunities and challenges that arise from adapting business processes. In the following, we consider these in the form of three questions:

- How can you ensure that the transition to SAP S/4HANA adds value to your business?
- How can the transition to SAP S/4HANA be made as quickly as possible?
- How does your company ensure that the transition to SAP S/4HANA is feasible?

Companies need to be able to find answers to these key questions so that you can successfully manage the change.

For the transition to SAP S/4HANA to result in added value for customers already running SAP ERP, a company must be able to improve its business processes. The following five steps show how companies can go about this:

1. Provide process transparency.
2. Evaluate the analysis results.
3. Define necessary process adjustments to realize identified added value potentials.
4. Implement the process adjustments.
5. Review to what extent the adjustments have led to a process improvement.

The starting point for any process improvement is to create transparency about how processes are performed in your company today. Most customers don't meet this requirement, or only to a limited extent. Ask yourself the following questions to gain visibility into your processes: Are there processes that aren't running at all or not running efficiently because of the system configuration? Does the process performance show any conspicuous features when comparing different company units; that is, are processes, for example, executed faster, with fewer errors in one area of the company compared to another area? Do additional findings result from a continuous observation of process key figures in the form of trends? Are there a large number of different process variants, some of which are used less or perhaps not at all, or yet others which have the shortest throughput times?

Based on the process transparency gained, it's necessary to evaluate it with the process owners. How can the results be explained, what are the causes of certain issues, and what process improvements can be realized? To evaluate added value potentials, comparisons with other companies are often useful in addition to internal company comparisons. In this step, it's necessary to scrutinize the lived process reality, because changes are often viewed critically by individual process participants. In the course of the change to SAP S/4HANA, it's also necessary to involve subject matter experts with detailed knowledge of SAP S/4HANA process innovation. The result of this step is an evaluation of the process transparency achieved in the first step and a quantification of the benefits. It makes sense to consider benefits and effort in equal measure and also to create a roadmap that defines at what point in time (before, during, or after the transition to SAP S/4HANA) adjustments should be implemented.

When evaluating the results of the process analysis, it was already fundamentally considered with which SAP S/4HANA innovations and with which adaptation effort which benefits can be achieved. In the next step, the definition of the process adaptations describes in detail which system-side adaptations lead to which adaptations of existing processes (and vice versa), which processes may become obsolete, and which completely new processes are possible at all. This step provides the basis for the next step, the implementation as well as the training and enablement of the process participants affected by the changes. Once this next step has been completed and the process changes are in productive use, it makes sense to review the success of the measures by performing a renewed process analysis. New findings lead to the need to reevaluate them and, if necessary, define further adjustments and implement them. This is a continuous process with incremental improvement steps. Coordination with all stakeholders on the progress of the analyses and adjustments and open communication are prerequisites for realizing added value potential.

Let's now look at the second dimension of time in addition to the dimension of added value. How can the transition to SAP S/4HANA be made as quickly as possible? Basically, SAP S/4HANA projects, as described at the beginning of this chapter, are complex and therefore also time intensive. The complexity and thus the time required can increase additionally if the goal is to adapt processes beyond the minimum required to achieve a benefit for the company: a larger number of process adaptations leads to a higher effort for process design, implementation, training, and roll-out. However, the use of business process transformation management can accelerate two key factors:

- **Time-to-insight**
  This describes the time required to analyze existing processes and quantify added value potential.
- **Time-to-adapt**
  This describes the effort required to adapt existing processes or to partially or completely reimplement processes.

Realizing added value for a company, especially if it's already using SAP ERP, means, as already explained in detail, creating the basis for this through process transparency. The time-to-insight factor is relevant here. To reduce time, but also effort, it's necessary to use data-based analyses in addition to manual process analyses based on surveys or workshops. These analyses can be carried out much faster and can also be highly automated with less effort. They thus complement manual process analyses and at the same time provide input for them. Standardized data-based analyses have another advantage; they make it possible to compare the process performance of one's own company with other companies, both within one's own industry and across industries.

In addition to the acceleration of the time-to-insight factor, the second factor is also relevant: time-to-adapt. For SAP customers who are already using SAP ERP, a key aspect here is reducing the time required to identify the measures necessary for process optimization. Which corrective measures can already be derived automatically from the data on the basis of the process analyses, and which can already be implemented in the run-up to the change, if necessary? Which innovations in SAP S/4HANA should be considered more closely against the background of the findings from the process analyses?

For both new customers and existing customers who opt for a new implementation, time-to-adapt means implementing as close to the standard as possible, taking into account SAP's best practices and a fit-to-standard approach.

Once a target process has been defined on the basis of a process analysis, the derived corrective actions, the review of SAP S/4HANA innovations, and SAP best practices, the transparency about it for all process participants is crucial for the speed of implementation. A central repository for processes (*process repository*), which can be accessed by all employees and where comments on the process or on individual process steps can be stored, reduces the risk of adjustments becoming necessary at a later point in time during implementation. These would lead to project delays. To avoid this, not only must the collaboration of all stakeholders be enabled, but the formal process release and the change process must also be regulated as part of the establishment of project governance.

If business process transformation is used to support the analysis and design of business processes, the implementation is carried out using ALM tools. An integrated tool chain, in which the process design is automatically transferred to ALM after it has been approved—including requirements that must be taken into account during implementation—also ensures that implementation takes place as quickly as possible. The implementation is based on the processes released with the process participants and process owners. Testing of the solution as well as training and enablement are also process-oriented. This significantly minimizes the risk that only individual functional tests are carried out in the project and that end-to-end business processes aren't tested, that unexpected errors occur after the go-live in productive use, or that the end users aren't able to operate the new functions.

Finally, we consider the dimension of feasibility using this question as a guiding principle: How does a company ensure that the change to SAP S/4HANA is feasible? Feasibility takes into account the answers to the previous two questions (How can companies ensure that the change to SAP S/4HANA creates added value for your company? How can the change to SAP S/4HANA be implemented as quickly as possible?) and looks at the following six areas:

1. Approach and planning of the transformation
2. Transformation goals
3. Management buy-in
4. Availability of employees with decision-making authority
5. Cooperation of all employees
6. Change management and enablement

As already mentioned in Section 10.1.2, the change scenario must be selected against the individual background of the respective company and should take into account the goals, target architectures, and initial situation. Of course, the result must also be implementable; that is, the necessary process adjustments must be taken into account in the time, personnel, and financial planning and must be reflected in the project plan. The selection of the right transformation approach for the company, as well as project planning, is significantly supported by the use of business process transformation management. The analysis and description of the current processes (also called *current state*), the target processes (also called *future state*), and the description of the necessary adjustments provide the basis for project planning.

The goals of the transformation are to be defined as one of the most important parameters as a basis for transformation planning and the evaluation of the change path by companies. They influence the definition of possible target architectures. In addition, they serve to focus on the target areas when analyzing the initial situation. Necessary process adjustments derived on the basis of clearly formulated goals create a sense of purpose for the change and are also a prerequisite for the next aspect.

To ensure the feasibility of the project, *management buy-in* is required. Management buy-in is when the company's management not only accepts the project as such, including the goals and also the necessary changes, but also actively supports them. Project managers who succeed in demonstrating and quantifying the added value of the planned changes have a much easier time gaining management support. If management is behind the project and its goals, it's also much easier to ensure the availability of employees and their decision-making authority because the project is given the necessary priority over competing projects in the company. The collaboration of all employees in the various specialist departments with the IT contacts is critical to success, as already described for the added value and time dimensions. Transparency and collaboration are only made possible by a shared view of the processes.

The final aspects of the feasibility dimension are *change management* and *enablement*. For the change to SAP S/4HANA to add value, you must be prepared to make adjustments to your existing processes. These can take the form of leveraging new SAP S/4HANA innovations, reducing in-house development, harmonizing or automating processes using SAP BTP, or a combination of these actions. In turn, company-wide defined processes are a prerequisite for the planned management of change processes from the initial state to a new target state, for change management, and for the enablement of employees, that is, the necessary training measures.

Business process transformation provides answers to the central questions and supports customers in realizing opportunities and overcoming challenges. Just as with the change to SAP S/4HANA, there are also opportunities and challenges in establishing business process transformation management. In the next section, we look at these as well as the success factors and dimensions of business process transformation management.

10.2 Business Process Transformation Management

Business process transformation management is a holistic approach to improving business processes with the aim of increasing the efficiency, effectiveness, and flexibility of organizations. Business process transformation may require a fundamental redesign of processes and underlying systems to ensure that they meet the changing needs and requirements of customers, employees, and stakeholders. In this context, business process transformation management involves the identification, analysis, and optimization of business processes, as well as the implementation of technologies to increase process effectiveness and efficiency. Successful implementation requires careful planning and implementation of various factors to ensure that the goals of the transformation are achieved. Various success factors and dimensions, such as the following, play an important role:

- Strategy and leadership
- Process responsibility and organization
- Process control
- Communication and changes
- Single source of information
- Skills and methods
- Tools and technologies

10.2.1 Success Factors and Dimensions of Business Process Transformation

In this section, we take an in-depth look at these various factors and explain their importance for a successful business process transformation. However, it's important to emphasize that not all factors are equally important for every company and that their implementation should depend on the individual situation and needs. These issues may be weighted differently depending on the context and organization. It's therefore important to consider the specific challenges and goals of the organization and develop a tailored business process transformation strategy.

Strategy and Leadership

Successful business process transformation requires clear alignment with the company's business strategy. The link between business process transformation management and business strategy is crucial to ensure that all activities are aligned to the same goal. A clear vision and strategy are therefore also essential. The clear vision gives the various managers a common understanding of what goals the company wants to achieve and how business process transformation management can contribute to this. Likewise, they help focus on what is important and ensure that all stakeholders are working in the same direction. The strategy then sets the roadmap for implementing the vision.

Successful implementation of a business process transformation strategy requires not only clear alignment with the business strategy and a clear vision and strategy but also effective change management. It's critical that all individuals and departments involved are informed about the goals, scope, and impact of the transformation and are actively involved in the change process. A solid change management program helps minimize resistance to change and supports the smooth introduction of new processes. An effective change management program in the context of business process transformation should fundamentally include topics such as goal definition and stakeholder identification, transparent communication, training and support, and resistance management. This program should not be seen as a one-off initiative, but as an ongoing process that accompanies and supports the transformation.

Another important success factor for business process transformation management is the creation of a culture of continuous improvement. By integrating continuous improvement into daily operations, the organization is able to make improvements to business processes on an ongoing basis and also improve performance. In addition to an environment that fosters continuous improvement, leadership commitment is also crucial. Leaders must demonstrate their commitment to transforming processes and take responsibility for the transformation. This creates an environment where employees are motivated and inspired to make improvements and optimize the company's performance.

In addition, management commitment is another key success factor for a successful business process management transformation. Management must support and promote process transformation to ensure that it receives the necessary attention and resources. By demonstrating commitment to business process transformation management, it creates a culture of continuous improvement and fosters buy-in from the people and departments involved. Without management commitment, it will be difficult to successfully implement business process transformation management.

In addition to the key factors mentioned so far, another factor for the success of the transformation is the involvement of employees in the process. They often know best which processes need to be improved and how this can best be achieved, so their involvement is essential. The business process management team and the employees involved must also have the necessary knowledge, skills, and experience to achieve the goals of the program. Targeted training and education for employees can help strengthen their skills and competencies in the area of business process transformation.

In addition to involving employees, continuous measurement and evaluation of the results as well as progress of the business process transformation management program is also very important. Regular evaluation can ensure that the program is on track and corrections can be made as needed. It's also important to ensure that adequate resources are in place to ensure the success of the program. Business process transformation management often requires investments in technology, training, and personnel. Appropriate resource allocation can help maximize the chances of success.

Another important success factor for business process transformation management is *risk management*. Business process transformation management can involve risks, including resistance to change, process disruption, or data loss. It's important to identify risks and develop plans to minimize or avoid them. Change management can also play an important role here by ensuring that all people and departments involved are aware of the risks and implement strategies to minimize them.

Process Responsibility and Organization

Process responsibility and organization refer to the structure and management of business processes within a company. In this context, process responsibility describes the assignment of specific roles and tasks relating to a particular business process to individual persons responsible. Process organization, on the other hand, is concerned with the way in which these processes are structured and coordinated to maximize efficiency and effectiveness. Both elements play a key role in creating an efficient and effective business process management system. For a detailed explanation of process responsibility and organization, see Chapter 11, Section 11.1.

Successful implementation of process ownership and organization in a company requires the interaction of several factors. One important component here is the creation of a *process management organization*, which provides the necessary structures and resources to manage and improve business processes in the company. Clear responsibilities and competencies, good cooperation between different departments, and a culture of continuous improvement are key success factors.

In addition, a process-oriented mindset and the implementation of process-oriented structures are required to ensure that processes in the company are also regarded as an important factor for business success. Targeted promotion of process-oriented thinking through training and education as well as the creation of structures that support the process-oriented mindset are important success factors here.

To carry out process management effectively and efficiently, clear roles and responsibilities, such as process owner, process expert, or process analyst, are essential. Defining these are crucial success factors for executing and monitoring processes, as well as ensuring that all stakeholders are informed and trained about their roles and responsibilities.

In addition to defining clear roles and responsibilities, process owners must also be able to take responsibility for their processes and be empowered to improve them. Creating incentives for process owners to take responsibility, providing resources to support process owners in performing their tasks, and creating freedom for process owners to make decisions and implement changes are important factors for success in this regard.

Another important factor for successful implementation in the company is the creation of a process transformation office (PTO). A central unit that focuses on the continuous improvement of business processes in the company can help to anchor process management as an important factor for business success in the company. Creating an organizational unit focused exclusively on process management, having senior management support the PTO, providing resources to achieve the PTO goals, and fostering a culture of continuous improvement in the enterprise are key contributors to success (for more details, see Chapter 11, Section 11.1).

Process Governance

Governance is a key dimension that should be considered in the context of business process transformation and includes the definition of guidelines and standards for process management and the definition of a framework for metrics and key performance indicators (KPIs) to measure processes and process improvements. These metrics should be able to quantify business objectives and measure process improvements. Regular monitoring of these metrics is essential to ensure that process performance goals are met and that action is taken to improve processes when goals aren't met.

The establishment of committees and decision-making processes is also an important part of process transformation. These committees must ensure that processes are developed and implemented with due regard for corporate strategy, compliance requirements, and customer needs. They should also be able to make decisions regarding prioritization of improvement projects, resource allocation, and risk management.

Finally, the creation of process standards is also an important factor in successful business process transformation. These standards must be clearly defined and visible to ensure that all employees have the same understanding of the processes and that process deviations are minimized. The creation of process standards also facilitates the implementation of best practices and the reuse of solutions in other business units.

Communication and Change

Change management is an important aspect of business process transformation management. Companies must be able to manage change successfully and analyze the impact on business processes and adapt business processes accordingly. Transparent communication, training, and a process management community can help to successfully implement changes and improve business processes sustainably.

Transparent communication of changes to business processes is essential to ensure that all stakeholders are informed about changes and can implement and follow them effectively. A clear and targeted communication can help to improve the acceptance of changes and optimize the performance of business processes.

Employee training and development is an important part of the successful implementation of business process changes. Employees must have the necessary skills and knowledge to work effectively with new processes. Ongoing training and development can empower employees to make continuous improvements to business processes.

A process management community can help to create a culture of collaboration and continuous improvement. By sharing knowledge and experiences, best practices can be identified, and problems can be solved more quickly. A strong community can also help to successfully implement change and improve business process performance.

Single Source of Information/Single Source of Truth

The *single source of truth* is a concept that is widely used in various industries and disciplines and helps to create a uniform and consistent source of information. In the context of process management, the single source of truth refers to the creation of a holistic source of information that contains all processes, including additional information and artifacts.

A consistent process architecture/structure is an essential component of the single source of truth. As a result, it's easier for companies to manage, update, and document processes. By creating a unified structure, companies can also ensure that process documentation is easily accessible and easy to find, leading to greater efficiency. In addition, this also helps to create a common language and understanding of processes.

By standardizing processes, companies can create synergies between different departments and teams by identifying and promoting common goals and processes. A single source of truth also reduces risk by ensuring that processes are performed in the same way across all departments or teams. This reduces errors and inconsistencies and minimizes the likelihood of errors in process execution. In industries where compliance and best practices are a high priority, such as finance or healthcare, a consistent single source of truth is especially important.

Skills and Methods

It's important to define the skills and knowledge required for the various roles in a process transformation and to ensure that everybody is trained accordingly. This includes knowledge of business processes management, project management, change management, and data analysis.

Training employees in process management methods and tools is crucial to the success of a transformation. Here, in particular, it's important to keep the entry threshold as low as possible to ensure acceptance by all employees. Depending on the role, the skills required differ. However, for a successful process transformation, it's crucial that the required skills per role are defined and that employees have the opportunity and freedom to learn these skills. For this purpose, regular training and Q&A sessions should be offered so that employees are always up-to-date, issues are clarified early on, and topics can be standardized.

Beyond skills training, it's important to establish standards and best practices for process transformation. This includes establishing policies and procedures that standardize the documentation, analysis, improvement, and monitoring of business processes and ensure that processes meet business objectives.

Tools and Technology

To effectively support a business process transformation, the selection and implementation of the right tool is crucial. The planning of the tool support plays a decisive role in business process transformation. The tool should meet the requirements of the business and facilitate holistic process optimization from design, implementation, and performance measurement to analysis and improvement. The integration of the process management tools into the existing IT infrastructure is also an important aspect. The tools should be able to collect data from different systems to provide a comprehensive and data-driven view of the process. Similarly, scalability and flexibility are other important aspects, as the demands on the process management tools may increase over time. Therefore, the tools should be able to grow with the changing needs of the business.

10.2.2 Opportunities and Challenges of Business Process Transformation Management

Business process transformation management offers several opportunities that companies can use to improve their business processes and increase their competitiveness. At the same time, however, business process transformation management also harbors risks that must be considered during implementation to avoid unforeseen consequences. In the following, we'll look at the opportunities and risks of business process transformation management and how companies can make the best use of opportunities and minimize risks.

Effective implementation of business process transformation management offers companies several opportunities that can have a positive impact on the efficiency, effectiveness, and flexibility of their business processes. Some of the most important opportunities offered by business process transformation management are as follows:

- **Business process improvement**
  Business process transformation management offers companies the opportunity to optimize and improve their business processes by eliminating redundant steps, removing bottlenecks, and streamlining inefficient processes. As a result, companies can streamline their business processes, save time and resources, and increase productivity.
- **Increasing customer satisfaction**
  Business process transformation management can help companies align their business processes to better meet the needs of their customers. By identifying weaknesses in processes and implementing solutions that address these weaknesses, companies can increase customer loyalty and satisfaction.
- **Increasing employee satisfaction**
  Business process transformation management can also help employees feel more productive and effective by removing obstacles and frustration factors in business

processes. Better collaboration, easier access to information, and faster response times can also increase employee satisfaction and motivation.

- **Improving the corporate image**
  By implementing business process transformation management, companies can improve their image and be perceived as innovative, efficient, and customer-oriented organizations. This can help to strengthen the trust of customers, business partners, and investors, as well as improve the company's reputation.

Although business process transformation management offers a wide range of opportunities, it also entails risks that must be considered during implementation. Some of the most important risks of business process transformation management are as follows:

- **High costs**
  Implementing business process transformation management can be costly, as it often requires changes to the IT infrastructure, process optimization, and employee training. Financing these changes can be a challenge and affect the profitability of the project.
- **Resistance from employees**
  Implementing business process transformation management often requires changing processes and procedures, which can be met with resistance from employees. If employees aren't sufficiently involved in the change or don't receive enough training to understand and implement the changes, this can lead to a lack of motivation and lower effectiveness.
- **Technological challenges**
  Business process transformation management often requires a significant shift of the IT infrastructure to support the new processes and systems. If implementation isn't carefully planned or technical requirements aren't adequately addressed, technical problems and errors can occur that hinder implementation or compromise the security of the systems.
- **Poor implementation**
  Poor implementation of business process transformation management can result in the intended benefits not being realized and instead unforeseen negative impacts occurring. A failed implementation can not only lead to high costs and frustration but can also affect the trust of customers, business partners, and investors.

It's important that organizations consider these risks when implementing business process transformation management and focus on careful planning and implementation to ensure that they can reap the benefits of business process transformation management without taking unnecessary risks.

10.3 Transition to SAP S/4HANA as an Entry Point to Business Process Transformation Management

Having provided a comprehensive overview of business process transformation management, we'll now look at how it interacts with SAP S/4HANA and why the move to SAP S/4HANA is an optimal time for companies to establish business process transformation management.

As described in detail in the previous sections, not only new customers but also existing customers must deal intensively with the business processes implemented in SAP ERP during the changeover for the following reasons:

- The design principles of SAP S/4HANA require a redesign of business processes against the backdrop of simplification elements and compatibility scope.
- The transition to SAP S/4HANA only offers the corresponding added value for companies if business processes are analyzed and optimized with SAP S/4HANA functions in mind.
- The target architectures described, which affect the operating model and the instance strategy, for example, make it necessary to adapt business processes.

However, business process transformation management supports customers in redesigning their business processes but also in answering the questions outlined in Section 10.1.3:

- How can you ensure that the transition to SAP S/4HANA adds value to your business?
- How can the transition to SAP S/4HANA be made as quickly as possible?
- How does your company ensure that the transition to SAP S/4HANA is feasible?

Business process transformation management enables opportunities to be realized and challenges to be minimized.

It's also clear when considering the timing of a transformation project that business process transformation management supports activities in all phases. The following are examples of the individual phases:

- **Preparation**
  Determination of the change path and the potential of the transformation.
- **Implementation**
  Process design and handover of process design to ALM.
- **Hypercare**
  Review of the extent to which processes or process adjustments lead to efficient workflows and early proactive identification of weaknesses.
- **Roll-out**
  Creation of necessary local process variants.

SAP doesn't make the use of business process transformation management mandatory. However, anyone who doesn't use the transition to SAP S/4HANA as an entry point into the topic of business process transformation management is missing an opportunity. Successful SAP S/4HANA transformation projects are business initiatives, not IT projects. For them to succeed, business process transformation management should be an integral part of any SAP S/4HANA transformation; to what extent depends on the project plan. The timing is ideal in any case, as the benefits for the project and each person involved in the process are immediately noticeable, and thus motivation is also understood as an opportunity and not as an additional requirement.

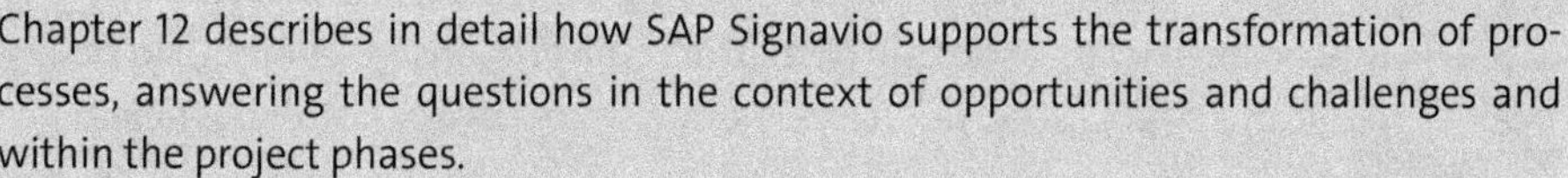

Note

Chapter 12 describes in detail how SAP Signavio supports the transformation of processes, answering the questions in the context of opportunities and challenges and within the project phases.

After the change to SAP S/4HANA, new business process requirements also require the adaptation of business processes under consideration of new product functions. Business process transformation is also required to ensure continuous process improvement after the transformation. Chapter 13 describes conceivable expansion stages of business process transformation management after the transition to SAP S/4HANA.

10.4 Summary

The introduction of SAP S/4HANA represents a unique opportunity to establish business process transformation management. The transition to SAP S/4HANA not only introduces a new technology but also changes the underlying business processes. A successful transformation requires a holistic view of the success factors and dimensions of business process transformation management, such as the identification of business processes and their alignment with business goals, the collaboration and involvement of all stakeholders, and the definition of the skills and knowledge required for process management. In addition, opportunities and risks must be carefully weighed up as part of the transformation process.

The implementation of SAP S/4HANA enables the optimization of business processes and thus creates the conditions for a successful transformation. Conversely, a successful transformation also requires the design of to-be business processes that are optimally aligned with the requirements of the new technology.

Overall, the transition to SAP S/4HANA offers an excellent opportunity for companies to optimize and transform their business processes. However, careful planning and implementation are essential to take advantage of the opportunities and successfully overcome the challenges. With the right approach, companies can optimize their business processes and become more competitive to survive successfully on the market in the long term.

In this chapter, we've looked at the basic principles of SAP S/4HANA and examined the opportunities that arise from integrating business process transformation management into the SAP S/4HANA implementation. We also looked at the success factors and dimensions of business process transformation management and the associated opportunities and risks. In doing so, we've outlined a holistic approach that considers the right skills, methods, and tools, which is crucial for a successful transformation.

We've explored how SAP S/4HANA can serve as an entry point to business process transformation management. In doing so, we've shown how the implementation of SAP S/4HANA can be used as a starting point for a comprehensive overhaul of business processes to optimize processes, reduce costs, and improve operational efficiency.

The introduction of SAP S/4HANA offers an excellent opportunity to increase the efficiency and productivity of the company. By taking a holistic view of business process transformation management and the associated success factors and dimensions, the transition to SAP S/4HANA can serve as a catalyst for successful business transformation.

Chapter 11
Fundamentals of Business Process Transformation Management

In this chapter, we explore the fundamental concepts of business process transformation management. We give you an overview of the most important aspects and show why they are important for a successful transformation.

Business process transformation management is an important component of corporate management, as it aims to make business processes more efficient, cost-effective, and end-customer focused. By monitoring and optimizing processes, companies can improve their business outcomes, increase customer satisfaction, and successfully tackle market challenges. In this chapter, we look at the basics of business process transformation management and why it's so important. In Section 11.1, we first present the importance of process organization and process roles. This is followed by a discussion of the topics of process architecture (Section 11.2) and variant management (Section 11.3). In Section 11.4 and Section 11.5 we introduce you to the interrelationships of process management with application lifecycle management (ALM) and enterprise architecture management (EAM).

Understanding these topics is critical for a successful business process transformation. It provides the basis for sustainably successful process design and is important for designing processes that meet business requirements and needs. In this chapter, we therefore take a closer look at the individual topics and highlight their importance for business process management.

11.1 Process Organization and Process Roles

Establishing a process organization involves distinguishing and defining three different categories of roles: executing, controlling, and consulting process roles. Each of these categories plays a critical role in ensuring the smooth and efficient functioning of the organization (see Figure 11.1).

Let's start with the executing process roles. These roles are directly related to the execution of specific tasks and responsibilities and are assigned or defined during process design. They include specific competencies and task areas that are required for the successful operation of the organization. A purchaser or buyer is an example of such a role. In this position, the person negotiates prices with suppliers and executes purchase orders. Similarly, the purchasing manager is an executing process role in that they must approve all purchase orders with a commodity value of EUR 1,000 or more. Another important executing process role in this context is that of the demand planner, who is responsible for planning in advance of purchase orders. The demand planner contributes to the optimization of the purchasing process and ensures that the organization is adequately supplied with the required resources.

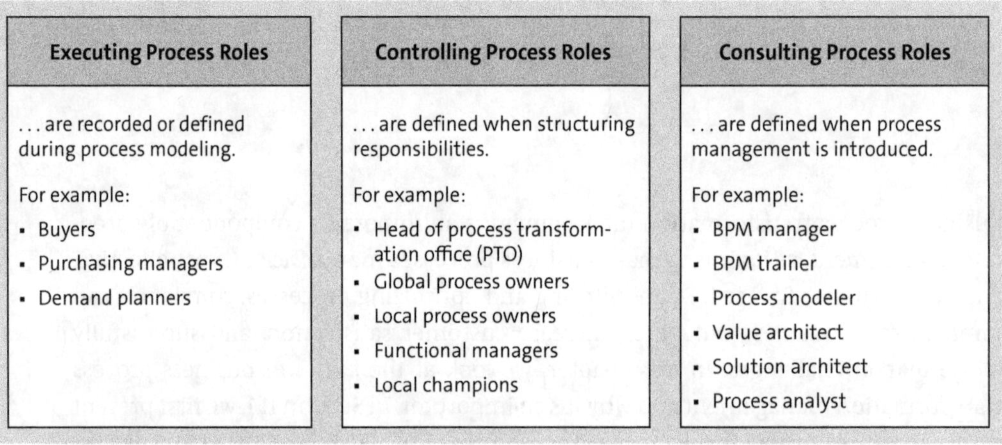

Figure 11.1 Role Types in the Process Organization

The controlling process roles are responsible for the management and coordination of processes within the organization. They are the ones who structure responsibilities and ensure that the processes run effectively and efficiently. Usually, these roles are divided into global process owner (GPO), local process owner (LPO), functional manager, and local champion. The GPO has overall responsibility for a particular process at a global level, while the LPO is responsible for managing the process at the local level. The functional manager, on the other hand, has oversight of a specific function or department within the organization. Local champions are specialized subject matter experts in their respective process area and have both extensive business process expertise and in-depth knowledge of process management methodologies. They use this knowledge to identify and implement process improvements and work closely with the LPO. It's important that the interaction of process responsibility and line responsibility is clearly defined in these roles to avoid conflicts and ensure efficient process execution.

It's critical for the effective and efficient management of business processes at the global and local level that the GPO and LPO work together. The GPO is typically responsible for designing, implementing, and optimizing a specific business process at the global level. This role develops and maintains process standards and best practices, ensuring that they are implemented and adhered to at the global level. The LPO, on the other hand, is responsible for implementing and optimizing these processes at the local level. They adapt the global process standards set by the GPOs to local conditions and needs and monitors compliance with these processes within their local organizational unit. Effective collaboration between GPOs and LPOs is based on clear communication, alignment, and mutual understanding of each other's roles and responsibilities. GPOs must ensure that LPOs understand the global process standards and best practices and how they can be adapted to local circumstances. At the same time, LPOs need to provide regular feedback to GPOs on the implementation and performance of processes at the local level and highlight any issues or opportunities for improvement, which might also apply to the global process. In addition, GPOs and LPOs can work together to promote continuous improvement and process innovation. By jointly analyzing process performance measurements and discussing best practices, they can identify and implement opportunities for process improvement. Overall, close collaboration between GPOs and LPOs leads to more efficient and effective business processes and helps achieve the organization's goals and strategies.

Collaboration between the functional manager, local champion, and LPO is also critical for effective process management in an organization. Each role has specific tasks and responsibilities, which, when effectively coordinated, help to optimize business processes. The functional manager is primarily responsible for the strategic direction and performance of their respective function or department. They set goals, develops strategies, and ensures that resources are used effectively. Local champions, who have both business process expertise and knowledge of process management, act as a link between the functional manager and the LPO. They support the functional manager by suggesting and implementing process improvements within his specialist function. At the same time, they help the LPO by providing information about specific aspects and requirements of their functional area that need to be considered for process optimization. The LPO is responsible for coordinating and optimizing processes at the local level. They work closely with the respective local champions to understand how processes are performed and how they can be improved across functions. The LPO may also advise the functional manager by offering insights into process flow and suggestions for improving process efficiency.

Overall, the functional manager, the local champion, and the LPO must work closely together to optimize the processes in their respective function or department but also across functions. Both the functional manager and the local champion tend to focus on optimization within their function, while the LPO wants to optimize processes across functions. They need to communicate openly and transparently, hold regular meetings to discuss progress and any issues, and work together to find solutions and implement improvements. Effective collaboration between these roles can help improve process performance and efficiency, increase customer satisfaction, and drive business success.

In Table 11.1 we've summarized the controlling process roles and their respective descriptions.

Roll	Role Description
Global process owner (GPO)	A GPO is responsible for monitoring, auditing, and optimizing a business process that is carried out across functions or countries. In doing so, they usually define a template process, which is adapted to the respective situation by the LPO.
Local process owner (LPO)	An LPO is responsible for monitoring, auditing, and optimizing a business process in a specific country across functions.
Functional manager	A functional manager is responsible for managing a specific function or department within a company. For example, the function or department may be IT, finance, human resources, or marketing. A functional manager works closely with an LPO to ensure that processes within and outside of their department are executed effectively and efficiently.
Local champion	A local champion is a subject matter expert in their respective functional area, such as purchasing or finance, who also has in-depth knowledge of process management methods. They work closely with the LPO to optimize processes that span their area. The local champion serves as a link between process management and the functional area and uses their expertise to improve processes and ensure efficient process execution.

Table 11.1 Overview of the Controlling Process Roles

In addition to the executing and controlling process roles, the consulting process roles are important to strengthen and improve the organization's process management approach. They are usually defined when process management is introduced in the organization and can be divided into different specialized roles, such as head of process transformation office (PTO), BPM manager, BPM trainer, process modeler, value architect, solution architect, and process analyst. They are usually organized in a *process transformation office* (PTO) and contribute to improving process efficiency and effectiveness by providing training, designing process models, proposing solutions, and performing process analysis.

The BPM manager is responsible for leading the organization's overall BPM initiative, while the BPM trainer trains employees in process management methods and tools. The process modeler is responsible for modeling the organization's processes, while the value and solution architects propose solutions to maximize the value of the processes. The process analysts, on the other hand, investigate the processes to identify weaknesses and suggest improvements.

In Table 11.2, you can find an overview of the consulting process roles.

Roll	Role Description
Head of process transformation office (PTO)	The heads of PTO are leadership experts in the field of process management and are responsible for the strategic direction and introduction of process management in the organization. They lead a team of consultants, including BPM managers, trainers, process modelers, value and solution architects, and process analysts. They are responsible for driving process efficiency, overseeing improvement initiatives, and ensuring effective collaboration between the other consulting process roles. Their primary goal is to maximize process performance and add value to the organization.
BPM manager	The BPM managers are responsible for defining, implementing, and monitoring standards and methods for process management throughout the company. This includes the development of methods and tools to improve the efficiency and effectiveness of business processes.
BPM trainer	The BPM trainers are responsible for training and sensitizing the employees in the areas of process business management.
Process modeler	The process modelers are in charge of modeling business processes. They work closely with process owners and other stakeholders to identify, define, and model the business processes. This includes the application of methods such as process analysis and optimization to identify weaknesses in the processes and take measures to improve them.
Value architect	The process value architects are responsible for designing and implementing processes within a company to maximize added value for the company and its customers. They work closely with the process owners, who are responsible for the content of the processes, while value architects bring with them the methods and tools to identify potential and sustainably optimize processes.
Solution architect	The solution architects are responsible for designing and implementing technical solutions for business processes. Solution architects work closely with process owners and value architects to ensure that technical solutions for business processes meet business goals and requirements.
Process analyst	The process analysts investigate, optimize, and document business processes. They identify weaknesses and opportunities to improve efficiency, and implement changes to improve business performance.

Table 11.2 Overview of Consulting Process Roles

Establishing a community within an organization where GPOs can share ideas is an essential aspect of a PTO. In this community, GPOs can share their experiences, challenges, and best practices, which contributes to improved process performance. Within this community, global experts are usually selected among other GPOs to play a particularly active role. These experts bring a wealth of experience and knowledge that they make available to improve processes and foster collaboration between the different process roles. They act as thought leaders, innovators, and facilitators between different stakeholders, and they share their knowledge and experience with the LPOs. Their role is to ensure that LPOs understand and follow the policies and procedures defined and established at a global level. They serve as a liaison between the GPOs and LPOs, facilitating communication and ensuring a steady flow of information. They help to ensure that LPOs can effectively manage and improve their processes. This exchange and collaboration is particularly important to strengthen process management in the organization overall. By sharing knowledge and experiences, GPOs and LPOs can work together to find solutions and identify best practices that can be applied across the organization. Including global experts in this community helps to improve process efficiency by providing new ideas and innovative approaches.

However, organizing such a community also requires effective leadership and coordination, which is the head of PTO. This role is responsible for leading the community, fostering exchange, and ensuring that the global experts can fulfill their roles effectively. The head of PTO oversees activities in the community, provides resources, and ensures that communication runs smoothly and effectively.

The role of a community consisting of GPOs and global experts in the context of process organization can't be overemphasized. The establishment of such structures enormously strengthens the possibility to improve knowledge sharing and experience transfer. It also fosters collaboration between different process roles and actively supports the optimization of process performance. The full potential of this community unfolds under effective leadership—a role ideally assumed by the head of PTO. A possible interaction is shown in Figure 11.2.

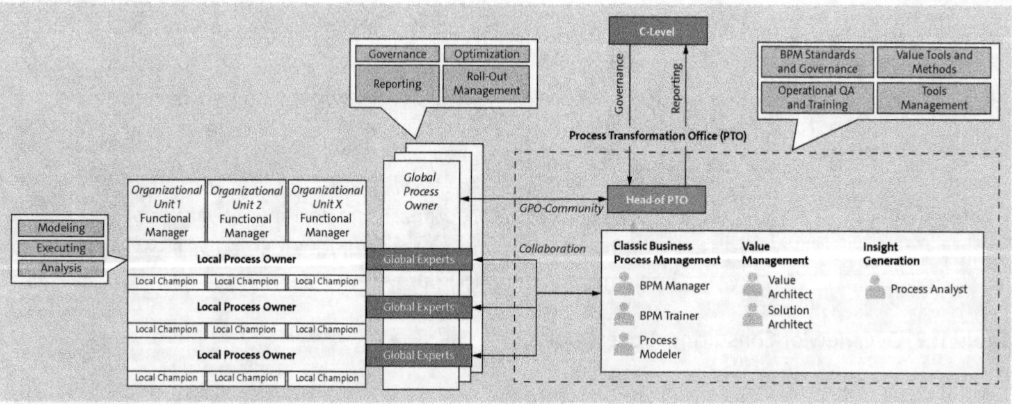

Figure 11.2 Sample Collaboration between the Individual Roles

In summary, the roles in a process organization serve to design and execute the processes efficiently and effectively. The executing process roles are responsible for performing specific tasks within the processes themselves, the controlling process roles are responsible for managing and improving processes, and the consulting roles are responsible for improving and strengthening the process management approach within the organization. By clearly defining and assigning these roles, the organization can ensure that its processes run smoothly and thus realize their full potential.

In addition to the interaction of the individual roles, there are three main design alternatives for a PTO (see Figure 11.3):

- Centralized PTO
- Hybrid PTO
- Decentralized PTO

11

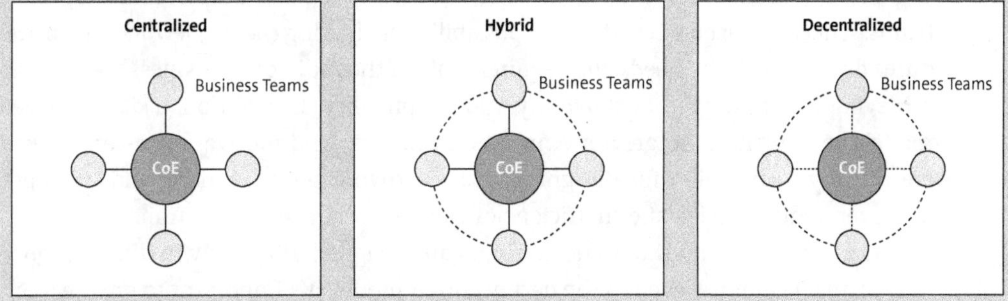

Figure 11.3 Options When Designing the Operating Model

Choosing the right operating model for a company is crucial. In the following, we describe the individual design alternatives and discuss the advantages and disadvantages of the three possibilities.

A *centralized PTO* is a facility or department within an organization that specializes in a particular topic or function and serves as a single point of contact for that topic or function. A centralized PTO can take on a variety of roles in a process management organization. Typically, it serves as a central point of contact for all business teams for process management issues and concerns, and it provides support and guidance to the various business units. The centralized PTO can also act as a think tank, developing new ideas, methods, and concepts to improve processes within the company. Another important aspect of the PTO is to disseminate best practices and know-how in the area of process management and to perform a governance function. For this purpose, it can provide training and education to improve employees' knowledge and skills in process management. In addition, a PTO can also act as a hub, fostering collaboration and sharing of experiences and ideas across the organization. By centralizing competencies, resources, and central contacts in process management, the centralized PTO can, on one hand, help to exploit synergies and increase efficiency within the company. On the other hand, however, a centralized model can also mean that, due to the large number of inquiries from the business teams, individual inquiries can't be processed quickly, which means that adjustments for process optimization can only be made later. In addition, such a model can lead to the methods and concepts in the company being less aligned with the needs and requirements of the individual business teams and thus being less flexible and adaptable. It can also lead to longer decision-making paths and less individual responsibility on the part of the individual business teams.

In contrast to a centralized model, in a *decentralized PTO*, the responsibility and competencies for process management are distributed among the individual areas or departments within the company, that is, the respective business teams. For instance, this could imply that every team develops its own ideas, methods, and concepts on how to conduct process analyses, for example, to identify weaknesses and potential for process improvement. The distribution of the company's own best practices and know-how in the area of process management can also take place at the level of the individual teams. A decentralized PTO offers the possibility of aligning the framework and procedures directly with the needs and requirements of the individual business teams, thus achieving a high degree of flexibility and adaptability. In addition, a decentralized model can contribute to greater employee ownership and motivation, as employees are directly responsible for designing and improving their methods and concepts. However, redundancies and inefficiencies can also occur in a decentralized model if, for example, similar methods and concepts are designed differently in different business teams. Therefore, even in the decentralized model, it's important to ensure a certain level of coordination and communication between the individual departments. This is usually done in a kind of virtual PTO, which is composed of one or more representatives of the respective business teams. Nevertheless, even in a decentralized PTO, there is usually a central team that manages the infrastructure, such as a process modeling or analysis tool.

In contrast to the centralized and decentralized model, the *hybrid PTO* combines elements of both approaches, whereby the PTO usually specifies the methods and concepts, but the adaptation is distributed among several points within the company called *single point of contacts* (SPOCs). Within a business team, a SPOC is a central point of contact for the topic of process management, which in turn is part of the PTO and can thus act as an evangelist within the respective business team. In this model, the PTO could, for example, offer training and continuing education to improve the competencies of employees in process management but also act as a think tank and develop new ideas and concepts. In this way, synergies could be leveraged and resources used effectively. Overall, a hybrid model offers the opportunity to combine the advantages of centralized and decentralized models to create a flexible and adaptable structure for process management in the company that leads to improved reach and increased scalability, as well as meeting the requirements of individual business teams.

11.2 Process Architecture

The *process architecture* is a systematic structuring and representation of all processes that are of importance to a company. It includes all business processes that are required to achieve the company's goals. The processes are arranged in a hierarchical order and their relationship between the processes is captured. The process architecture ensures that all relevant activities and processes are recorded, regardless of the departments or functions responsible for their execution. It provides a holistic view of the company's processes and enables effective management, optimization, and control of the entire process landscape.

To be successful in the long term, a holistic understanding of processes is therefore of great importance for companies. Ever-increasing customer requirements and steadily growing global competition are leading to permanent pressure on operational procedures and consequently on the underlying processes. In this context, business processes must be optimized from a holistic point of view instead of driving optimizations within individual silos. For this reason, we differentiate different views of business processes: the *functional view* and the *process-oriented view* (see Figure 11.4).

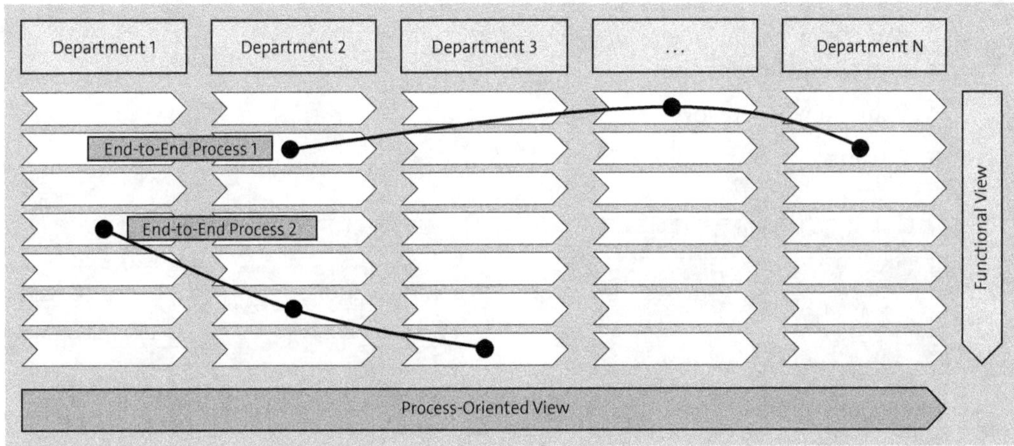

Figure 11.4 Functional versus Process-Oriented View

The functional view typically describes the subprocesses within a department, such as finance, sales, or purchasing, and focuses on the fulfillment of specific functional tasks. Accordingly, both the control and the optimization of the subprocesses within the individual departments are driven forward.

The process-oriented view focuses on an overarching end-to-end value chain that combines subprocesses from different specialist areas into a holistic picture called end-to-end processes. Examples are the order-to-cash or purchase-to-pay processes. In contrast to the control and optimization of functional subprocesses within individual departments, end-to-end processes are controlled and optimized within the process-oriented view along the value chain.

Introducing end-to-end processes means striking an efficient balance between functional expertise and process integration, but not replacing functional structures and processes. However, introducing and establishing end-to-end processes is simpler in words than in action. Without a uniform taxonomy, that is, a process architecture as well as an established governance structure, the individual subprocesses are merely a jumble of processes. This leads to the following problems, among others: employees, for example, in purchasing, can't find the processes relevant to them, and processes may be defined multiple times by different areas or depending on the end-to-end process.

Most companies use a four- to six-level process hierarchy, usually managed centrally by the BPM governance team or the PTO. For simplicity, we describe a four-level process architecture that you can adapt to your individual requirements.

As shown in Figure 11.5, we've divided the process architecture into a functional view and a process-oriented view. Each of these views comprises four levels, which together provide a comprehensive understanding of the process architecture:

- Map
- End-to-end or function
- Value step or category
- Process

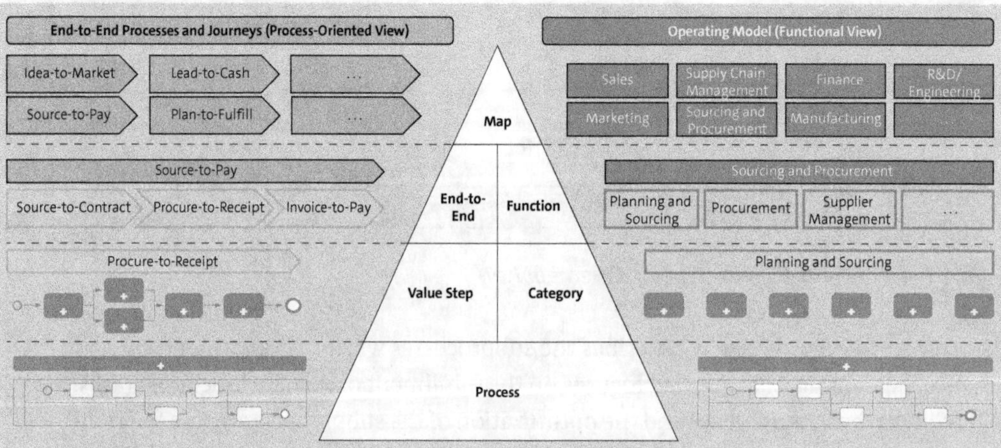

Figure 11.5 Process Architecture

We start with the right side of the pyramid—the functional view. This view reflects the operating model of a company and is therefore usually divided into the individual departments and functions of a company, such as finance, sales, or purchasing.

As there are a large number of subprocesses that can be assigned to the respective departments and functions, we recommend introducing at least one further structuring level—category. This structuring level divides the processes within a function into different categories; for example, the purchasing function may be divided into the planning and procurement, purchasing, and supplier management categories.

Here, you can find the individual subprocesses, such as supplier selection, supplier onboarding, and supplier evaluation, which all fall into the category of supplier management.

It's important that the assignment of the subprocesses to the categories or, at a higher level, to the functions, is clearly defined and that no subprocess can be assigned to more than one function or category. Accordingly, when defining the functions and categories, make sure that they are clearly distinguishable.

In summary, the functional view of the process architecture is structured as follows:

- **Level 1**
  Summary of all departments or functions.
- **Level 2**
  Detailing of a function into different categories.
- **Level 3**
  Listing/assignment of all subprocesses to a category.
- **Level 4**
  Modeled subprocess.

This functional structuring enables easy navigation to the respective relevant subprocesses, especially for the executing process roles.

On the left side of the pyramid—the process-oriented view—you can find the respective end-to-end processes of a company. For reasons of consistency, we recommend a four-level structure here as well. Accordingly, we start here at the top level with the listing of all end-to-end processes of a company. Similar to the category on the right side of the pyramid, we also recommend introducing at least one further structuring level on the left side of the pyramid, which divides the end-to-end process into logical sections and thus significantly reduces the scope and complexity of the end-to-end process. The next level then defines a *value step*, in which the respective subprocesses from different specialist areas/functions are linked together and placed in a logical sequence.

In summary, the process-oriented view of process architecture can be structured as follows:

- **Level 1**
  Summary of all end-to-end processes.
- **Level 2**
  Detailing of an end-to-end process into value steps.

- **Level 3**
  Logical connection/flow of all subprocesses within a value step across departmental boundaries.
- **Level 4**
  Modeled subprocess.

This process-oriented structuring creates a high level of transparency, especially for process owners, but also in the context of holistic process optimization projects. This makes it possible to reduce interactions (an optimization of a subprocess in department A can lead to a deterioration of a subprocess in department B because the latter depends on the output of the subprocess from department B) and to increase the efficiency of the processes.

Many other process architectures that are widespread in practice are based on some kind of process segmentation. Because such structuring can quickly become very complex, we don't recommend structuring processes on the basis of process segmentation, but we do recommend performing such segmentation at the subprocess, category, or value step levels in any case. Process segmentation divides processes into different categories based on their degree of standardization or differentiation. In this way, it enables the targeted design and alignment of processes. Segmentation helps identify efficient standard processes and optimized tasks that can be standardized, while differentiating processes can be designed to be more flexible and adaptable to individual requirements. This supports the targeted allocation of resources and enables effective process optimization and control in line with corporate objectives.

A possible segmentation of the processes can be done, as shown in Figure 11.6, into the following four categories:

- **Key differentiators**
  Key processes that are developed specifically for a company to differentiate itself from the competition and deliver a unique value proposition (UVP).
- **Differentiating processes**
  Processes that leverage an improved industry standard with company-, business unit-, or region-specific characteristics to outperform the competition in terms of speed, quality, or the like.
- **Harmonized core**
  Processes that deviate only slightly from the industry standard without requiring extensive adaptation, as these processes don't represent direct differentiators for end users, for example, supply chain management (SCM) processes.
- **Standard processes**
  Standard processes in back-office processes (group-wide) that are designed for exact repetition, correctness, automation, and maximum efficiency, for example, in finance, compliance, and reporting.

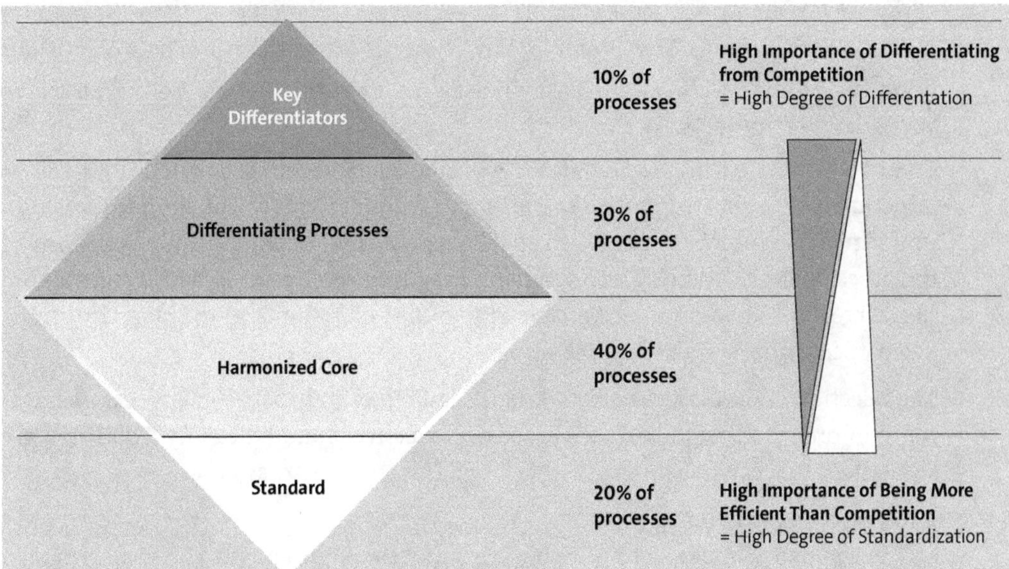

Figure 11.6 Process Segmentation

11.3 Variant Management

In process management, we often talk about *standard processes*, also called template processes, and the associated *process variants*. Standard processes form the basis for the execution of business processes and offer a standardized approach to make business processes more efficient and consistent. Process variants, on the other hand, allow these standard processes to be adapted to specific requirements. This may be necessary, for example, if a company operates in different regions or has to meet specific customer requirements.

The importance of process variants in the context of process management lies in their ability to offer companies flexibility and adaptability. By using process variants, companies can make their processes more efficient and better serve their customers by addressing their specific needs. In addition, the use of process variants can help companies better monitor, control, and optimize their processes.

An example of a process variant that differs due to geographical location, for example, could be found in a logistics company. This company could have a standard delivery process that is the same in all regions worldwide. But due to differences in geography and legal requirements in different countries, it may be necessary to customize the delivery process in certain regions. For example, one region might have special requirements for packaging and transportation, while another region might have special regulations for handling hazardous materials. To meet these requirements, the logistics company can develop process variants for each region that meet the specific requirements in those regions. This example shows how process variants can help companies make their processes more efficient and legally compliant by addressing geographic specifics and requirements.

To drive process harmonization, it's important that the company establishes a clear structure for its standard processes and process variants. This can be supported using a uniform process architecture, which enables a consistent definition and monitoring of the processes and variants, and a standardized governance process, which controls the initial creation of variants and enables the return of variants to the standard in the event of changing standard processes.

The selection of the correct process variants is an important aspect of process management in order to design business processes effectively and efficiently. Following are some steps that can help in selecting the right process variants:

1. **Identify requirements.**
   The company should identify its business process requirements and goals. This can be done by reviewing the business requirements and processes.
2. **Review standard processes.**
   The company should then review its standard processes to determine whether they meet the requirements or need to be adapted.
3. **Identify deviations.**
   After reviewing the standard processes, the company should identify deviations that require an adjustment of the processes. This may be due to regional legislation, customer requirements, or specific business needs, for example.
4. **Evaluate process variants.**
   Based on the identified deviations, the company should evaluate different process variants to find the process variant that best meets the requirements. Aspects such as efficiency, costs, and compliance should be considered.
5. **Implement and monitor.**
   After identifying the appropriate process variant, the company should implement and continuously monitor it to ensure that it continues to be effective and meet requirements.
6. **Perform a roll-out of standard process changes.**
   Over time, standard processes may also change. Here, it's important to apply these changes to the process variants if relevant.

To sum up, process variants in the context of process management are an important component for adapting processes to the specific requirements of customers and companies and for making process management more efficient and flexible. When selecting the right process variants, a systematic approach is important in which requirements are identified, standard processes are reviewed, deviations are identified, and process variants are evaluated and finally implemented and monitored.

11.4 Interplay with Application Lifecycle Management

To improve the efficiency and quality of business processes in the long term and contribute to the success of the company, the management of processes is essential. Ensuring the technical implementation of previously defined processes represents an essential component to success.

For this purpose, it's crucial that there is a defined and established process for the implementation of processes in the company, which creates a framework for interaction between the different roles and persons responsible in the company and ideally controls it in a system-supported manner. If this process is realized in your company, you can achieve a more effective and efficient implementation on one hand and a better tailored software application to support your business processes on the other. Likewise, you can ensure that the development and operation of software applications are well coordinated toward the same goal of supporting the defined business processes. This can help improve the quality of software applications, shorten development time, and reduce costs.

The decisive factor for success is the timing after the design of new processes (versions) and the coordination between the various functional managers. Many process innovations are already discarded before implementation if there is no defined process for handover to IT after the process has been defined.

Now, we'll briefly go into the general purpose and the subject areas in *application lifecycle management* (ALM). ALM is an approach that supports the entire lifecycle of software applications from conception to continuous operation. ALM covers the planning, monitoring, and control of activities and resources required for the development, testing, and operation of software applications. For the link with process management, we'll focus on the development phases of ALM.

To explain the potential of the interplay and the interactions between both worlds, a concrete scenario of an example company is explained next. As part of the continuous process improvement for our example company, the process for ordering direct materials in the company was analyzed. During the analysis of the process, it was noticed that the time from the request for the material to the receipt of the material takes considerably longer in some countries than in others. The currently defined process describes that the order is created by the buyer, then released by the purchasing management, and then sent to the supplier. Once the goods have arrived at the warehouse, the receipt of the goods is confirmed by the warehouse clerk. The supplier invoice is then posted by the accounting department. The process model for the described process can be found in Figure 11.7.

The process analysts get to the bottom of the cause of the delayed lead times and see in the data-driven process analysis that in most cases the blocker lies in the approval of the order. After interviews with different senior people in purchasing, it turns out that some of them aren't keeping up with the approval process given the large volume of orders. The analysts take this insight as an opportunity to rethink the approval process in the company and agree to only approve orders with a purchase value of more than USD 1,000.

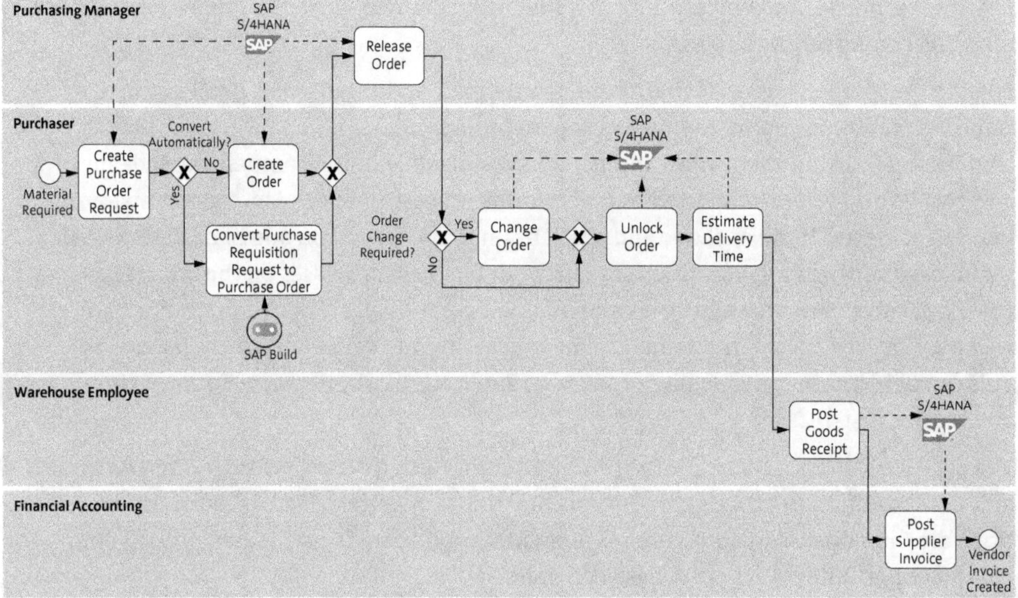

Figure 11.7 Actual Process Model for Ordering Direct Materials

The globally valid process model for the ordering process of direct materials is adapted accordingly and supplemented by the decision that purchase orders only have to be approved if their purchase value exceeds USD 1,000. Because the process adaptation is a technical change and the process has also not yet been released, this process version is saved as a draft and isn't yet visible to all process participants.

In the next step, the approval process is triggered to approve the process and check for conditional technical changes. This is controlled by the SAP Signavio Process Governance component in a company-specific workflow. As described in Chapter 8, you can use this component to define different workflows and adapt them to the needs of your company. In our current example, we consider an approval process that consists of two approval steps (formal and functional), the steps to hand over the process to ALM for implementation, and the steps after the technical implementation has been done.

In our example, the process is formally approved by the BPM manager of the PTO. The person globally responsible for the process is then asked to check the content of the process and to approve or reject it. Once the process has been approved, it's possible to check what is required for implementation in the company. The first decisive factor here is whether the changes to the process are purely organizational, or whether the changes to the process also affect the supporting systems. In our example, the implementation of the new process version also requires a change in the underlying SAP system, as certain purchase orders no longer need to be approved.

To implement the change accordingly, it's important to make sure that all needed changes are transparent to the system owners. In SAP Signavio Process Governance, the previously started workflow can now be used to assign the process change to be implemented to the responsible persons. Based on the process changes, they can define the technical requirements necessary to optimally support this process.

The technical requirements can now be incorporated into the planning for implementation and then implemented accordingly. To ensure transparency, it's important that not only the technical requirements are derived from the process changes but also that the objects providing information about the status of the implementation are linked to the process. This is the only way to ensure that after the technical implementation of new processes or process versions, the training and publication of these processes can take place.

Once the technical changes have been implemented and the new process version has been tested, the implementation is marked as **Complete**. This information is relevant not only for the IT managers but also for the process owners because the next steps to roll out the process can now be taken. Depending on the integration between SAP Signavio and the respective ALM tool, this transfer of the implementation status can be automated as a task of the SAP Signavio Process Governance component or has to be transferred manually by a user.

After the completed implementation has been made transparent in SAP Signavio, the person responsible for the implemented process receives the notification that the process has been implemented from the technical side and can be executed in daily operations. However, before this happens, the process has to be made transparent for the process participants accordingly. For this purpose, the process owner decides whether special training or other information must be created and carried out for the new process version. Once the process participants have been trained accordingly, the process owner can initiate the implementation of the process into daily operations. This publishes the process in SAP Signavio so that it's available to all process participants in the company and is correspondingly transparent for execution and future analysis.

In summary, it's important that with the approval of a new process version, the handover to the appropriate person responsible for the implementation of the process is successful, and after the technical implementation of the process, the way back to the person responsible for the process is found. For this purpose, SAP Signavio enables you to define a workflow supporting your collaboration with SAP Signavio Process Governance and to control and monitor this process. Figure 11.8 summarizes the described process with the steps from process management and ALM.

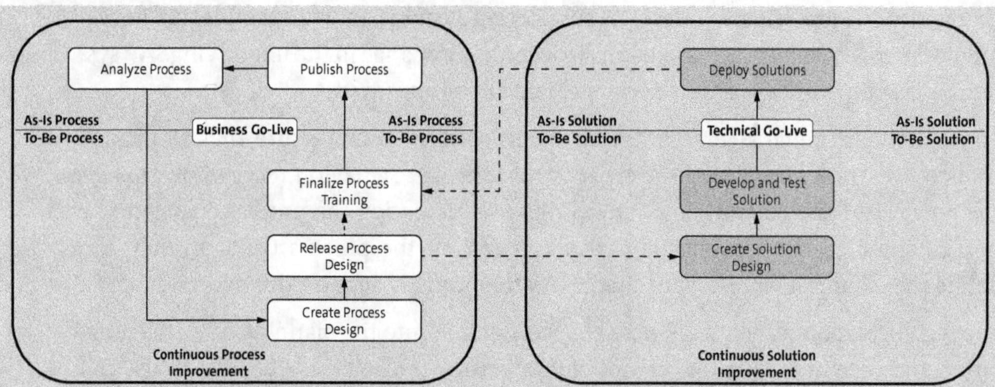

Figure 11.8 Interaction of Business Process Management and ALM

11.5 Interplay with Enterprise Architecture Management

In today's business world, managing processes plays a crucial role in optimizing business operations and achieving business goals. However, it's not just about the pure management of processes but also about monitoring and shaping the overall business. This is where *enterprise architecture management* (EAM) comes into play.

EAM describes a discipline for designing and monitoring the overall architecture of an enterprise, including the capabilities, IT systems, data, and infrastructure required by the business. It helps achieve a company's business goals by ensuring alignment between business processes and the IT landscape. Through the interplay of process management and EAM, a company can gain a holistic view of its business operations and drive optimization of its business processes at all levels.

In this section, we look at the interplay between process management and EAM and show how the two disciplines are linked and can support each other to achieve successful enterprise management. We first address the different artifacts for the types relevant to the interplay, through which a connection between the two disciplines can be achieved. We then present the different scenarios in which the combination can benefit.

In the context of EAM, capabilities play a central role in the planning and overview of the enterprise architecture. These capabilities represent the business competencies a company needs to achieve its business goals. These can come from different business areas, such as production, finance, or marketing. Based on this, decisions can then be made about the direction of the business and necessary changes, as well as application planning based on prioritization of the supported capabilities.

In the implementation of EAM capabilities, the selection and integration of IT systems and tools, referred to as applications, plays a critical role because the selected applications not only must meet business requirements but also provide efficient support for the organization's business capabilities. Here, in addition to factors such as scalability, integration capability, and cost, both support and adaptability to business processes must be a primary consideration. Optimal support for the enterprise can be achieved through careful alignment of capabilities, applications, and processes.

Throughout this book, we've already emphasized the importance of processes several times and provided a comprehensive overview of the various process management disciplines, from process analysis to process design to the use of process definitions during system implementation. Processes represent the sequence of activities required to achieve a specific goal. Aligning and linking these processes with capabilities from your EAM discipline can help abstract the goals and required capabilities of a process. In addition, it can help to make more accurate assessments of capabilities because linked processes and their performance can be better analyzed.

In a further integration stage, processes and individual process steps can refer to applications from EAM. This linkage enables a transparent view of business activities and facilitates the detection of system breaks. In addition, continuous optimization of business processes and improvement of business competencies can be achieved through monitoring and analysis. Likewise, in the event of a system failure, it's possible to quickly assess whether there are alternative ways to execute a process.

The linking of process management and EAM results in a variety of advantages:

- **Improving processes in line with the existing and future IT architecture**
  By linking processes and EAM, it's possible to ensure that business processes are in line with the existing and future IT architecture. Improvements to processes can thus be integrated directly into the IT architecture to ensure optimum support. This enables efficient and targeted design of business processes and IT architecture planning.
- **Accelerated root cause analysis through transparency in business processes and IT landscape**
  By linking processes and EAM, a transparent overview of the business processes and the IT landscape can be ensured. Root cause analysis of system errors or faults can be accelerated because the link between processes and IT systems is directly visible. This means that problems in the business process or IT landscape can be quickly identified and efficiently remedied.
- **Holistic business continuity management by linking processes and IT architecture**
  Holistic business continuity management can be realized by linking processes and EAM. The monitoring of business processes and IT systems enables early detection and reaction to disruptions or failures. In this way, continuous availability of business processes and IT systems can be ensured to guarantee business continuity.

- **IT portfolio and roadmap planning as well as investment decisions based on business process insights**
  By linking processes and EAM, a comprehensive overview of the underlying processes can be gained. These insights can be used to perform portfolio and roadmap planning in an efficient and targeted manner. Investment decisions can also be made based on business processes and their performance to ensure optimal business support.

Let's now look at how process management and EAM can be linked by using SAP Signavio. To integrate the information from EAM, we use the dictionary in SAP Signavio. For this purpose, we use one dictionary for capabilities and one for applications from the EAM tool. SAP Signavio can receive this information through an interface so that an appropriate EAM tool that provides this information can perform the transfer. As soon as the data is available in the SAP Signavio Process Manager dictionary, it can be linked in the processes. It's recommended to link capabilities at the process level via a process attribute by default, while linking applications at the process step level. This method ensures that a finer assignment of process steps to applications is made and that successive process steps can be mapped with different applications.

SAP Signavio also enables you to read out the information of the defined processes again via an interface. Especially for EAM tools with their own process objects, it's advisable to transfer the process definitions and the link with capabilities and applications to the EAM tool. This ensures that the linkage in both tools is available to the various users and that they can use it optimally in the use cases described.

In this section, we've examined the importance of process management and EAM and presented the advantages of linking the two disciplines. The linkage makes it possible to monitor and optimize business processes transparently, as well as ensure that business activities can be carried out continuously. By linking them, different scenarios, such as data-driven portfolio and roadmap planning, can be realized. However, it's important to emphasize that linking process management and EAM is an ongoing process and requires constant monitoring and adjustment. Regularly reviewing the linkage and adapting the processes and architecture can ensure that the enterprise is always up to date and meets the changing business requirements for the business.

In conclusion, linking process management and EAM is a valuable combination for corporate management to optimize business processes, secure business operations, and run the business in accordance with goals and requirements.

11.6 Summary

In this chapter, we've taken an in-depth look at the fundamentals of the topic of business process transformation management and highlighted important aspects such as process organizations, process roles, process architectures, variant management, and the interplay of process management with ALM and EAM. In summary, successful transformation requires an early examination of the fundamentals described in this chapter, which form the foundation for long-term transformation. The goal is to build an efficient and flexible process architecture that can adapt to the changing needs of the business and form the basis for different transformation scenarios, such as an SAP S/4HANA implementation. With the concepts taught in this chapter, you're well prepared for this journey.

Chapter 12
Implementing Business Process Transformation Management in the Transition to SAP S/4HANA

If you're considering a transformation to SAP S/4HANA or have already started, you want to make sure that this investment provides you with added business value. You can only achieve this if you also optimize your business processes in the course of the transformation. In this chapter, we show you how to do this.

In Chapter 10, we described the basic design principles of SAP S/4HANA, possible transition paths, and the opportunities and challenges of switching to SAP S/4HANA. We explained why business process transformation management should be an integral part of a switch to SAP S/4HANA and why the switch is the ideal entry point for customers who haven't yet comprehensively established business process transformation management in their organization. In Chapter 11, we described theoretical principles of business process transformation management. In this chapter, we describe which components are used when and how in the change to SAP S/4HANA by explaining the issues, challenges, measures, results, and deployment scenarios of SAP Signavio. We base this on SAP Signavio's methodology for end-to-end business process transformation, which we describe first in Section 12.1. After we've shown in detail how to use SAP Signavio in the individual phases (Section 12.2 to Section 12.6) and activities, we present customer case studies in Section 12.7.

12.1 Methodology for End-to-End Business Process Transformation

SAP S/4HANA requires a redesign of business processes. This is necessary due to the fundamental design principles of SAP S/4HANA and because it's the only way to realize added value for companies when switching to SAP S/4HANA. Therefore, you don't need to focus exclusively on necessary system adjustments, but identify processes where added value can be achieved through a redesign, taking into account the corporate strategy. Customers who successfully transform to SAP S/4HANA consider the dimensions of corporate strategy, processes, systems, and people in equal measure, as already described at the beginning of Chapter 10.

The corporate strategy describes the long-term overarching plan of a company with the ultimate goal of ensuring economic success. The strategy should also serve as the basis for evaluating which business processes should be redesigned when transitioning to SAP S/4HANA. It also forms the basis for defining an enterprise architecture. In Chapter 11, Section 11.2 and Section 11.5, we've already explained the importance of enterprise architecture management (EAM) in detail. It helps companies achieve their corporate goals by ensuring alignment between business processes and IT architecture. Derived from the corporate strategy, individual business goals can be defined, which in turn determine which capabilities must be part of the enterprise architecture and which processes are of particular importance to the company. The Open Group Architecture Framework (TOGAF) supports customers in the design, planning, implementation, and maintenance of enterprise architectures. Within the framework, capabilities are assigned to one of three classifications: strategy, core, or supporting. Necessary adaptations of existing capabilities or processes and the new implementation are orchestrated by application lifecycle management (ALM).

[»]

The Open Group Architecture Framework (TOGAF)

TOGAF Architecture Development Method (ADM) is a framework that describes the procedure for developing an enterprise architecture. In TOGAF, three domains are modeled: business architecture, information system architecture (consisting of application architecture and data architecture), and technology architecture. The business architecture considers, among other things, the business capabilities required for an enterprise. For more information, see *http://s-prs.de/v917415.*

Figure 12.1 shows that the corporate strategy is the basis for both business process transformation management and EAM. It also shows how an integration with ALM accelerates implementation. From the product point of view, business process transformation management is supported by SAP Signavio solutions, EAM by SAP LeanIX software, and ALM by SAP Solution Manager or SAP Cloud ALM.

Based on the corporate strategy and the business models that a company wants to support, not only the capabilities but also the applications are described in the context of EAM, and the value streams and customer journeys are described in the context of business process transformation management. These two are in turn elaborated on in a process design, taking into account the applications described in the enterprise architecture. The solution design specifies in detail how the process design is implemented. Once implementation has been completed, a process analysis is performed that is used as the basis on which a process redesign is performed, or additional applications have to be taken into account. Figure 12.2 shows how the interplay enables companies to implement their strategy.

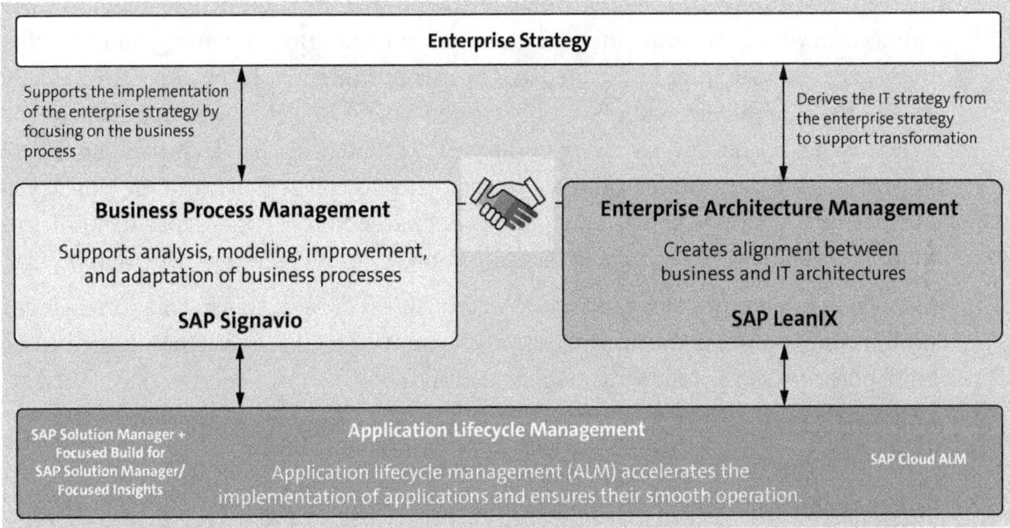

Figure 12.1 Interplay of Business Process Transformation Management, Enterprise Architecture Management, and Application Lifecycle Management

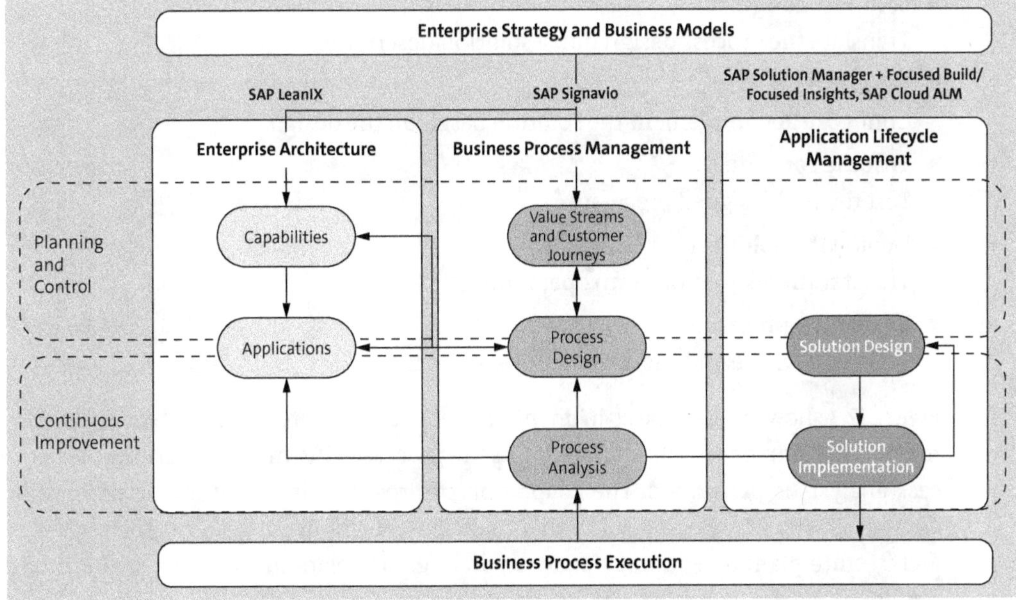

Figure 12.2 From Corporate Strategy to Implementation

The interplay of business process transformation management and ALM is of particular relevance, as this is the only way to ensure that process adaptations can be translated into system adaptations during the transition to SAP S/4HANA and continuously thereafter. Without the use of business process transformation management in close connection with ALM, companies lack the tools for analyzing, designing, and adapting business processes in an integrated and consistent manner. In practice, this usually means that companies had often only carried out technical release upgrades to the extent that is absolutely necessary in the past. By doing so, when transitioning to SAP S/4HANA, little or no added value for the company is realized. Because the benefits of business process transformation management extend far beyond the actual transformation, we address key aspects of the continuous improvement phase in Chapter 13.

SAP Signavio's methodology for end-to-end business process transformation includes both business process transformation management and ALM activities. It comprises eight phases with the following main objectives:

- **Analyze the process**
  Perform business process analysis.
- **Enhance the process**
  Enhance (existing) business processes.
- **Process design**
  Create a process (re-)design.
- **Solution design**
  Translate the process design into a solution description.
- **Build the solution**
  Configure and implement the solution based on the design.
- **Test the solution**
  Test the (re-)designed processes.
- **Deploy the solution**
  Transfer the adaptations into operation.
- **Enable the process**
  Train and enable end users to use the solution.

Figure 12.3 shows these phases in an *infinity loop*; that is, after deployment, an enablement of the employees follows. As soon as the process adjustments are used, a new process analysis is performed. The adaptation of processes is a continuous process. In addition to the process and solution dimensions, the figure also shows the phases of architecture planning: strategy analysis, architecture planning, and the creation of a roadmap.

In Part II of this book, we described the individual solutions of SAP Signavio Process Transformation Suite in detail. These solutions support the individual activities of the SAP Signavio methodology for end-to-end business process transformation. Figure 12.4 shows which phases are supported by which solution components.

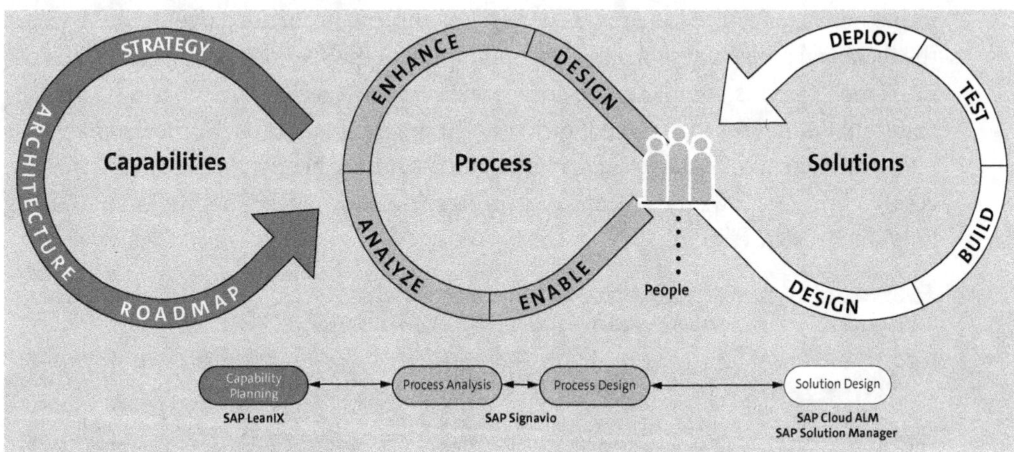

Figure 12.3 SAP Signavio Business Process Transformation Method

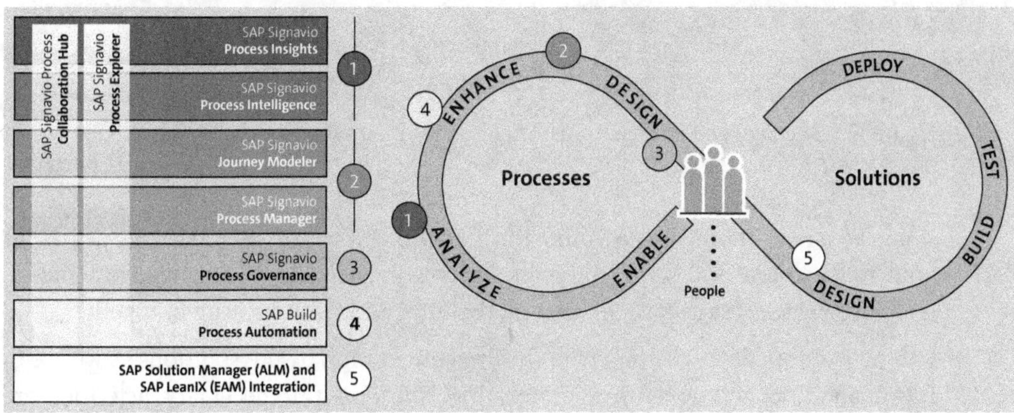

Figure 12.4 Support of the SAP Signavio Method by SAP Signavio Process Transformation Suite

We now look at the methodology in the context of the transition to SAP S/4HANA. The starting point for business process transformation management are processes that are executed with SAP ERP, a non-SAP ERP system, a combination of both, or even with multiple systems in many cases. Customers who want to improve their business processes usually don't know where the weak points—and thus the potential for improvement—lie. So, customers first need to analyze the processes to understand how they can be improved. Based on the insights gained, the currently executed processes can be adjusted to immediately add value, or the insights are taken into account when designing new processes. The basis for designing the new processes is the *SAP standard process content*. When customers rethink their business processes, they can replace enhancements with SAP standard and improve processes in general or accelerate a new implementation. Once the process design is complete, it's handed over to ALM, where the technical process design and implementation of the solution takes place. When the implementation is complete, the company is able to execute the process in the new solution. Employees are trained, process changes are made, and a process analysis is performed again to understand if adjustments that have been made improved the process. Figure 12.5 shows the SAP Signavio process transformation method when moving to SAP S/4HANA.

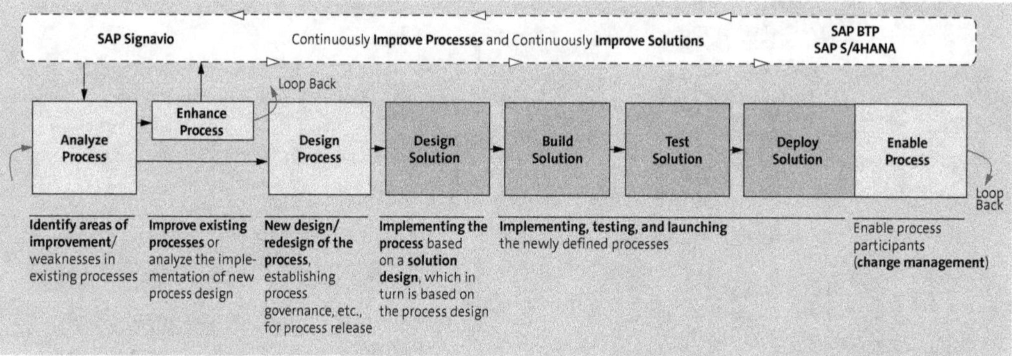

Figure 12.5 SAP Signavio Process Transformation Method When Transitioning to SAP S/4HANA

If you've already familiarized yourself with the *SAP Activate* methodology as part of your move to SAP S/4HANA, you're probably wondering how SAP Signavio's end-to-end business process transformation methodology and SAP Activate fit together.

SAP Activate supports customers in the implementation of SAP solutions as well as in the adaptation and extension of existing solutions. All necessary activities are described in the form of a roadmap, which is made available to customers via the Roadmap Viewer from SAP. These activities are presented chronologically in the discover, prepare, explore, realize, deploy, and run phases and are grouped in terms of content into workgroups such as application design and configuration, integration, and extensibility. These workstreams include all types of activities and not only those in the context of process transformation. SAP Signavio's process transformation methodology, on the other hand, describes all activities related to the analysis, design, and continuous improvement of business processes. As already described, the activities are divided into the analysis, enhance, process design, solution design, build, test, deploy, and enablement phases. The activities support the individual phases and tasks of SAP Activate, accelerate them through data-driven analyses, and digitize the results. In this way, they also create the basis for continuous process improvement.

Another question that arises about the background of a decision already made or still to be made regarding the transition path is whether the SAP Signavio end-to-end business process transformation method is relevant for all transition paths. In Chapter 10, Section 10.1.2, we explained the three change paths and pointed out that the decision as to which transition path is the right one depends on, among other things, how many processes need to be transformed:

- **Solution-oriented transformation**
  In a solution-oriented transformation, most of your processes are reused, and processes are often only redesigned where necessary due to simplification and the scope of compatibility. The transformation is executed as a system conversion.
- **Selective transformation**
  In a selective transformation, the number of processes you reconsider is significantly higher because you want to take advantage of many more opportunities in SAP S/4HANA from the start to improve your processes, you want to eliminate enhancements or process variants that are no longer needed or used, and you harmonize processes (possibly because you want to consolidate your existing SAP ERP systems when you switch to SAP S/4HANA). The implementation is executed as a selective data transition.
- **Process-oriented transformation**
  In a process-oriented transformation,you rethink most of your processes and reuse only a few processes (possibly none at all if you're a new SAP customer). The transformation is executed as a new implementation.

The exact number of processes that are redesigned versus those initially reused varies from transformation project to transformation project and is one of the key findings of the analysis phase. Figure 12.6 therefore shows examples of how the ratio is represented in the different transformation paths.

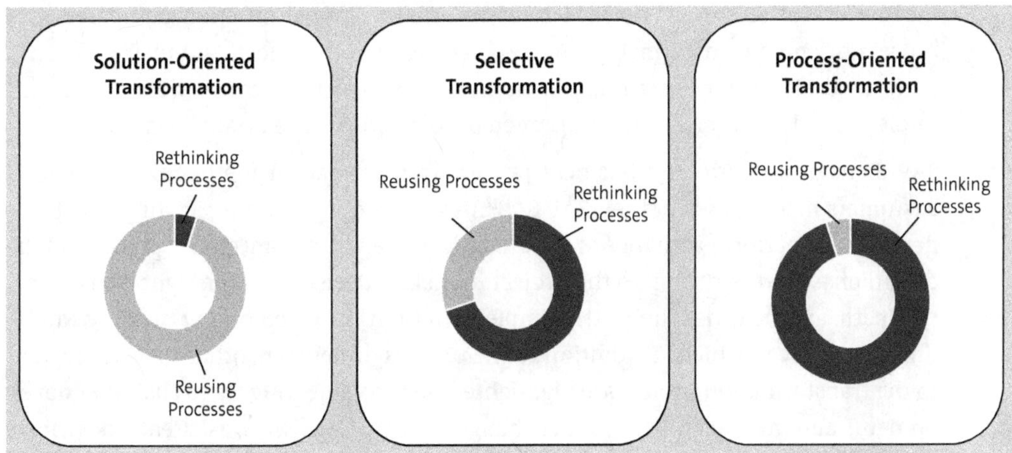

Figure 12.6 Ratio of Rethought and Reused Processes

Regardless of the chosen migration path and scope of the initial process redesign, SAP Signavio is necessary for a successful transformation. This is also true because the process adaptation isn't completed with the transition to SAP S/4HANA (see Figure 12.7). In the first step, the ratio of processes to be reused and processes to be redesigned is very different. Over time, more and more processes are adapted in the continuous improvement phase, which also takes into account new requirements and product innovations.

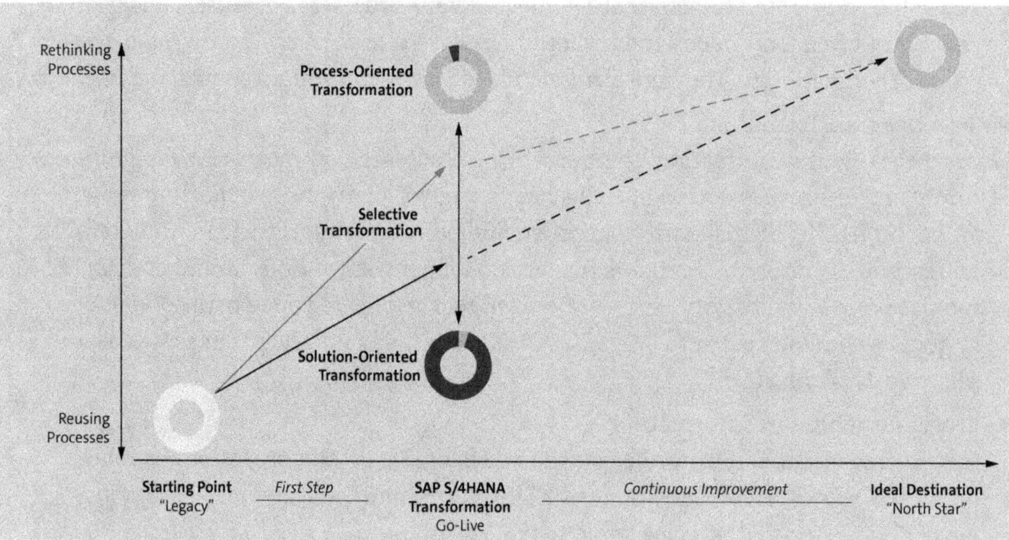

Figure 12.7 Continuous Process Redesign

For customers who have opted for a solution-oriented transformation, the focus is on changes based on simplification items and the compatibility scope, as well as selectively other adaptations with very high added value. Figure 12.8 illustrates this by showing the use of SAP Signavio in a solution-oriented transformation compared to a process-oriented transformation. As described earlier, the selective transformation is performed either as a new implementation or as a system conversion; that is, it's implemented as either a process-oriented or solution-oriented transformation.

SAP Signavio's end-to-end business process transformation methodology supports customers in all phases of their SAP S/4HANA transformation project. In practice, it's necessary to perform activities of the analysis phase, the enhancement phase, and the design phase in the run-up to the project to make a decision on the right transformation path, among other things. The implementation is carried out by means of a preliminary study in which, in addition to the process dimension, other aspects relevant to the transformation project are highlighted, such as adjustments to customer developments and interfaces. Because the change to SAP S/4HANA is, as already described, the ideal time to get started with business process transformation management, a prestudy offers the opportunity to get to know SAP Signavio.

In the following sections, we describe the issues, goals, and challenges of the transformation method, as well as how SAP Signavio provides answers with the solutions presented in Part II. The focus is on the methodological approach.

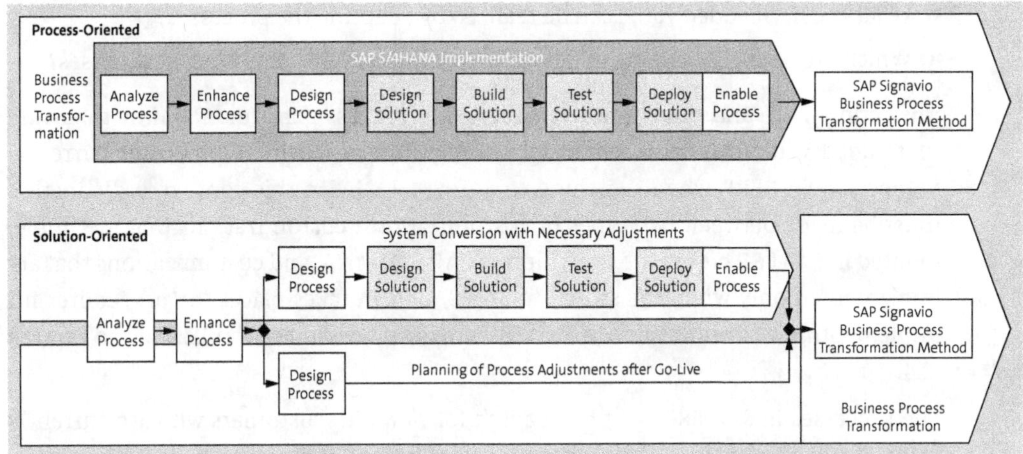

Figure 12.8 SAP Signavio Process Transformation Method for Solution-Oriented and Process-Oriented Transformations

12

12.2 Analyze the Process

The first step of the end-to-end business process transformation method aims to create transparency about current process performance.

12.2.1 Issues, Challenges, and Goals

In Chapter 10, Section 10.1.3, we discussed the following three issues in detail:

- How can you ensure that the transition to SAP S/4HANA adds value to your business?
- How can the transition to SAP S/4HANA be made as quickly as possible?
- How does your company ensure that the switch to SAP S/4HANA is feasible?

To find answers to these questions and to redesign the right processes when transitioning to SAP S/4HANA, *process transparency* is the most important prerequisite. This creates the basis for answering a wide range of questions:

- How is the performance of selected end-to-end processes?
- Are there differences in process performance between different organizational units or due to different process variants?
- Which process variants are used and how often?
- How does the process performance in my company compare to competitors in my industry?
- Which processes give my company a competitive advantage?

- What financial added value can be realized by adapting the processes?
- Which process adjustments make sense to eliminate the identified weaknesses?

Answering these and other questions related to process performance aims to ensure that added value can be achieved by redesigning the business processes currently mapped in SAP ERP when switching to SAP S/4HANA. In this context, a distinction must be made between two types of customizations: customizations that are implemented in SAP ERP before the transition to SAP S/4HANA, and customizations that are implemented only when the switch to SAP S/4HANA takes place. The procedures for implementing customizations in SAP ERP are described in Section 12.3 and in Section 12.4.

The processes in SAP ERP can't be adapted for new SAP customers who are currently using ERP solutions from competitors. In the subsequent design phase, SAP best practices or standard content form the basis for fit-to-standard workshops. The analysis of processes in the currently used IT solutions with the help of SAP Signavio Process Intelligence is conceivable in principle, but, in practice, it's very seldom the case because effort and benefit aren't in a meaningful relation to each other. Because SAP ERP solutions are fundamentally different from competing solutions, the analysis results don't provide much added value anyway. We therefore don't address this scenario any further.

Existing SAP customers who are running an SAP ERP system and have decided to switch to SAP S/4HANA Cloud Public Edition are carrying out a new implementation, just like new SAP customers. Here, an analysis of the actual processes can be useful. The motivation and simultaneous prerequisite for the change to the public cloud is the standardization of the processes based on what's available in SAP S/4HANA Cloud Public Edition. As a result, you wouldn't perform an analysis of SAP ERP in this scenario for the purpose of making adjustments during or after the transition to SAP S/4HANA Cloud Public Edition because you're moving toward the standard anyway. An analysis of the existing SAP ERP system can nevertheless be useful for the following reasons:

- To identify potential for process improvement and derive which added value can be realized by a new implementation, taking these findings into account
- To analyze process variants because business processes are harmonized in the course of the new implementation with the aim of reducing complexity
- To create transparency about the as-is processes—complementary to the digital discovery assessment—and assurance that the functional scope of SAP S/4HANA Cloud Public Edition is sufficient

From the perspective of existing SAP customers converting to SAP S/4HANA or seeking a new implementation, process transparency is the foundation for the following goals:

- Creating added value in the redesign of business processes
- Maximizing the benefits of SAP S/4HANA

- Minimizing complexity by reducing the number of process variants and document types
- Creating the basis for the modeling of processes and creating a process repository
- Comparing with already modeled processes

The first objective mentioned can be described as a value-added analysis in which it's necessary to define in advance which process areas are to be analyzed. This isn't necessary for the other objectives.

Customers generally have no comprehensive transparency about their processes, which means that they aren't in a position to answer important questions across the company. Against the backdrop that existing IT resources are already stretched to capacity and now the SAP S/4HANA project is added to this workload, transparency is a challenge because, in addition to a shortage of resources, the time available for analyses is also limited. This means that companies are facing a number of challenges:

- No evaluation options are available, resulting in ignorance of process performance.
- Process performance data, if any, is out of date.
- Process performance can't be compared with competitors (cross-company) or within the company (intra-company).
- No overview is available of which process variants exist, how often they are used, and which are most efficient and error-free.
- There is no organization dealing with the topic of process management and therefore no dedicated resources with the corresponding know-how.

12.2.2 Procedure

As described in the previous section, the analysis of business processes pursues various goals. These can be grouped into two types:

- **Value-added goals**
  To create specific process transparency to generate added value when redesigning business processes (*value-added analysis*).
- **Benefit-maximizing and complexity-reducing goals**
  To create basic process transparency to maximize the benefits of SAP S/4HANA and reduce complexity (*benefit-maximizing and complexity-reducing analysis*).

In practice, both happen at the same time, as the goals complement each other.

For the first type of objectives, it's first necessary to define which processes are to be analyzed and to determine whether there is potential for improvement. Processes can be complete end-to-end processes, modular subprocesses, or even a subprocess step. As described in Section 12.1, the corporate strategy forms the basis for identifying the processes to be analyzed. Based on the corporate objectives and the associated value drivers, the business processes to be analyzed and the relevant process indicators are identified.

SAP Value Lifecycle Manager can be used to supplement the definition of a process analysis scope based on the corporate strategy. Part of SAP Value Lifecycle Manager is the *Move the Needle* tool. It enables customers to compare their company with competitors using predefined key figures based on publicly available standardized financial ratios. The result is a calculation of the full value-added potential. Business capabilities and processes supported by SAP S/4HANA play no role in this consideration, nor do the added values customers can realize by using SAP S/4HANA.

This is where the next form of top-down value-added calculation comes in. SAP Value Lifecycle Manager enables the creation of a business case with another capability of SAP Value Lifecycle Manager. The calculation of a business case is supported by an SAP-internal benchmarking database to refine the insights already gained by comparing data points from other companies. Value drivers, which are determined by the selection of products, business capabilities, or processes when the business case is created, are also analyzed here. To validate the manual entry of the performance of the value drivers, it's possible to import system-related data into the business case. This is done by the SAP Signavio Process Insights, discovery edition. This solution analyzes the most important cross-industry metrics. SAP Signavio Process Insights significantly extends the analysis content and also offers the possibility to flexibly evaluate the already determined key figures. In Chapter 14, we'll give you an overview of both solutions and explain how they get you started with business process transformation management.

Figure 12.9 shows the interaction of a top-down calculation based on key performance indicators (KPIs), which are based on external benchmarks, and process performance indicators (PPI), which are provided by an analysis of the SAP ERP system.

SAP Value Lifecycle Manager

SAP Value Lifecycle Manager is available to all customers with a support contract. You can find more information about this product here:

- Information for getting started with SAP Value Lifecycle Manager: *http://s-prs.de/v917416*
- The learning journey for the Move the Needle tool: *http://s-prs.de/v917417*
- The learning journey for a business case: *http://s-prs.de/v917418*
- The learning journey for integration process discovery and the business case: *http://s-prs.de/v917419*

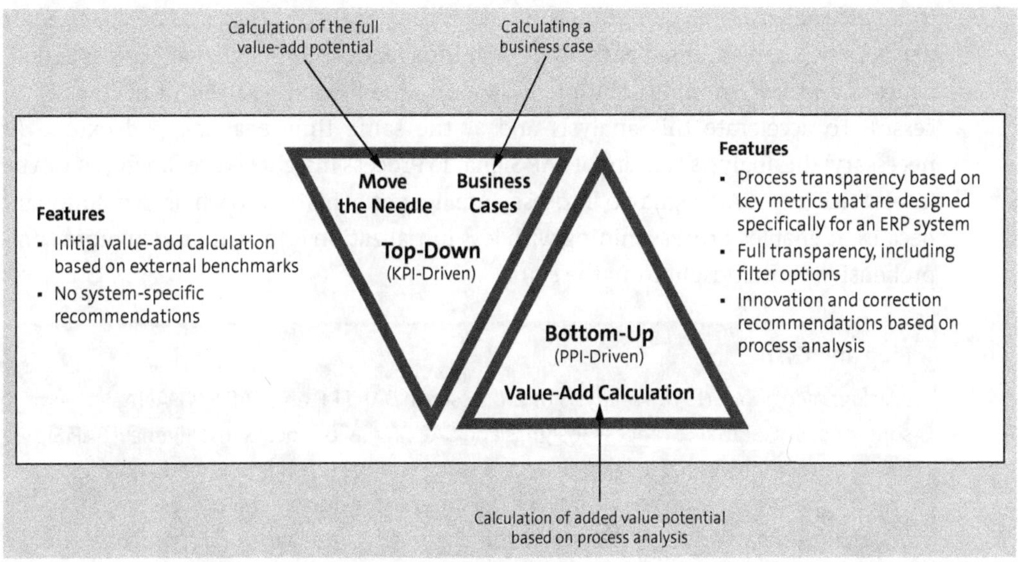

Figure 12.9 Top-Down and Bottom-Up Added Value Calculations

The analysis of value-added potential often takes place before the start of the project, as the investment costs make it necessary to determine value-added potential. In the form of an SAP S/4HANA preliminary study, the decision for a change path is made here in addition to a benefit analysis, and further analyses, such as the readiness check described in Chapter 10, are performed. Likewise, an initial analysis of the necessary process enhancements or process redesigns is already underway. The SAP S/4HANA pre-study thus includes activities from the analysis, enhance, and design phases.

By using SAP Value Lifecycle Manager, it has already been possible to identify added value potential. Nevertheless, there is a need to look at the assumed value potential in more detail to determine necessary adjustments and, based on this, the implementation effort. SAP Signavio Process Insights and SAP Signavio Process Intelligence support this analysis. Both solutions also create the basis for SAP S/4HANA benefit-maximizing and complexity-reducing customization.

SAP Signavio Process Insights supports the analysis of SAP ERP systems based on predefined analysis content. After activation, this content is available in the form of process flows, process key figures, corrections, and innovation recommendations. For value-oriented analysis, SAP Signavio Process Insights offers a value driver–oriented entry point. In addition to the analysis of the number of objects for process flows and PPIs, the calculation of the financial added value is also supported. For the benefit-maximizing and complexity-reducing analysis, an entry via lines of business is possible. Extensive filter options allow a detailed view and thus also validation and exploration of the findings.

If the analysis content isn't sufficient with regard to the processes to be analyzed or needs to be supplemented with additional information, SAP Signavio Process Intelligence is a process mining solution. It supports the flexible analysis of business processes. To accelerate this analysis and at the same time evaluate and extend (if necessary) the analysis content of SAP Signavio Process Insights more flexibly, it can be integrated into SAP Signavio Process Intelligence. This approach is revolutionary because it enables process mining with less initialization effort and much more comprehensive content right from the start.

[+]

Plug and Gain

SAP Signavio supports the analysis of processes in SAP ERP and SAP S/4HANA with two solutions: SAP Signavio Process Insights and SAP Signavio Process Intelligence. SAP Signavio Process Insights enables process transparency based on predefined analysis content with almost no effort. However, a flexible extension of the process analysis by additional process steps, an adjustment of the analysis interface, or a comparison with existing process models isn't possible. With the *plug-and-gain* approach, you can use SAP Signavio Process Intelligence for a more flexible analysis by generating event logs based on the SAP Signavio Process Insights analysis content; those logs are then analyzed in SAP Signavio Process Intelligence. Plug and gain thus not only increases the benefit of SAP Signavio Process Insights but also enables customers to perform process mining with a drastically reduced time effort. In addition to flexibility, additional analysis content is provided such as the transaction code analysis. It's also possible to enrich the analysis by enhancing the event logs with data that isn't part of the SAP Signavio Process Insights analysis content.

The data-supported analysis enables a fast and objective process analysis. It provides the basis for a discussion with the stakeholder on the business unit and IT side, in which the findings obtained are discussed and validated in workshops: How can the results be explained? What enhancements and adjustments are necessary to eliminate the identified weaknesses? How great is the added value, and how costly are the adjustments? The answers to these and other questions form the basis for a cost-benefit analysis, prioritization, and classification of whether the adjustments are made in the enhance or design phase.

The following results are obtained in the analyze phase:

- Calculation of the total value-added potential using the Move the Needle tool and the SAP Value Lifecycle Manager business case through a top-down approach to value-added calculation
- Refinement of the top-down view based on the PPIs by a bottom-up analysis based on the KPIs of the Process Discovery report
- Identification of processes with the greatest potential for performance improvement

- Execution of an analysis of process flows and further key figures of SAP Signavio Process Insights
- Use of the plug-and-gain approach to visualize the analysis content from SAP Signavio Process Insights in SAP Signavio Process Intelligence as well as additional specific plug-and-gain content
- Analysis of further business processes not covered by the content from SAP Signavio Process Insights with SAP Signavio Process Intelligence
- Creation of an overview of which processes offer the greatest potential, with regard to two factors:
  - Which processes create the most added value in terms of corporate strategy and the top-down, outside-in view based on Move the Needle or the business case view?
  - Which processes enable the reduction of complexity by reducing process variants, document types, and master data?
- Supplementation of the overview for the identified added-value potential with an initial rough estimate of the effort required, preparing a cost-benefit analysis, and classifying whether the adjustments are to be implemented in the enhance or in the design phase

12.3 Enhance the Process

The second step of the end-to-end business process transformation method aims to define how the identified added value potentials can be realized by extending existing processes.

12.3.1 Issues, Challenges, and Goals

In the analyze phase, the existing business processes are examined, weaknesses are identified, and potential for improvement is highlighted. As described in Section 12.2.2, a distinction is made between two types of adjustments:

- Adjustments that can already be made in SAP ERP during ongoing operations, as they involve little implementation effort, are classified as noncritical and don't require SAP S/4HANA innovations
- Adjustments that are only made in the course of process design because they are too extensive or require SAP S/4HANA innovations

In the enhance phase, the implementation effort for the adjustments that can be implemented in SAP ERP is first estimated. Taking into account the benefits that can be achieved in the analysis, the adjustments are prioritized and a roadmap is created for when they will be implemented. To do this, the following factors must be clarified:

- **Effort**
  With what effort can the identified weaknesses be eliminated? This includes both the technical adaptation effort and the necessary functional retests, as well as clarification of who on the IT and business side must be involved in the changes. As a result, it's also possible to determine the feasibility of the adjustments and whether enhancements to SAP ERP processes are possible during ongoing operations and not just during the conversion to SAP S/4HANA. The more detailed consideration of the analysis results and the required effort can also lead to the result that potential improvements identified in the analysis phase no longer bring any or the same benefits and are therefore no longer pursued.
- **Responsibility**
  Once the necessary activities for the technical implementation have been identified, who is responsible for individual partial adjustments or overall can be clarified.
- **Time**
  Once work packages and responsibilities have been defined, the question of how much time all those involved in the adaptation will need can be finally clarified.
- **Cost-benefit ratio**
  If effort and benefit have already been determined in the analysis phase, the initial prioritization is updated to take the findings into account.
- **Clustering**
  In preparation for the creation of a roadmap, an evaluation is carried out to determine which adaptations can be sensibly carried out together, for example, to reduce test efforts.
- **Roadmap**
  Finally, a roadmap is created that describes the chronological order in which the adjustments will be made, taking into account personnel availability.

After budget approval, implementation can begin. Through this, the following goals are realized:

- **Realizing added value quickly**
  Process transparency makes it possible to realize targeted improvement potential and thus also demonstrate the benefits of process changes.
- **Complexity reduction**
  Cleaning up document types and process variants that are no longer or not frequently used leads to a reduction in complexity. During the conversion to SAP S/4HANA, this can also reduce testing efforts and focus on the frequently used process variants.
- **Interaction**
  As already described, the number, scope, and added value of process changes are often very small; the focus is on operating the systems. Enhancements on a limited scale are ideal preparation for the necessary interaction between all stakeholders in IT and in the individual business units during the changeover to SAP S/4HANA.

Although the overall effort is usually low and potential benefits can be realized, some challenges can also arise during this phase:

- **Willingness to change**
  In the past, the focus has been on operating systems (keeping the lights on) rather than continuous process adjustments (innovation). Now, when adjustments are to be made based on data-driven process transparency for the first time in a long time, this can lead to resistance from stakeholders.
- **Know-how**
  Although customizations are usually straightforward, the necessary technical know-how, for example, for a configuration adjustment, is no longer necessarily available due to employee turnover and the fact that the number and scope of customizations in recent years have been small.
- **Resource availability**
  Not only the know-how but also the resource availability can be a challenge because, in parallel, the ongoing operation has to be supported with a very often thin personnel cover, and further preparatory activities of the SAP S/4HANA project have to be supported. The activities of the enhance phase can still be ongoing if the conception has already begun.

12.3.2 Procedure

The process adjustments in the enhance phase are based on the results of the analysis phase. In this phase, the benefits of the process adjustments were analyzed, but the implementation effort was only estimated very roughly to decide whether the process adjustments can already be implemented in SAP ERP or whether they are part of a more comprehensive process redesign. In a first step, therefore, the necessary adaptation effort is determined. The following results are analyzed with regard to the implementation effort:

- **Correction recommendations**
  SAP Signavio Process Insights already provides concrete information on which adjustments can be made in SAP ERP in the form of correction recommendations. These recommendations are based on selected PPIs. Necessary steps for the adaptation are described, and the implementation effort is classified into the categories of low, medium, and high based on empirical values. A validation can be carried out on this basis.
- **Innovation recommendations**
  In addition to correction recommendations, SAP Signavio Process Insights also provides innovation recommendations at the process flow level. These also include recommendations on which bots can be used to improve processes through SAP Build Process Automation or SAP Fiori apps. An individual assessment of the implementation effort is required.

- **Process flows and PPIs**
  In the case of improvement potential identified on the basis of process flows or individual PPIs, it must first be determined which adjustments are necessary. The blockers identified for process improvement in the process flows as well as the individual PPIs should be analyzed using the filter options, if not already done in the analysis phase. Complementary to the analysis in SAP Signavio Process Insights, an analysis based on the plug-and-gain approach can also be performed in SAP Signavio Process Intelligence. In addition to the data-driven system-based analysis, dialogue with the end users should always be sought to validate planned adaptation measures.

Once the necessary adjustments and efforts have been defined on the basis of the correction recommendations, the innovation recommendations, the process flow analysis, and the PPIs, the responsibilities and time requirements can be clarified.

Before a roadmap is drawn up, the individual activities under consideration are grouped together, and the ratio of benefit to expense is reassessed. After a final decision has been made as to which adaptations are to be implemented, budget approval is given if necessary. In practice, the following implementation of the adjustments can take place in parallel with the design phase that now follows.

The following results are achieved in the enhance phase:

- Definition of the necessary adaptation activities as well as the required effort
- Clarification of the resources required for implementation
- Timing for the realization
- Update of the benefit-cost analysis and an assessment of which adjustments should be made, including prioritization
- Grouping of adaptation activities
- Creation of a roadmap

12.4 Process and Solution Design

An essential step on the way to a successful SAP S/4HANA transformation is the process and solution design phase. In this phase of the end-to-end business process transformation methodology, the foundations for the transformation of business processes are laid and future solution development is planned. In this phase, numerous questions and challenges arise that must be carefully analyzed and addressed to ensure a successful change.

12.4.1 Issues, Challenges, and Goals

When designing the new business processes or solution, customers have to deal with various issues and overcome some challenges. One of the first aspects that companies have to deal with is identifying the business processes that are to be integrated into the new system. Both existing and future business processes must be considered. Companies must also decide which processes should be optimized within this initiative and which processes can be retained based on the influence of the analysis results from Section 12.2.

Another important aspect is the design of business processes with regard to the corporate strategy. The aim here is to design the business processes in such a way that they contribute to achieving the company's goals. To this end, the business processes must be designed so that they are efficient and enable rapid implementation.

Building a process architecture is an important step in understanding, optimizing, and managing the processes in an organization and should be a central part of the design process and solution phase. The following steps can help:

1. **Identifying the business processes and building the process architecture**
   Identify all business processes in your company, from product development to delivery, and cast them into a structure.
2. **Prioritizing the processes**
   Prioritize the processes according to their importance for the company and their value creation potential.
3. **Designing a process**
   Design an optimized process that eliminates the weak points and increases efficiency.

In the following section, we explain the three steps in detail.

12.4.2 Procedure

As part of the SAP S/4HANA transformation, it's crucial to build a solid process architecture that serves as the foundation for the redesign of business processes. This section is dedicated to the structured planning and design of the process architecture as well as the prioritization and segmentation of the processes to be transformed. It also explains how to develop a final modular process design to maximize the efficiency and flexibility of the enterprise.

Structure of the Process Architecture

Identifying and scoping the business processes is the first step in developing a process architecture. It's important to identify all business processes that exist in the company. Various methods can be used for this, such as interviews with employees or data analysis.

The company's business strategy and business models should also be considered when identifying business processes. The company's business strategy and business models determine which activities and processes are required to achieve the company's strategic goals. By analyzing the business strategy and business models, companies can derive their key business processes. In addition to the processes that are relevant for the implementation of the business strategy or the business models, the identification of business processes should also take into account the activities and tasks that aren't necessarily required for the implementation of the business strategy or the business models. In doing so, it's helpful to be guided by the various departments and functions of the company to ensure that all relevant processes are identified.

First, the end-to-end processes should be identified or derived from the business models that support the company in its core areas, for example, make-to-order or engineer-to-order models. An end-to-end process consists of a series of steps necessary to achieve a specific outcome. However, these steps can be broken down into smaller, modular subprocesses that can be executed independently and affect the overall outcome of the end-to-end process. Examples include production order creation, production planning, or accounts receivable. In the following, we describe in more detail how modular subprocesses can be derived from an end-to-end process:

- **Step 1: Identify the overall result of the end-to-end process.**
  Before you can identify the modular subprocesses, you need to understand the overall outcome of the end-to-end process. What is the goal of the process? What is the expected outcome? Once you've identified the overall outcome, you can proceed to identify the individual steps required to achieve that outcome.
- **Step 2: Identify the steps necessary to achieve the overall result.**
  Now identify the steps that are necessary to achieve the overall result. Write down each of these steps, and note what happens at each step.
- **Step 3: Group similar steps.**
  Once you've identified the individual steps in the end-to-end process, group similar steps together. These steps should be closely related and perform a similar function. Group these steps into modular subprocesses.
- **Step 4: Identify the dependencies between the modular subprocesses.**
  After you've identified the modular subprocesses, identify the dependencies between them. Which modular subprocesses must be completed before other subprocesses can begin? Which subprocesses depend on each other?

Subsequently, all identified end-to-end processes and all modular subprocesses can be sorted into the process architecture described in Chapter 11, Section 11.2 (see Figure 11.4).

When developing a process architecture, it's important to consider and identify all relevant processes, regardless of whether they are part of the end-to-end process or not. An example of a subprocess that isn't normally part of an end-to-end process is the execution of inventory. Inventory execution may be a subprocess that doesn't interact directly with the customer or supplier, but still plays an important role in keeping the business running. If this process isn't performed effectively and efficiently, it can impact inventory management, sales, and customer satisfaction. Therefore, all relevant modular subprocesses should be identified and included in the process architecture.

The incorporation of reference content or reference architectures, such as SAP Signavio Process Explorer (see Figure 12.10) or content from partners, into the development of a process architecture can be very helpful as it allows the company to leverage best practices from other companies. Reference content or reference architectures are typically already established models or frameworks that represent best practices and standards in specific industries or functional areas. They provide a structured method for identifying, analyzing, and improving processes that have already been successfully implemented in other companies.

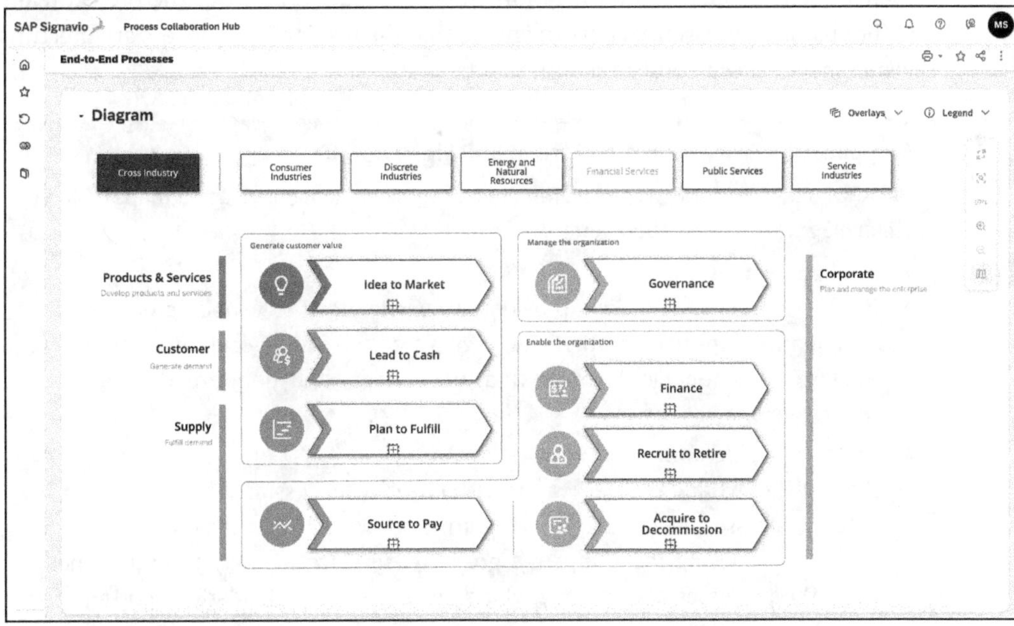

Figure 12.10 SAP Signavio Process Explorer

To incorporate reference content or reference architectures into a company's process architecture, these models and frameworks should first be carefully evaluated and analyzed. Here, it's important to consider the specific requirements and needs of the company to ensure that the selected models and frameworks cover the relevant aspects. The company should then adapt the selected reference content or architecture and apply it to their specific business processes/business model. This could potentially indicate that certain parts of the reference content or architectures need to be modified or supplemented to meet the needs of the business.

The goal is to create a comprehensive and structured overview of the business processes in the company, which serves as a starting point for the further development of the process architecture. With this overview, companies can better understand their processes, recognize weaknesses, and identify potential for process improvement.

If a customer already has a process architecture, it must be understood and checked regarding whether it's still up-to-date and relevant for the planned project. Usually, an existing process architecture is taken as a starting point for a new project to avoid redundancies and possible inconsistencies. However, it's possible that the existing process architecture is outdated or incomplete. In this case, the process architecture should be revised and updated before it's used as a starting point for the new project. To do this, the processes included in the process architecture should be reviewed, updated, and adjusted to ensure that they meet the requirements of the new project. It's also important that the affected employees are involved in the revision of the process architecture. Open communication and collaboration between the project team and the customer is essential here to ensure that the result of the process architecture revision meets the customer's requirements.

[»]

Mapping of a Process Architecture in SAP Signavio

In SAP Signavio, process architectures can be represented via navigation maps or value chain diagrams, as described here:

- **Navigation maps**
  These visual representations provide an overview of the various processes and their relationships to each other. These maps can serve as navigation assistance for employees and enable them to find, understand, and optimize the processes within the organization.
- **Value chain diagrams**
  These diagrams are another way to visualize the process architecture. Here, the various processes are depicted in the form of value chains to show how they contribute to achieving the company's goals. These diagrams help to better understand the processes and identify which processes are particularly important and should be optimized.

To better organize the processes within the process architecture, processes can be separated from each other using *folders*. In this way, the various processes can be categorized and grouped to simplify their management and maintain an overview.

A well-structured process architecture is an important component of a successful company. By using process management tools, designing navigation maps or value chain diagrams, and organizing processes in folders, companies can better understand and optimize their processes.

In addition to the pure representation of the process architecture, user-specific entry pages via process maps are also important. This implies that each user group sees an individual page tailored to their needs and tasks when logging in to SAP Signavio. For example, employees from Germany can see a page listing the processes relevant to them, while employees from the US see a page tailored to them.

Another important feature of SAP Signavio is the ability to assign specific process maps to process owners. For example, the process owner for the make-to-order process can see a special page that lists only information about this process, while the process owner of the purchase-to-pay process has a separate page that lists only information about this process.

This feature is managed through user groups and audiences. A user group can be created to group all employees from a specific country or with a specific role. An audience can be created to identify a specific group of users that defines which specific entry page should be displayed to a user.

By using user-specific entry pages and process maps in SAP Signavio, efficiency is increased, and workload is reduced. Employees no longer have to dig through irrelevant information but receive only the information relevant to their work. This leads to higher productivity and a more efficient workflow.

Prioritize and Segment Processes

When prioritizing processes, companies should evaluate the importance of the individual processes for the company and their value creation potential. There are different criteria according to which prioritization can be carried out. One important factor in prioritizing processes is their strategic importance for the company. As mentioned earlier, companies should always think about their business strategy and prioritize the processes that best contribute to the implementation of this strategy. For example, if a company's strategy is to serve its customers quickly and effectively, it should prioritize the processes that are directly related to customer satisfaction. Another criterion may be the value creation potential of the process. Companies should evaluate the processes that can make the greatest contribution to revenue and profit or that deliver the greatest cost savings. The risks associated with each process should also be considered. Critical processes, where errors could have serious consequences, are prioritized higher. It's also critical to consider the opinions and experiences of employees and customers, as they often have important insights into the effectiveness and efficiency of processes. Once prioritized, organizations should focus their resources and efforts on the most important processes and continuously improve them to increase their effectiveness and efficiency. In Chapter 11, Section 11.2, we've already presented the theoretical basis for process segmentation. As described, we recommend performing such segmentation at the level of subprocesses, categories, or value steps.

A possible segmentation of the processes can be made into the following four categories, as described in Chapter 11, Section 11.2:

- Key differentiators
- Differentiating processes
- Harmonized core
- Standard processes

These categories have a significant influence on process design, which we'll discuss in more detail next.

Mapping of Process Segmentations in SAP Signavio

Segmenting processes in the process map/architecture can help companies better understand and optimize their processes. This helps identify which processes are of strategic importance to your company and which aren't. Depending on how important a process is for your company, the process design can vary greatly.

The differentiation of the process categories—key differentiators, differentiating processes, harmonized core, and standard processes—can be recorded in SAP Signavio using, for example, attributes.

Segmenting processes in SAP Signavio is an important step in ensuring that the company's processes are optimally designed. By defining attributes at the process level, for example, as a dropdown attribute, companies can ensure that their processes are tailored to their specific requirements.

Another methodology that can influence prioritization and process design is the concept of a *customer journey*, which we'll describe in more detail here.

Customer behavior has changed significantly in the recent years. Customers today have a variety of ways to buy products and services. They use different channels to do so and expect a seamless and personalized experience. To meet these expectations, companies must have a deep understanding of their customers' needs and expectations. Customer journeys are an important tool to gain this understanding.

Customer journeys have become a central part of process management in recent years. They offer the opportunity to create a detailed analysis of the interactions between the company and its customers. By linking customer touchpoints with internal processes, companies can gain deep insights into their customer relationships and identify and prioritize optimization potential.

A customer journey is the totality of all experiences a customer has when interacting with a company, from the first contact to the completion of a purchase or other type of interaction. It includes various touchpoints, that is, points at which a customer encounters the company. These can be physical (e.g., a retail store) or digital (e.g., a website or app).

Customer journeys are important for several reasons. They provide insights into customers' needs, preferences, and behaviors. They also enable companies to identify and address weaknesses in their processes. In addition, they help to improve the customer experience (CX) and increase customer loyalty.

The implementation of customer journeys (see Figure 12.11) requires a methodical approach. The following steps can help:

1. **Define customer journeys**
   The various customer journeys must be identified and defined. This can be done through conversations with customers, employee feedback, and data analysis.
2. **Identify touchpoints**
   The next step is to identify the various touchpoints in each customer journey. This includes all points at which customers encounter the company.
3. **Measure mood**
   Customer sentiment should be measured at every touchpoint. This can be done through surveys, feedback tools, or analysis of customer feedback on social media.
4. **Link internal processes**
   The identified touchpoints should then be linked to the company's internal processes. This enables a more detailed analysis of the customer journey and helps to identify weak points.
5. **Identify and prioritize optimization potential**
   Finally, the collected data should be analyzed to identify and prioritize optimization potential. This can be done through data analysis and interpretation as well as through discussions with employees and customers.

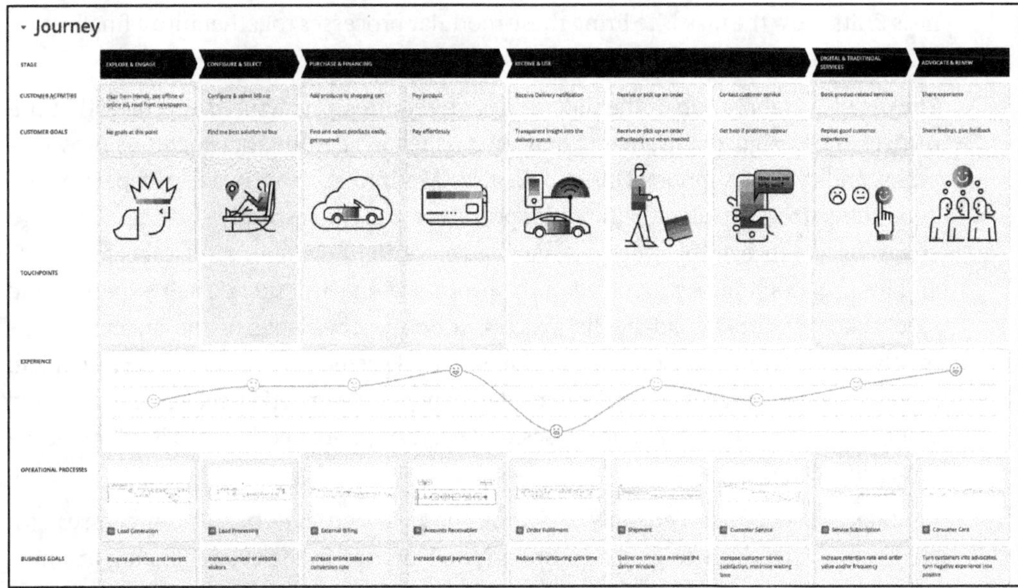

Figure 12.11 View of a Customer Journey in SAP Signavio

Mapping a Customer Journey in SAP Signavio

In SAP Signavio, customer journeys can be represented in different ways to gain a better understanding of customer interactions.

One approach is the use of *customer journey maps*. These represent the individual phases of the customer journey and include the various interactions between the customer and the company as well as the associated emotions and needs of the customer. These representations help companies to gain a deeper understanding of their customers' needs and can serve as a basis for identifying optimization potential. Overall, customer journeys in SAP Signavio provide an effective way to understand and optimize customer behavior and interactions with the company. By analyzing the customer journeys in detail, companies can improve the CXs and increase customer loyalty.

Creation of a Final Modular Process Design

Achieving a successful modular process design requires a structured approach. In the first part of this chapter, we therefore showed how to derive the end-to-end processes starting from the corporate strategy and how these can in turn be subdivided into modular processes. Once the basic set of modular processes has been defined, these can be segmented so that, depending on how the process is segmented, the design should be closer to the system standard or to the respective requirements of the business units. Now the task is to bring these modular processes together into a final design that best meets the needs of the business.

The detailed elaboration of the process design requires a structured approach that prioritizes the design requirements and establishes a clear link between business processes and system support. This should also take into account possible dependencies and interactions between the modular processes and the respective business areas. To ensure acceptance of the process design and smooth implementation, process experts and managers should be involved in the process at an early stage. Their expertise and feedback are crucial to adapt the process design to the needs of the business units involved and to the requirements of the technical systems. In this section, we explain how the detailed process design can be developed, what criteria need to be considered when designing the modular processes, and what steps are necessary to implement the final process design.

Modular process design enables organizations to design their business processes efficiently and effectively to gain a competitive advantage in the ever-changing business world. By applying a modular approach, organizations can divide their business processes into individual, reusable, and adaptable components. This facilitates adaptation to new requirements, adoption of best practices, and continuous process improvement. Modular process design consists of creating detailed process models that contain all business-relevant information as well as the respective steps that are carried out in a system-supported manner. It's important to find the right balance between customization and standardization to both preserve the uniqueness of the company's processes and benefit from best practices. Such attention to detail is critical, as it enables the underlying systems and technologies to be understood in detail and taken into account when creating the target process design. This leads to closer integration of business processes and IT systems and ensures that the solutions developed both meet business requirements and are technically feasible and sustainable.

An important aspect of modular process design is the close link between business process transformation management and ALM. By integrating these two disciplines, companies can obtain a holistic view of their processes and the supporting technologies, leading to better alignment between business and IT strategies. In this context, business process transformation management provides the methods and tools to analyze, model, and optimize business processes, while ALM ensures the management of the entire lifecycle of the applications and systems required to support these processes.

In this context, the described segmentation of business processes into different categories (key differentiator, differentiating, harmonized core, and standard), and plays a crucial role in specifying the basic assumptions for process design. By applying this segmentation, organizations can target their resources and efforts to the processes that provide the greatest business value and best support the business strategy. This creates a synergistic relationship between business process transformation management and ALM that allows business processes to be optimally designed according to their strategic importance and role in the overall portfolio (see Table 12.1).

Process Category	Significance for the Process Design
Key differentiator	The process design follows the recommendations of the process experts, as they have the most know-how about the requirements and specifics of these processes. The needs and requirements of the company are put in the foreground, and the processes are designed to provide maximum competitive advantage. The experience and knowledge of the process experts is used to develop innovative solutions tailored to the specific requirements of the company.
Standard processes	The focus is on standard reference processes. These are less critical for differentiating the company and offer less scope for competitive advantage. Process design is based more on proven, standardized procedures to maximize efficiency and cost savings. Deviations from standard processes only occur if there is a good justification for them, for example, to meet specific legal requirements or to respond to special market conditions.
Differentiating processes and harmonized core processes	Here, the focus is on a balanced mix of customization and standardization. Process design addresses both the individual needs of the company and proven best practices. The aim is to design processes in such a way that they support the corporate strategy as well as being efficient and cost-effective.

Table 12.1 Segmentation of Processes and Their Importance for Process Design

Segmenting processes into different categories helps to effectively control the basic assumptions for process design. Companies can thus focus their resources and efforts specifically on the most important processes and find the right balance between customization and standardization. This enables more effective implementation of the corporate strategy and helps to achieve sustainable competitive advantages in the ever-changing business world.

Now that the segmentation of business processes and its impact on the basic assumptions for process design have been explained, it's time to look at the *process workshop*, which plays a central role in the development of the target process design. The process workshop is where the relevant stakeholders come together to identify areas for improvement and develop a new process design that supports the strategic goals of the organization. We'll present the various aspects of the process workshop to provide you with a comprehensive understanding of how to design and conduct an effective workshop.

The process workshop is an essential part of process optimization, as it provides a platform for collaboration and the exchange of knowledge and experience between the various stakeholders. Preparation is therefore critical to success and involves several aspects. These include the selection of participants, where care should be taken to ensure that all relevant stakeholders affected by the processes are represented. Prior to the workshop, objectives should also be clearly defined to ensure that all participants know what is to be achieved. For example, these objectives may be to gain a comprehensive understanding of business requirements, identify areas for improvement, develop solutions, or prioritize the implementation of process changes. The workshop ensures that the relevant stakeholders, such as process owners, process experts, IT experts, and representatives from various departments, work together to develop the target process design. By weighing different options, a process design is created that both meets the strategic goals of the company and takes into account the needs of the various stakeholders. The workshop also enables potential challenges or risks associated with the proposed changes to be identified at an early stage and appropriate mitigation measures to be planned.

To make the workshop effective and to make informed decisions, it's important that the participants have the right input. Information about the company's current processes and workflows, the results of the previously conducted process analysis, and other sources of information, such as customer feedback, employee surveys, or industry standards, can serve as the basis for the discussion. Reference materials, such as SAP best practice processes or other (industry-specific) reference processes from partners, can serve as a guide for designing processes and help identify best practices.

Next, we consider the conduct of the process workshop itself. During the workshop, it's important to create an open and collaborative atmosphere in which all participants can freely express their opinions, ideas, and concerns. The facilitation of the workshop should aim to guide the discussions to ensure that all relevant issues are addressed and that the established objectives are achieved. Before the workshop is actually conducted, the roles and responsibilities of the participants should be clearly defined. Each participant should know their specific role in the workshop to effectively contribute to the achievement of the workshop objectives. Roles can include process owners, business and IT experts, and managers. Responsibilities vary depending on the role and may include identifying improvement potential, developing solution approaches, or approving the target process design.

In the first phase of the workshop, the results of the process analysis are presented and discussed (see Figure 12.12). This gives the participants the opportunity to develop a common understanding of the current situation and the identified potential for improvement. Subsequently, solution approaches and ideas for the target process design can be developed and discussed. Consideration should be given to the different segmentation categories of the processes to ensure that the right approaches and reference materials are used.

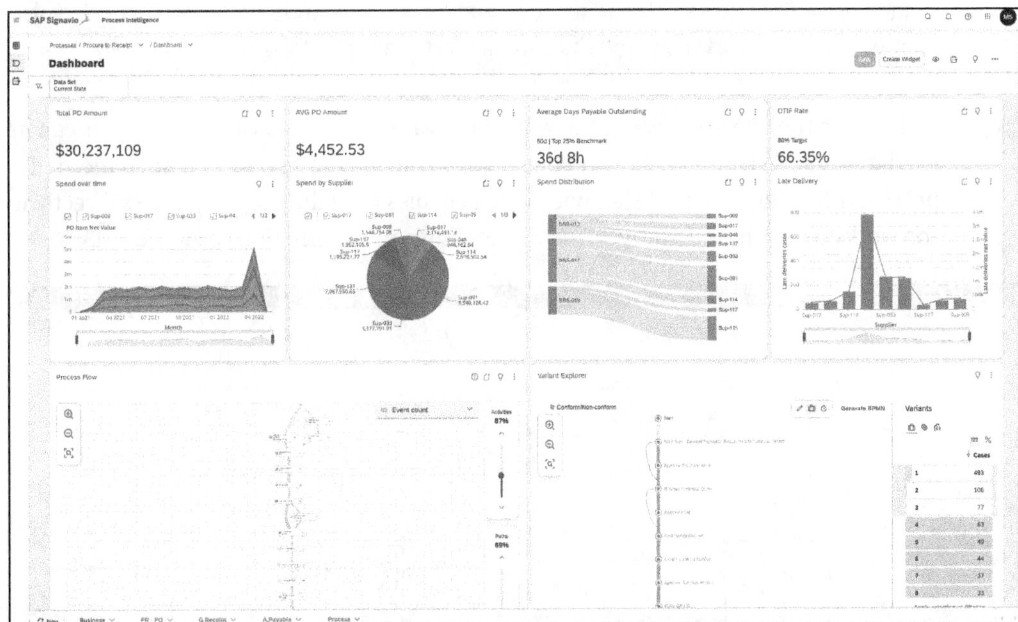

Figure 12.12 Process Analysis and Optimization Potentials in SAP Signavio Process Intelligence

Defining Your Own Subsets for Notations in SAP Signavio

Defining custom subsets for notations in SAP Signavio is a powerful feature that helps modelers focus on specific aspects of their process landscape. By customizing the available diagram elements, they can focus their process models on the most relevant details and thus improve the clarity and readability of the models.

A use case for creating custom subsets could be to distinguish between different levels of process modeling. For example, end-to-end process diagrams that provide an overview of process sequences might require only basic elements such as start and end events, tasks, collapsed subprocesses, pools/lanes, and connectors. Modeling complete processes or technical process steps might require a richer set of BPMN elements to represent technical details and exception behavior.

With SAP Signavio, administrators can easily define and manage these subsets for notations in SAP Signavio Process Manager. The process starts with the selection of the modeling language under **Setup • Define Notations/Attributes**. Here, a new subset can be added by clicking on **Add Subset.** After entering a name for the subset, you can select or remove the desired diagram elements. By default, all elements for the new subset are selected.

You can also copy and edit existing subsets, which allows for quick and efficient adaptation of modeling options to different modeling scenarios.

Note that defining subsets for notations is a privileged operation that requires an administrator account. This ensures that notation standards remain consistent throughout the workspace and that changes are made carefully and deliberately.

To increase the effectiveness of the workshop, various methods and techniques can be used to support the participants in developing the process design. One such method is process modeling with SAP Signavio, which enables modeling of the process directly in the workshop with its intuitive process modeling environment (see Figure 12.13).

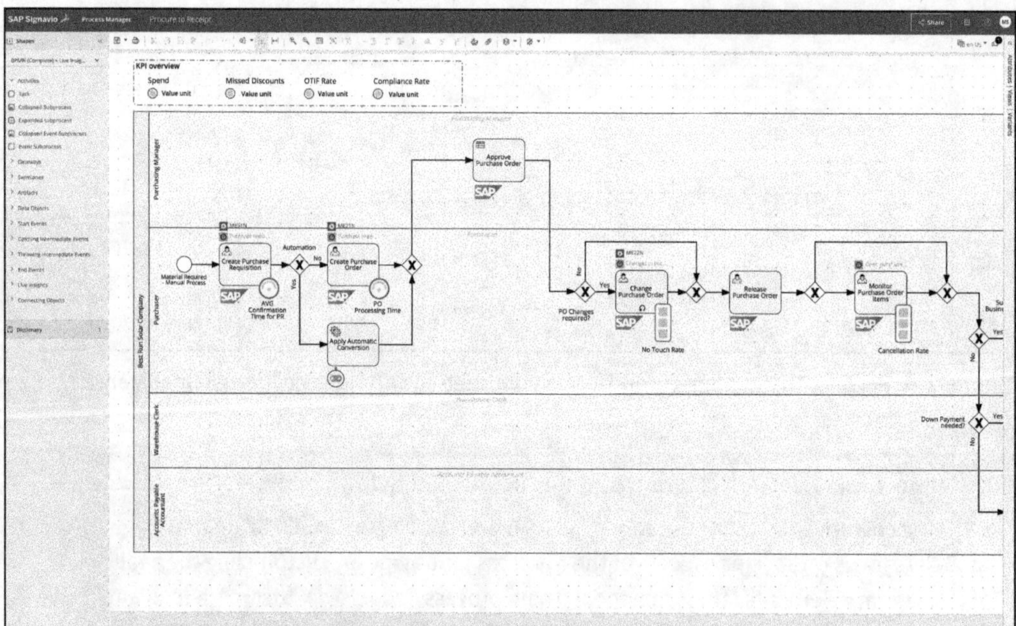

Figure 12.13 Process Modeling in SAP Signavio Process Editor

In practice, process models are usually rudimentarily captured or defined in the workshop and later prepared accordingly. As described in the previous box, SAP Signavio offers the possibility to limit the objects for modeling the processes so that only certain objects are used across all models and areas, and the same facts aren't mapped with different objects.

When revising the process models after the workshop, further comments and insights from the workshop can also be incorporated. Modeling will usually accompany the workshop in combination with other techniques such as brainstorming sessions or group discussions. However, the selection of appropriate methods and techniques depends on the specific requirements and objectives of the workshop.

[«]

Modeling Conventions in SAP Signavio

The modeling conventions in SAP Signavio play a crucial role in the quality assurance and consistency of process modeling. They ensure that all process diagrams follow a consistent format and adhere to certain rules regarding BPMN, naming, process structure, and diagram layout. This is particularly important as it ensures the comprehensibility and comparability of process models across different departments and projects. Likewise, it facilitates communication and collaboration between the various stakeholders.

SAP Signavio therefore offers a function for managing existing modeling conventions. This allows you to view the active and inactive conventions and to adjust them if necessary. This can be done on a global level for the entire workspace as well as specifically for certain functions, such as the **Check** button in the toolbar and the automatic check in the save dialog.

In addition, you can define custom modeling conventions and rules. This allows you to address specific business or project requirements and reflect them in the modeling conventions. New modeling conventions can be created or an existing convention can be copied and adapted. This allows a high degree of flexibility in the design of the modeling conventions.

Another aspect of modeling conventions in SAP Signavio are user-defined rules. These can be added to complement existing modeling conventions. These rules have to be checked manually, as they can't be checked automatically by the system. In addition, you can define mandatory attributes. This feature reports if a diagram contains empty mandatory attributes and thus supports the quality assurance of your process models.

Overall, the modeling conventions in SAP Signavio provide a robust and flexible platform to ensure quality and consistency in process modeling. They enable a high degree of adaptability to specific business requirements and help to increase the comprehensibility and comparability of process models.

The process design should also capture business requirements that will later serve as the basis for implementing the processes. These requirements should be clear and concise, describing the desired outcomes, functions, and performance characteristics of the new processes. Business requirements can come from a variety of sources, including customer expectations, regulatory requirements, business goals and strategies, competitive requirements, and market conditions. It's important that these requirements are considered in the process design to ensure that the new processes meet the business requirements and deliver the expected benefits.

In addition, technical requirements can also be taken into account in the process design that will later be of importance in the implementation. These requirements can include, for example, the integration of existing IT systems, the automation of manual process steps, or the use of specific technologies.

It's of great importance that both business and technical requirements are captured and documented in the process design immediately as they arise during discussions. This ensures that the new process design both meets the business requirements and is technically feasible. Requirements capture and documentation can also help ensure that no important requirements are overlooked or forgotten during implementation. In combination with structured requirements management, such as that available in Focused Build for SAP Solution Manager, requirements can also be effectively managed and prioritized later.

After the target process design has been developed, it's important to define the business process value and the KPIs. The *business process value* describes the business benefits to be achieved by implementing the process changes. Identifying the KPIs makes it possible to measure the performance of the new processes and evaluate their success. Defining the business process value and KPIs allows priorities and resources to be established for implementing the process changes and ensures that there is stakeholder acceptance and support for implementing the new process design. At the end of the workshop, the participants should jointly review and discuss the solution approaches developed and the target process design.

After completion of the workshop, the solution approaches and the target process design should first be documented and summarized. Subsequently, the developed process design is shared with relevant stakeholders and decision-makers in a structured feedback and review process.

[»]

Comment Function in SAP Signavio Process Collaboration Hub

The comment function in SAP Signavio is an essential feature for collaboration and communication between different stakeholders during business process transformation. It enables the exchange of feedback, the formulation of questions, the conduct of discussions, and the clarification of issues in the direct context of the process models (see Figure 12.14). This promotes transparency and accelerates decision-making processes because discussions are held directly where the relevant information is available.

For example, a typical use case might be a team working on a complex process model. A team member could add a comment to a specific element of the diagram to ask a question or address a problem. Other team members could then respond to that comment, sharing their perspectives and suggestions and leading a discussion that way. You could also mention specific people to draw their attention to the comment. Once a decision has been made, the original comment could be marked as clarified. This documents the discussion and allows for future reference. In addition, deleting comments is possible, but note that this process can't be undone. Finally, SAP Signavio offers a notification function for comments. Users are notified about various activities related to comments, such as adding a comment to one of their diagrams; clarifying, rejecting, or reopening one of their comments; being mentioned in a comment; or replying to one of their comments. This ensures that users are always aware of relevant discussions and developments.

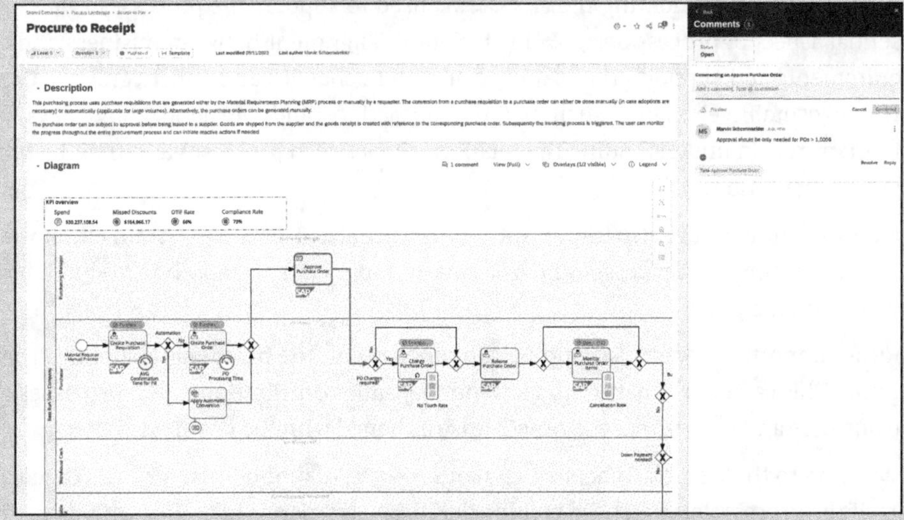

Figure 12.14 Comment Function in SAP Signavio Process Collaboration Hub

In this feedback and review process, stakeholders and decision-makers are asked to provide their feedback and comments on the target process design. In doing so, they can point out potential issues, improvements, or change requirements that may not have been identified or addressed during the workshop. Validation of the target process design should also consider compliance requirements and regulations. After feedback has been collected and evaluated, the target process design should be adjusted and updated accordingly. In doing so, it's important to find a balance between the different requirements and priorities of the stakeholders and to ensure that the revised process design meets the goals and needs of the company.

As part of the process design and transformation, a large number of *process artifacts* must be defined and maintained. These comprise a wide range of documents and materials that are created, updated, and used throughout the lifecycle of the process.

The core artifacts, which are usually defined and assigned during a process workshop, include process models, requirements documents, process descriptions, and roles and responsibilities. These form the basis for a detailed understanding of the processes, requirements, and responsibilities within the process.

KPI definitions play a central role in measuring and monitoring process performance. They enable performance targets to be set and progress to be tracked to support continuous improvement.

Another important category of process artifacts relates to the way tasks are executed within the process. Classifying tasks according to how they are executed, for example, whether they are automated or manual, helps in understanding the degree of automation and efficiency of processes. This in turn can provide information on targeted improvement measures.

Identifying and documenting the IT systems used to support the processes is also an essential aspect of process design. This facilitates alignment between business and IT requirements and supports the planning of system changes and enhancements. Capturing executable elements, such as SAP transactions and SAP Fiori apps, plays an important role in understanding the relationship between processes and their technical implementation.

Training documents are also important process artifacts. They help to train employees in the use of the new processes and systems and to ensure a smooth introduction of the process changes.

Another important aspect is mapping the processes to the business capabilities they support. This can help to better understand the value contribution of the processes to the business and ensure that the processes are aligned with the business strategy.

In addition to these core artifacts, several other elements should also be considered as part of the process design. These include defining governance roles, documenting the scope of the process and its criticality, identifying and assessing process risks, and developing control measures. Documenting and analyzing business decisions within the process can help improve the decision-making process. Education and training materials, known as *enablement content*, are important to promote understanding and acceptance of the processes among employees and to ensure that they have the necessary skills and knowledge to implement the processes effectively.

Finally, the identification of value drivers and the use of data-driven process analytics. By understanding the key value drivers of a process and using data to analyze process performance, companies can make informed decisions to optimize their processes and achieve their business goals.

It's important to emphasize that the previously mentioned process artifacts should not be viewed in isolation. Rather, they should be viewed as part of a cohesive system designed to deepen understanding of the processes, monitor their performance, and support continuous improvement. Effective maintenance of these process artifacts requires close collaboration between the various stakeholders, including process owners, process experts, IT experts, and process analysts.

In conclusion, the quality of the process design depends significantly on the quality and completeness of the process artifacts. The more complete and accurate these artifacts are, the better companies can understand, monitor, and improve their processes. Therefore, the maintenance of process artifacts should be considered an integral part of process design and transformation.

Once all relevant artifacts for the process have been defined, and a joint decision exists on the future direction of the process, the process must be finally approved in its latest version (see Figure 12.15). The *process approval* is a crucial control and quality assurance feature. It provides the opportunity to review the designed or revised processes and ensure that they meet the requirements of the company and the applicable regulations. The involvement of relevant stakeholders plays a central role in this process. They have the opportunity to contribute their expertise, identify possible weak points, and thus contribute to the continuous improvement of the processes.

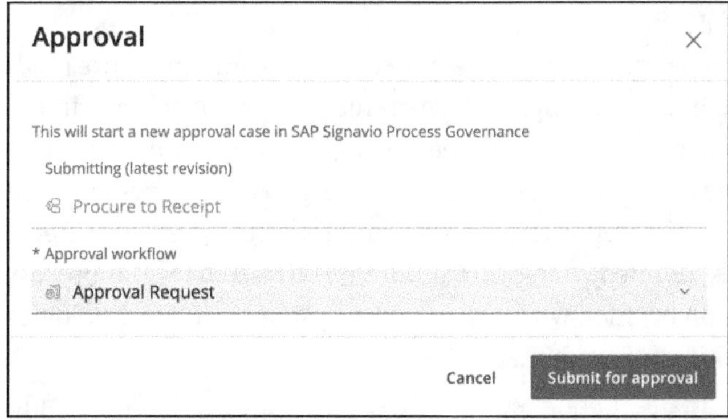

Figure 12.15 Starting an Approval Process from SAP Signavio Process Collaboration Hub

Effective process approval also requires a clear understanding and definition of roles and responsibilities within the approval process. It should be clear who grants approval, who reviews the processes, and who is responsible for carrying out the approved processes. This transparency contributes to the efficiency and effectiveness of the approval process.

Another critical consideration in the approval process is weighing the added value of the new process design against the approximate effort required to implement the requirements. It's important that the benefits of the new or revised process outweigh the resources required to implement it. This isn't just about direct costs. Factors such as the impact on staff workflows, training requirements, and the possible need to adapt IT systems must also be considered.

The added value of the process as well as the approximate effort of the requirements should (as described previously) ideally be defined before approval. This enables an objective assessment and helps to ensure that only processes that make a real contribution to achieving the business objectives are approved. It's important that this assessment is based on solid data and careful analysis to make an informed decision.

SAP Signavio Process Governance provides excellent technical support for the process approval flow. The low-code/no-code platform makes it possible to define individual workflows that support the approval process. The intuitive operation and customizability of the system allows the approval process to be adapted to the specific needs and requirements of the company. For example, approval workflows can be created that send automatic notifications to the appropriate people when a process is due for approval. In addition, if a process is rejected, the system can be configured to automatically send feedback to the responsible party and return the process for revision. This minimizes delays and improves the overall efficiency of the approval process. In addition, SAP Signavio provides the ability to visualize and monitor the approval process. This provides a clear overview of the status of each process and contributes to the transparency and traceability of the approval process. Furthermore, by using the integrated analysis functions, trends can be identified and possible improvements in the approval process can be uncovered.

Finally, the comparison of the newly defined target process with the architecture roadmap plays an important role in process approval. The target process developed in the workshop and refined by the team must now be checked against the company's architecture roadmap. This involves verifying that the process design and requirements are consistent with the roadmap. This step is critical to ensure that the intended systems have been used in the newly defined process. It should be verified that the process design and requirements can be met with the current systems and technologies defined in the enterprise architecture roadmap.

Furthermore, it must be clarified whether the newly defined process requires additional applications or capabilities that weren't previously included in the planning. Such new requirements could have a significant impact on the company's roadmap and budget and must therefore be carefully examined and discussed with the relevant stakeholders. The challenge is to strike a balance between process requirements and architecture planning that allows the benefits of process improvements to be realized without compromising the architecture roadmap. This requires careful consideration and close collaboration between process designers and those responsible for enterprise architecture.

Once the target process design has received the approval of the relevant stakeholders and decision-makers, the transfer of the process design into the solution design can take place. An important part of this phase is the use of the *business process model connector for SAP Signavio solutions* between SAP Signavio Process Manager and SAP Solution Manager 7.2. The connector enables an efficient and seamless transfer of the process designs and models from SAP Signavio to SAP Solution Manager, where the further detailing, requirements, and specific documents are created. From a methodological point of view, the connector takes a central role in the design process. It provides a link between process design and solution design, enabling a seamless integration of the two phases. The connector also enables the import of process diagrams, dictionary entries, and other content types from SAP Solution Manager into SAP Signavio Process Manager, as well as the export of folders, scenarios, processes, tasks, and user-defined attributes from SAP Signavio into SAP Solution Manager (see Figure 12.16).

Note that SAP Signavio is the leading system for processes. After changes have been made and approved in SAP Signavio, these changes are transferred to SAP Solution Manager 7.2 using the connector. It's important to ensure careful coordination between the teams working in SAP Signavio and with SAP Solution Manager 7.2. This ensures that the changes are communicated and implemented correctly, and that the integrity of the solution design is maintained throughout the transfer and implementation.

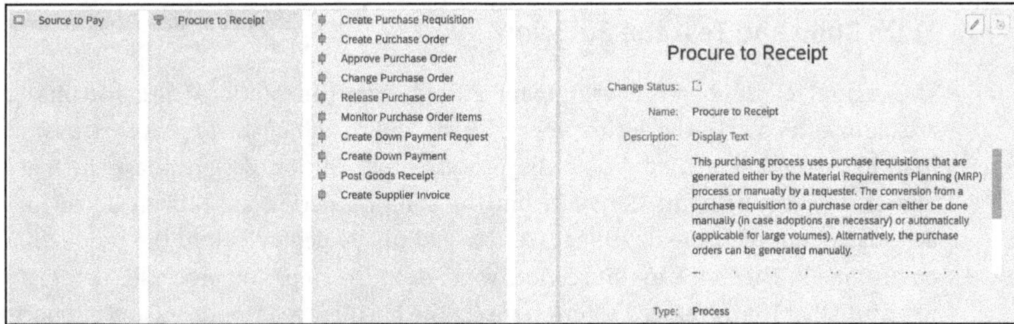

Figure 12.16 Transferred Process Structure in SAP Solution Manager 7.2

The solution design phase is a crucial step in ALM and immediately follows the approval of the target process design by the relevant stakeholders. In this phase, the approved process design is transformed into a detailed solution design that serves as a blueprint for subsequent implementation. This is a critical interface between business process transformation management and ALM, as it's in this phase that the theoretical process models and concepts are translated into concrete technical requirements and design specifications.

In the solution design phase, the process models, requirements documents, and other process artifacts created in the previous phases are thoroughly analyzed and translated into specific technical requirements. This also includes consideration of the IT systems to be used to support the processes and identification of any required adaptations or enhancements to these systems.

Another important aspect of this phase is the creation of design specifications. These documents describe in detail how the processes are to be implemented at the technical level. They also serve as guidance for the implementation team. They contain information about the selected technologies, architectures, and frameworks as well as detailed instructions for programming and configuring the systems. This phase also defines the requirements for data modeling and data migration. The design of the data structures and flows is a critical aspect for effective implementation of the processes. This includes planning how existing data will be integrated into the new systems and processes. The phase ends with the release of the solution design for the implementation phase. Here, the solution design is thoroughly reviewed once again to ensure that it correctly and completely maps the business process requirements and that it's technically feasible.

This phase requires close collaboration between the business departments and the IT teams. Only through a joint effort and a deep understanding of the business processes and the technical capabilities can an effective solution design be created that supports the successful implementation of the business process transformation.

12.5 Build and Test the Solution

This section focuses on the development and testing of the solution. These two phases are an integral part of the entire process of the end-to-end business process transformation method, and they follow the process and solution design phase. In these phases, the requirements derived from the solution design are broken down into actionable units that are developed, tested, and finally deployed into the production environment. Therefore, in this section, we present the steps and processes necessary to ensure that the developed solution meets the business requirements and supports the defined target process. The process of developing and testing a solution requires precise planning, careful implementation, and thorough testing to ensure that the solution is error-free and works as expected.

12.5.1 Issues, Challenges, and Goals

Various issues and challenges arise during the phases of developing and testing a solution as part of an SAP S/4HANA transformation. In the first phase, build solution, the company focuses on the actual development of the solution. Questions may arise, such as these: How can the processes defined in the previous process and solution design be broken down into actionable steps and activities for developing the solution? How can development be efficiently designed to meet the predefined processes and enable seamless integration with existing workflows?

In the second phase, test solution, the focus is on ensuring that the developed solution meets the business requirements and works smoothly. This raises questions such as these: How can the developed processes be mapped completely and correctly in the test phase? What test cases and scenarios need to be developed to ensure that the solution meets the requirements of the predefined processes? How can the test processes be designed so that they correspond as closely as possible to the real production processes and the expected business scenarios?

12.5.2 Procedure

The phase begins with release planning. First, an overview of the requirements to be implemented is given, and a framework for the implementation of the solution is created. Then, the solution architect breaks down the requirements into actionable units ready for development and testing. The actionable units are developed and individually undergo functional testing to ensure that they function correctly and have no bugs. Error corrections are made as needed to ensure the quality of the deployable units. After all deployable units have been developed and tested, an acceptance test is performed to verify the overall functionality of the solution.

After a successful acceptance test, training materials can then already be created that relate to the outcome of individual requirements. Training materials relating to the entire process flow, on the other hand, are only created after successful integration testing (at the beginning of the deploy phase). Once all requirements have been implemented, a process-oriented functional integration test finally follows to ensure that all parts of the solution work together correctly.

This phase is an important step in the overall business process transformation. It ensures that the developed solution meets the business requirements and is technically sound. Careful planning, development, and testing ensure that the solution is effective, efficient, and delivers the intended benefits to the business. In addition, this phase ensures that users are able to make the most of the new solution by creating training content to educate and support users before the process goes live.

This phase is also the point at which the theoretical concepts and plans from the previous phases are put into practice. This transition from theory to practice is a critical step that must be performed carefully to ensure that the developed solution meets the business requirements and delivers the desired process optimizations.

Now, we'll describe the individual steps of this phase in detail. High-level release planning is a key aspect of the implementation phase and plays a crucial role in the success of the project. It involves the strategic planning and coordination of all activities required for the development, testing, and delivery of the solution. In addition, effective release planning enables the management of risks and the efficient use of resources, along with ensuring that project goals are achieved. In this phase, the project lead assumes a central role and is responsible for coordinating and monitoring release planning and ensuring that all project stakeholders understand and fulfill their roles and responsibilities. The project leads ensure that the necessary resources are available and that all activities are in line with the project goals and deadlines.

Focused Build for SAP Solution Manager provides the project lead with powerful features to accomplish these tasks. With Focused Build, the project lead can effectively manage and monitor release planning while ensuring the necessary transparency and control. Focused Build's features help the project lead plan and monitor the implementation of requirements, coordinate development and testing, and plan and monitor solution deployment. For example, Focused Build enables effective planning and monitoring of sprints and makes it possible to track progress in real time.

Focused Build for SAP Solution Manager 7.2

SAP Solution Manager 7.2 is a comprehensive application management and administration tool that helps companies manage their SAP and non-SAP applications throughout the entire application lifecycle. As a central platform, SAP Solution Manager provides functions for project management, test management, change and release management, incident management, and many other areas.

A special feature of SAP Solution Manager 7.2 is its strong integration into the system landscape. This enables a close link between business processes and IT infrastructure, helping companies to efficiently design and optimize their business processes.

In addition to the standard functionality, SAP Solution Manager 7.2 offers *Focused Build for SAP Solution Manager*, a dedicated solution for system implementations. Focused Build provides a standardized and highly automated approach to all aspects of implementation, including project management, requirements management, development management, test management, and deployment management. With Focused Build, organizations can efficiently and effectively manage their implementation projects and bring them to a successful conclusion.

Following the planning, which is carried out by the project management, the solution architect enters the scene. Solution architects act as key intermediaries between business requirements and technical implementation. The role is responsible for understanding the business requirements defined in the design phase and translating those requirements into technical specifications. Individuals in the solution architect role must therefore have a deep understanding of both the business and IT sides and be able to connect the two.

This role has the task of breaking down the requirements into smaller, actionable work packages that can be assigned to and worked on by the developers (see Figure 12.17).

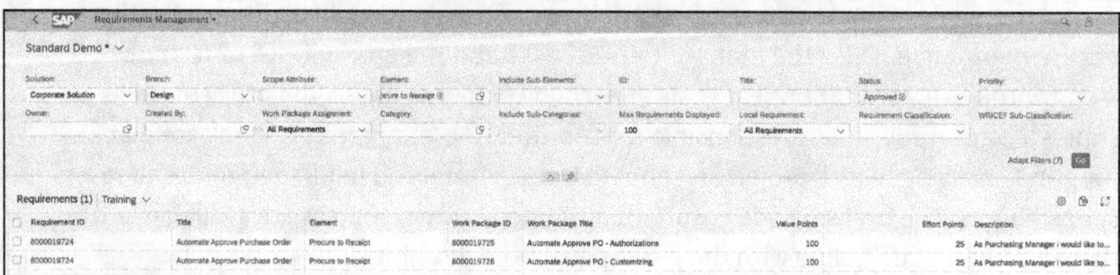

Figure 12.17 Defining and Specifying Requirements with SAP Solution Manager 7.2

Breaking requirements down into smaller units makes it easier to manage and track these units, which in turn leads to more efficient and effective implementation. This step is crucial to ensure that requirements are divided in a way that makes them implementable in the real working environment.

In addition, a technical document is created for each implementable work package. This documentation is important to ensure that developers have a clear understanding of what is expected of them and to enable accurate tracking and verification of the implemented requirements.

Once the requirements have been broken down into actionable units and are adequately documented, the next phase is the development of these units, in which the business requirements are translated into technical solutions. This is where the developers come in, whose main task is to translate the specifications provided by the solution architects into functional code and configurations. Developing actionable units typically involves a number of activities, including writing and testing the code, debugging, and retesting. This process requires a high level of technical expertise. It's important to emphasize that the development of deployable units is an iterative process. This means that developers not only write the code but also continuously test and improve it to ensure that it meets the requirements and works optimally. This process of continuous review and improvement helps ensure the quality of the developed solution and minimizes the risk of errors or problems.

After successful implementation of the individual work packages, the individual function tests follow. This process is a systematic check of each implementable unit to ensure that it functions as intended and meets the specified requirements. Single function testing, also known as *unit testing*, is a type of software testing in which individual components of a system are tested in isolation. The main purpose of these tests is to ensure that each component or unit functions correctly. Unit tests play a crucial role in maintaining code quality and minimizing bugs or defects in the software. When bugs or defects are discovered during unit functional testing, it's the developers' responsibility to fix them. This process includes identifying the cause of the error, developing a solution, and retesting the component to ensure that the error has been fixed.

The subsequent *acceptance test* ensures that the developed solution is tested at the requirement level to ensure that it meets the defined business requirements and is acceptable for the requirement. Acceptance testing is typically performed by process experts or actual end users, as they have the best knowledge of the business requirements and daily operations. During the acceptance test, the implementable units that were previously successfully developed and tested in the single function tests are retested. The process experts conduct the tests by using the developed solution in a real environment or an environment as similar as possible to the real conditions. The goal is to ensure that the solution meets the business requirements and provides value to the users. If errors or defects are identified during acceptance testing, they are reported back to the development team. The developers fix the defects, and the deployable units are retested to ensure that the fixes were successful, and the solution now meets the requirements.

Once the acceptance test has been successfully completed, the first training documents can be can be created at the level of the implemented requirements. Training content is critical to ensure that end users can effectively use the new solution. This content is created to provide users with the knowledge and skills they need to successfully use the solution in their daily workflows. For example, SAP Enable Now can be used to create and deliver training content, which also offers integration with SAP Signavio, and can link training materials directly to process steps or processes.

After the acceptance tests have been successfully performed at the requirements level, the next crucial step is the functional integration test at the process level. This test is usually performed by the key user and ensures that all implemented requirements function correctly not only individually but also in their entirety and in their interaction. In addition, the test can already provide information about whether the developed solution meets the requirements of the real business environment and whether it's capable of delivering the expected benefits.

Creating test cases for integration testing at the process level is a challenging task that requires a deep understanding and precise knowledge of the business process (see Figure 12.18).

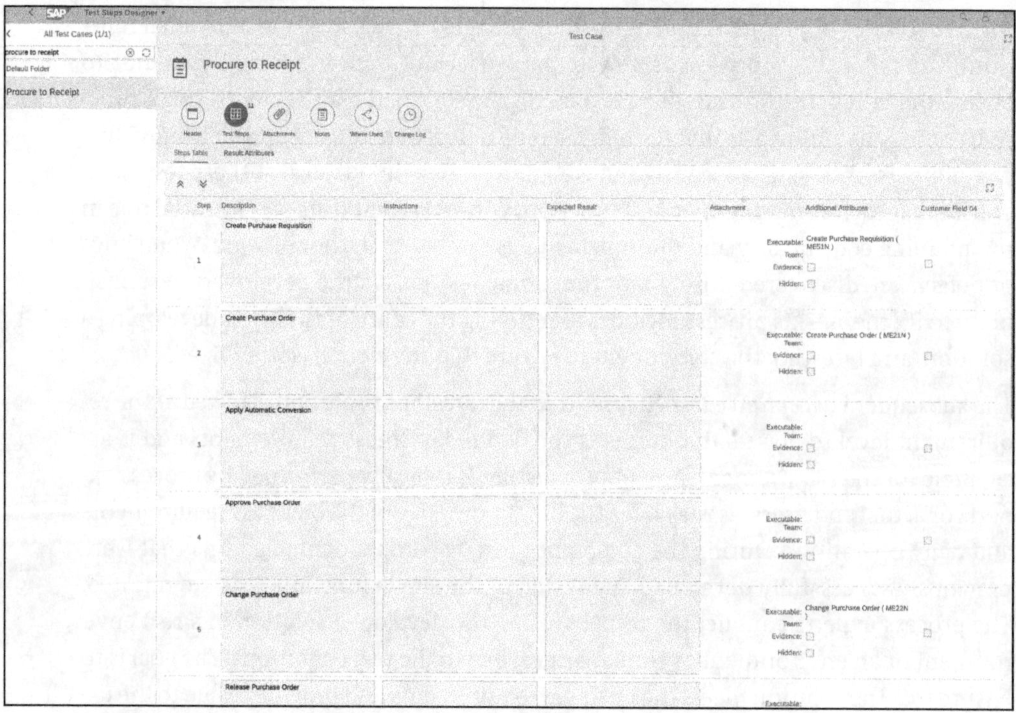

Figure 12.18 Creating a Test Case Based on the Defined Process Flow in SAP Solution Manager 7.2

An effective approach is to use the process design as a starting point and break it down into its individual steps. For each of these steps, you should first consider which functions and components of the solution are needed. Based on this, you can then develop test cases that verify that these functions and components work correctly.

This approach allows you to approach test case creation in a systematic and structured way. By breaking down the process into its individual steps, you can ensure that all relevant aspects of the process are covered in the tests. At the same time, it allows you to focus on the specific requirements of each step and ensure that the tests accurately reflect these requirements.

To move from the business process to the test cases, it's helpful to think of the process as a sequence of events that cause specific state changes in the solution. Each test case should then be designed to verify that the expected state changes actually occur when the corresponding event occurs. In this context, it's important to also consider possible unexpected or undesired state changes and verify that the solution handles them correctly. However, it's important to note that not all sequence flows in the process design can be useful in combination. Therefore, deriving test cases from the process design always requires some manual adjustment and review. In addition, you should always check which steps of the process are system-based and can therefore be tested. Nevertheless, the process design can provide a valuable basis for creating the tests and can speed up this process considerably.

The build and test phase is where the carefully crafted and detailed requirements and design elements finally transform into a working solution. Every step in this process—from release planning, requirements breakdown, and development and testing of actionable units, to acceptance and integration testing—helps ensure the quality of the solution and supports successful business process transformation. Although this process may seem technical and structured, it relies heavily on the collaboration and commitment of all stakeholders. Effective communication between the business and IT teams, deep understanding of the business processes and technical capabilities, and continuous review and improvement are essential elements that determine the success of this phase.

With the integrated tools, such as SAP Signavio and SAP Solution Manager 7.2, and a careful and methodical approach, it's possible to ensure that the solution meets the business requirements and works properly from a technical point of view. At the end of this process, the solution is released, paving the way for the next step in the business process transformation: training the process and deploying the solution.

SAP Cloud ALM

With the goal of further improving the process and application landscape for enterprises, SAP has developed a future-oriented strategy that includes the transition from SAP Solution Manager 7.2 to *SAP Cloud ALM*. This transition offers companies new opportunities to optimize their business processes and benefit from the advanced features of the cloud-based application. An important aspect of this transformation is that SAP Signavio will be further integrated with SAP Cloud ALM.

SAP Solution Manager 7.2 has long been a central application for ALM in SAP environments. It offers comprehensive functions for managing implementation projects, monitoring system landscapes, troubleshooting, test management, and much more. With the introduction of SAP Cloud ALM, the focus is shifted to the cloud-based solution to provide customers with more modern features and an improved user experience.

SAP Cloud ALM offers companies the ability to manage and monitor their applications in the cloud. It enables centralized management of projects, monitoring of systems, fault management, test automation, and comprehensive analytics. With an intuitive user interface (UI) and seamless integration with other SAP cloud services, SAP Cloud ALM provides a modern and user-friendly solution for ALM.

An important part of the long-term strategy is the interplay of SAP Signavio and SAP Cloud ALM. The interplay enables companies to make their business processes even more efficient and create synergies between process and application management. By integrating SAP Signavio with SAP Cloud ALM, companies will gain a comprehensive end-to-end view of their business processes. They will be able to analyze and optimize existing processes and model new ones to increase the efficiency and agility of their organization.

12.6 Deploy the Solution and Enable the Process

Business process transformation is a dynamic journey that goes far beyond implementing a solution. It's not just about deploying a technology but also about how that technology is used to make business processes more efficient and effective. An important part of this journey is implementing the developed and tested solution into the production environment and activating the process. This is what this section will be about as we describe the final phase of the end-to-end business process transformation methodology: deploy and enable.

In the world of SAP S/4HANA implementations, solution deployment is a critical step. This is not only a technical task, but it also requires a methodical approach to ensure that the solution integrates smoothly with the existing IT landscape and delivers the desired business outcomes. But the journey doesn't end with implementing the solution; an equally important part is rolling out the new process and way of working. This is about ensuring that the people using the solution have the necessary skills and knowledge to get the most out of it. In this section, we address the challenges and issues that arise when deploying the solution and activating the process. We also explain how we address these challenges and the methods and tools we use to do so.

12.6.1 Issues, Challenges, and Goals

Deploying a solution raises several questions: How can we ensure that the solution is transferred smoothly from the test environment to the production environment? How do we ensure that data integrity is maintained during the process? How do we minimize potential disruptions to business operations? How do we deal with any issues that may arise during deployment? Creating content for process enablement also presents challenges: How do we ensure that the content created is relevant and effective? How do we adapt the content to the different learning styles of the users? How do we ensure that content stays current as processes change? How can SAP Enable Now help us create and manage this content?

Ensuring transparency and training in the new process are also challenges that need to be overcome: How do we make the new process understandable for users? How do we overcome resistance to change? How can we ensure that the training is effective and that users are using the new processes correctly? How can we measure the success of the training and identify opportunities for improvement? Each of these questions requires careful consideration and planning. In the next section, we discuss the approaches we take to answer these questions and address these challenges.

12.6.2 Procedure

The deployment of the solution, or go-live, is the moment of truth in any implementation project. At this point, theory meets practice, and the developed solution must prove itself in the real business world. Given the complexity and importance of this step, a systematic approach is essential.

SAP Solution Manager 7.2 offers a range of functions that support us in this critical step. It's not just a technical tool, but a central platform that enables end-to-end management of applications and facilitates the interaction of the various elements of the project. At the beginning of the deployment, SAP Solution Manager 7.2 coordinates the various aspects of the process. This includes managing technical resources, coordinating the team, and monitoring progress. These functions help keep track of everything and ensure that all elements are on track. In Table 12.2, you can see an overview of the functions.

Challenge	Function in SAP Solution Manager
Transfer of the solution from the test environment to the production environment	SAP Solution Manager 7.2 provides change control management functions that support this process. These functions make it possible to track, manage, and document changes and ensure that the solution is correctly transferred to the production environment.

Table 12.2 SAP Solution Manager Functions for Solution Deployment

Challenge	Function in SAP Solution Manager
Ensuring data integrity	Features such as data consistency management can ensure that data remains consistent and accurate throughout the process.
Minimize disruptions to business operations	SAP Solution Manager 7.2 includes functions for IT service management. These functions enable potential problems to be identified and remedied at an early stage, thus minimizing the impact on business operations.
Collect feedback and make improvements after deployment	Process improvement and quality management functions make it possible to measure the success of the deployment, identify opportunities for improvement, and make the necessary adjustments.

Table 12.2 SAP Solution Manager Functions for Solution Deployment (Cont.)

In summary, SAP Solution Manager 7.2 supports us not only in the technical implementation of the deployment but also in the methodological management of the process. By combining these two aspects, we can ensure that the deployment is successful and achieves the desired business results.

Creation of Process Enablement Content

Business process publishing requires more than just a well-implemented solution. It also requires effective training and support materials that help users understand and effectively use the new processes and technologies. Creating such materials can be challenging, especially given the diversity of learning styles and constant changes in business processes. SAP Enable Now provides extensive opportunities to accomplish this.

SAP Enable Now and the Integration with SAP Signavio

SAP Enable Now is a comprehensive knowledge management and end-user training platform that helps companies effectively prepare their employees for new software solutions or process changes. With SAP Enable Now, companies can quickly and easily create customized learning content and guidance to help users effectively use new solutions. An interesting aspect of SAP Enable Now is its capability to integrate with other SAP products, particularly SAP Signavio. By integrating SAP Enable Now and SAP Signavio, companies can effectively link their business processes and training content.

In SAP Enable Now, users can link training content to the individual process steps or the entire process model. This is done by using user-defined attributes introduced in SAP Signavio. Linking training content allows users to access the relevant training content directly from the business process. The integration of SAP Enable Now and SAP

Signavio thus enables a close link between business processes and training content. This facilitates user training and helps ensure that users can use the new processes and solutions effectively.

Creating process enablement content starts with a clear overview and understanding of the process models created in SAP Signavio. These models serve as the basis for the content you want to create and can be imported directly into SAP Enable Now via the integration between SAP Signavio and SAP Enable Now. They provide the necessary understanding of the processes and tasks in the process and allow us to create specific training content tailored to the respective roles and responsibilities of the users (see Figure 12.19).

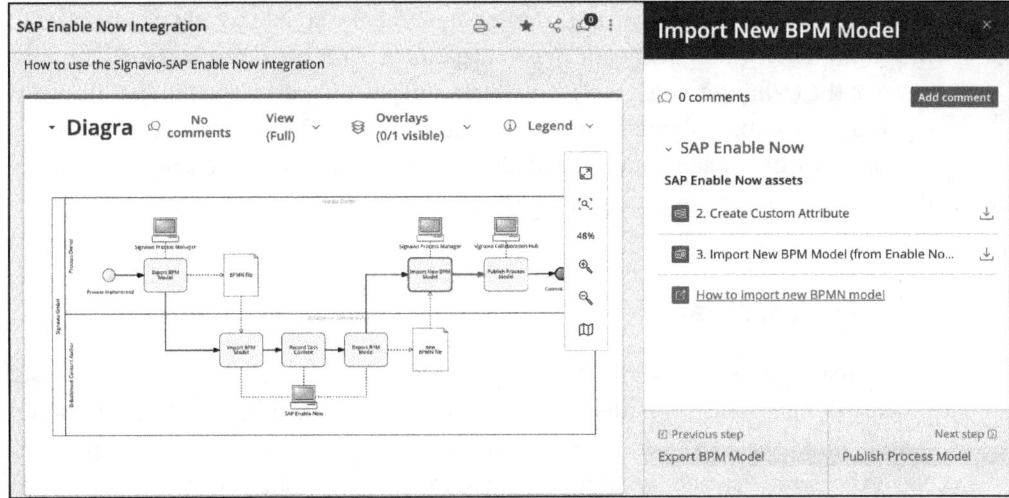

Figure 12.19 Example of Training Content in SAP Enable Now with Process Reference in SAP Signavio

SAP Enable Now allows us to create a variety of content types to meet the different learning styles and needs of users. This can range from written instructions and manuals to interactive tutorials, videos, and webinars. The key is to design content that is relevant to users and easy to understand.

Another important aspect of creating process enablement content is keeping it current. Because business processes can change over time, it's crucial that training content is also kept up to date. SAP Enable Now supports this by enabling efficient updating of content. When a process model is updated in SAP Signavio, you can quickly and easily update the corresponding training content in SAP Enable Now. Finally, it's important that the created content is easily accessible to users. Especially through the integration with SAP Signavio, the created enablement content can be visualized directly on the process models and made available to end users.

Ensure Transparency and Training of the New Process

Transparency and comprehensibility are key components for the successful implementation and use of new processes. They are critical to ensuring that users understand, accept, and effectively use the new processes. Therefore, it's important to ensure clear communication and training. The process models we created in SAP Signavio and the enablement content in SAP Enable Now are the basis for transparency. They ensure that users have a clear understanding of how the new processes work and what is expected of them. They provide a reference they can fall back on when they have questions or uncertainties.

User training is another crucial factor. This is not only about showing users how to execute the new processes but also about teaching them the why behind the processes. This helps them understand the benefits and purpose of the new processes, which promotes adoption and effective use. Training should be offered in a variety of formats to meet different learning styles and needs. Regardless of the medium chosen, it's important that the training content is clear, understandable, and tailored to the specific needs of the users. Training should also be an ongoing process. This means that it takes place not only during implementation, but also afterwards. This helps to keep users up-to-date and ensure that they can still use the processes effectively when changes occur.

12.7 Practical Examples

SAP Signavio is not only an application suite with solutions for process analysis, process design, and process automation, but it also offers a holistic transformation methodology that can be effectively applied to SAP S/4HANA transformation projects. Nevertheless, the initial situation is different for every company; every transformation program is characterized by different influencing factors and is therefore always set up differently. Some of the key influencing factors are as follows:

- **Size and complexity of the company**
  One important factor is the size and the complexity of a company. A medium-sized company usually has shorter decision-making paths and less coordination effort to establish process management in the organization than a large international corporation. This also has an impact on which roles are required and how these are defined in concrete terms.
- **Maturity level of the organization to date**
  Some companies have already established process management but haven't yet combined the topics of process analysis, process management, and process improvement. Other companies have already established enterprise architecture methodologies and built capability maps from an enterprise architecture perspective but see processes as merely realizing level 4/level 5/level 6 capabilities. Still

other companies have only various documentation from previous IT projects, but neither a consistent definition of end-to-end processes nor roles and responsibilities that address continuous process improvement.

- **Approach to the SAP S/4HANA transformation**
  As already described in Section 12.1, there are various forms of transformation to SAP S/4HANA. The process-oriented transformation, with its focus on rethinking and redefining processes in general, differs in key respects from the solution-oriented transformation, with its focus on migrating technically to the new SAP S/4HANA platform. In the context of a new SAP S/4HANA implementation or a system consolidation, there is no other option at all than to rethink the processes and take a process-oriented approach. In contrast, a customer operating an SAP ERP system that is only a few years old and can still be easily maintained is more likely to take the path of a solution-oriented transformation.
- **Focus on added value versus technical innovation**
  In some companies, a business case is required for every initiative, including an SAP S/4HANA project. In addition, once defined added values are continuously tracked, it must always be possible to see whether initially defined added values have also been achieved or at what point the goals weren't reached. Other companies see the transformation to SAP S/4HANA more as an investment in being technologically up to date in the future. They want a lean, manageable ERP system so that they can continue to introduce and use innovations at any time in the future. Such goals are more difficult to calculate in a business case.

In this section, we describe how SAP Signavio has supported various transformation projects in the past and will continue to do so in the future. The companies mentioned are fictional, but the examples are very closely based on real practical examples. We compare three different case studies:

- **Company A: Process-oriented transformation with a focus on process redesign**
  In this example, we focus on the methodology of a process-oriented transformation. We describe in detail how the company defined its transformation targets, how the transformation approach has been derived, and how journey models provided an additional view on the previously setup process hierarchy. All findings are described in the context of the actual implementation project.
- **Company B: Process-oriented transformation with a focus on the realization of initially defined added values**
  In this example, the company starts by identifying added values and calculating a business case, taking into account PPIs measured directly in the system. The analysis results are then used in further steps to find the root cause for the poor process performance and to derive improvement opportunities. Identified added value is realized, and users in the company are enabled to work with the new processes.

- **Company C: Solution-oriented transformation with realization of quick results and changes in selected areas.**
  The company essentially plans to adopt existing processes but would still like to realize selected improvement opportunities in selected areas as part of the transformation project.

12.7.1 Company A: Process-Oriented Transformation Focused on Process Redesign

Company A is a global leader in sports and leisure apparel, footwear, and accessories with strong roots in Europe. With a history that goes back more than seven decades, the company has built a remarkable presence in various sports and lifestyles, always with an emphasis on innovation and design. Company A's products are known for their quality and variety. They include everything from high-performance gear for professional athletes to fashionable clothing and footwear for everyday use. At the same time, the company has always placed great emphasis on sustainability and is committed to minimizing the environmental impact of its products and production processes.

However, Company A faces several challenges. One key aspect is the pursuit of an omni-channel model. This model intends to provide a seamless CX across all touchpoints, whether online, in stores, or via mobile apps. The goal is to provide customers with a unified journey, regardless of how and where they interact with the company. In the modern customer landscape, this is a huge challenge. Customers expect a personalized, convenient, and cohesive experience regardless of where they shop. This requires a significant investment in technology and data management to integrate and leverage customer data across all channels. In addition, internal processes must be adapted to ensure that staff can provide consistent information and service across all channels. The omni-channel model also requires deep knowledge of customer preferences and buying behavior. Company A must be able to provide its customers with what they want, when they want it, and how they want it. This requires sophisticated segmentation and personalization to offer relevant products and services.

In addition, Company A needs to adapt its physical infrastructure. Traditional stores may need to be redesigned or modernized to harmonize with digital channels and provide a seamless experience. This can incur significant costs and requires strategic planning and execution. Another critical aspect is logistics. To provide a seamless omnichannel experience, Company A must be able to ensure fast and reliable deliveries across all channels. This requires a robust and flexible supply chain that can adapt to changing conditions and requirements. In summary, Company A faces the challenge of transforming its customer relationship through the omni-channel model.

In this section, we'll look at the concrete activities and results of their transformation at Company A. The company is still in the middle of their transformation, so some phases have already been completed, and other phases are still ongoing. There, we provide a corresponding outlook. The sequence of the different phases isn't always sequential, so some phases have been grouped together. Short-term improvements or quick fixes weren't made at Company A, so the enhance process phase isn't found in the further description.

Analyze Process, Process Design, and Solution Design

Defining goals was a critical step in Company A's business transformation and an essential part of the analyze process phase. It provided clear direction and helped align all stakeholders on the same path. In defining the goals for SAP S/4HANA implementation, Company A followed these steps:

1. **Understand business strategy.**
   Before defining the transformation objectives, Company A analyzed its overall business strategy in detail. This provided a framework for the defined objectives. Company A has a strategy that aims to improve the CX, increase efficiency, and drive innovation.
2. **Obtain stakeholder engagement.**
   Stakeholder engagement was another important step. This involved workshops, interviews, and surveys to understand the perspectives and requirements of the various stakeholders. At Company A, for example, these included salespeople, marketing teams, IT people, and, of course, the customers themselves.
3. **Define goals.**
   Based on the understanding of the business strategy and stakeholder requirements, Company A defined its transformation goals. These were specific, measurable, achievable, relevant, and time-bound (SMART) to maximize their effectiveness. Examples of goals defined by Company A include improving customer satisfaction by 20% in the first year after implementing SAP S/4HANA, reducing delivery times by 15%, and increasing the innovation rate by 10%.
4. **Communicate and align goals.**
   Once the goals were defined, Company A communicated them and ensured that all stakeholders understood and were aligned. This was done through regular updates, meetings, and training.
5. **Monitor and adjust the targets.**
   Finally, Company A regularly reviewed the goals and adjusted them as needed. This ensured that they remained relevant and aligned with the business strategy.

Through this structured approach to goal definition, Company A has ensured that its SAP S/4HANA implementation is effectively contributing to the achievement of its business goals.

Company A chose an innovative approach by combining the benefits of process modeling and customer data analytics. Prior to the new implementation of SAP S/4HANA, the company built a customer journey in SAP Signavio. Company A's customer journey identified several key touchpoints with customers. These include the website, mobile apps, physical stores, customer service channels, and social media. Each of these touchpoints was analyzed and linked to the corresponding internal processes. For example, customer interaction was linked to website processes such as inventory management, order fulfillment, and customer care.

SAP Signavio enabled Company A to visualize and analyze the processes behind these touchpoints. This helped the company identify bottlenecks and opportunities to improve process efficiency and effectiveness. SAP Signavio also enabled the integration of customer sentiment data, which gave the company a better understanding of how customers perceived interactions at various touchpoints. This understanding helped Company A prioritize its internal processes and redesign relevant processes to provide a better CX. One example of a touchpoint Company A identified was post-purchase customer service. Analysis of sentiment data revealed that customers often had difficulty finding the information they needed or resolving issues quickly. As a result, the company prioritized its customer service process for redesign as part of the SAP S/4HANA implementation.

Another process area that the sentiment analysis was designed to improve was logistics. The analysis of customer touchpoints showed that delivery delays and problems led to a negative CX. Accordingly, Company A used these findings to prioritize its supply chain processes for redesign as well, which should result in faster deliveries and fewer problems.

Company A, after analyzing the customer journey, conducted a comprehensive process scoping exercise to determine the scope for SAP S/4HANA implementation. This important phase began with an in-depth examination of end-to-end processes, which were then in turn broken down into modular subprocesses. This allowed for a granular look at each process and how they fit into the big picture.

Following scoping, Company A decided to segment its processes as well. Segmenting the processes after scoping was crucial to determine which processes should be rethought or, on the other hand, standardized. The goal was to find the optimal balance between customized and standardized processes that would provide both operational efficiency and competitive advantage. Company A recognized that not all processes are created equal. Some processes offer unique competitive advantages and contribute significantly to differentiation in the marketplace. These processes require a tailored approach to ensure that they meet their specific needs and align with their strategic objectives.

At the same time, there are processes that, although important to the operation of the business, don't provide a clear competitive advantage. These processes can be standardized by leveraging the best practices and capabilities of SAP S/4HANA. This leads to improved efficiencies, reduced costs, and a focus on the company's core competencies.

Process segmentation also helped Company A to use its resources efficiently. By clearly identifying which processes needed to be rethought and which processes could go to the standard of SAP S/4HANA, the company was able to focus its investments and efforts on the areas that would provide the most benefit and value.

Here, the segmentation was made into the already-known categories:

- **Key differentiator**
  This category includes the processes that make Company A unique and give it a competitive advantage. They are at the heart of the company and reflect its core values and competencies. These processes require custom design and implementation in SAP S/4HANA to ensure that they deliver the desired results. Examples may include the following process areas:
  - **Product design and development**
    This process is crucial for creating innovative and trendy products that underline Company A's brand value and strengthen its position in the market.
  - **Sustainability initiatives**
    Company A places a high priority on sustainability. The processes related to creating environmentally friendly products and minimizing the environmental impact of its operations can be considered key differentiators.
  - **Customer loyalty and retention programs**
    Customized programs that address customer needs and preferences strengthen the relationship between Company A and its customers and foster brand loyalty.
- **Differentiating process**
  These processes aren't necessarily unique to Company A, but still help differentiate the company. They might include elements that go beyond the industry norm and help the company stand out in the marketplace. These processes may need to be customized to fit the company's specific needs. Examples may include the following process areas:
  - **Omni-channel marketing strategies**
    Company A's ability to deliver coherent and engaging messages across all channels can be considered differentiating.
  - **Personalized online experience**
    Providing a personalized online shopping experience based on individual customer behavior and preferences can also be considered a differentiating process.
  - **Customized product lines**
    The ability to develop and provide customized product lines for specific customer groups or regional markets can also be considered a differentiating process.
- **Harmonized core**
  This category includes processes that are consistent across all of Company A's business units. They are nondifferentiating but important to the efficient operation of

the business. These processes are typically standardized and harmonized to ensure consistency and efficiency. Examples may include the following process areas:

 - **Warehouse management and logistics**
   These processes are essential to the operation of Company A and must be consistent across all business units.
 - **Financial management**
   Accounting processes, financial reporting, and budgeting are other examples of the harmonized core.
 - **Human resources management**
   Processes such as recruitment, staff development, and employee appraisals are also harmonized to ensure a consistent corporate culture and efficiency.

- **Standard process**
  These are processes that are typical for Company A's industry and operations. They don't require any customization and can be taken directly from the standard SAP S/4HANA solution. Examples may include the following process areas:
  - **Procurement**
    Processes for the purchase of materials and services are usually standardized.
  - **IT support**
    Processes for routine support and maintenance of IT systems and infrastructure can largely be taken from the standard.
  - **Legal compliance**
    Processes to comply with industry regulations and legal requirements also usually don't require company-specific customization.

The results of the customer journey analysis informed the prioritization of processes and helped determine which processes should be rethought as part of the SAP S/4HANA implementation. For example, processes that impact CX, such as order processing or customer service, could be considered key differentiators and adapted accordingly.

The process design phase was an essential step in the new implementation of SAP S/4HANA at Company A. In this phase, the processes that were identified as key differentiators and as differentiating were rethought. In contrast, for the harmonization of the core and the standard processes, the SAP standard processes were taken as a starting point, and only necessary adjustments were made.

Redesigning the differentiating processes and key differentiators was a significant step. Company A wanted to ensure that these processes were optimally tailored to the company's specific needs and strategic goals. For example, the product development process, a key differentiator, was completely redesigned to foster innovation and creativity, respond more quickly to market changes, and reduce time to market.

Another example was all the processes surrounding the redesign of the customer loyalty program, a differentiating process. Company A redesigned this process to improve customer engagement, enable personalized offers, and increase customer satisfaction. This process was also rethought to integrate seamlessly with Company A's omnichannel marketing strategies.

In contrast, the SAP standard processes were taken as a starting point for the harmonization and standard processes. These processes are generally not differentiating and don't require a customized solution. For example, the warehouse management process, a harmonized core process, was adapted based on the SAP standard processes to optimize inventory management, reduce delivery time, and improve customer satisfaction.

Similarly, the standard procurement process could have been adapted based on the SAP standard processes to improve efficiency, reduce costs, and optimize the supply chain. Company A made a conscious decision not to analyze the processes running in SAP ERP so as not to carry any "ballast" from the old system. This meant that rather than relying on existing processes or constraints, the company focused on developing optimal processes tailored to its specific needs and goals.

This approach enabled Company A to take full advantage of SAP S/4HANA and develop processes focused on optimizing performance, achieving strategic goals, and improving the CX. It led to better alignment of processes with business strategy, improved efficiency, and a stronger competitive advantage in the marketplace. After rethinking or adapting processes, Company A moved on to the next phases of SAP S/4HANA implementation.

Build, Test, Deploy, and Enable

After the processes have been rethought or adapted, Company A now plans to move on to the next phases of the new SAP S/4HANA implementation. Here, the company would like to consistently orient themselves on the designed processes and build all subsequent steps on the processes:

1. **System configuration and customization**
   Based on the approved processes to be transferred to SAP Solution Manager 7.2 via the business process model connector in the future, SAP S/4HANA is to be configured and customized to support Company A's new processes and requirements. This includes setting up system parameters, integrating third-party tools, and customizing UIs to increase usability.
2. **Data migration**
   In this phase, the plan is to migrate data from the existing SAP ERP system to SAP S/4HANA. This involves cleansing and consolidating data to ensure that it's transferred consistently and correctly to the new system.

3. **Training and change management**
   Company A plans to invest in training employees to familiarize them with the new processes and SAP S/4HANA system. Change management initiatives are planned to promote acceptance of the new processes and systems and to prepare employees for the changes.
4. **Go-live and support**
   Once all preparations have been completed, the company plans to go live with SAP S/4HANA. During this phase, ongoing support and maintenance services are planned to ensure that processes run smoothly and that any problems or challenges can be resolved quickly and efficiently.
5. **Continuous improvement**
   After implementing SAP S/4HANA, Company A intends to focus on continuous improvement to further optimize processes and systems and increase performance, efficiency, and customer satisfaction, including the ongoing use of the customer journey map as well as process analytics capabilities.

With this structured approach, Company A plans to take full advantage of SAP S/4HANA and optimize its processes and systems to achieve its strategic goals and strengthen its position in the market.

[»]

How Did SAP Signavio Help Company A?

SAP Signavio has played a central role as a process management tool at Company A, particularly in the redesign and improvement of business processes as part of the SAP S/4HANA implementation, as well as in the customer journey analysis conducted in advance.

At the beginning, Company A displayed the customer journey map in SAP Signavio and connected the touchpoints with the internal processes there. SAP Signavio Process Manager enabled the company to design and model its processes. In particular, it was beneficial that the SAP Signavio solution enabled a clear and consistent representation of processes, which facilitated effective communication and discussion about processes across departmental boundaries. As part of the process design, SAP Signavio Process Collaboration Hub played an essential role. SAP Signavio Process Collaboration Hub is a platform that facilitates sharing and collaboration between different stakeholders in the company. In the case of Company A, SAP Signavio Process Collaboration Hub enabled open and effective communication between process owners, subject matter experts, and other relevant stakeholders during process redesign. Through SAP Signavio Process Collaboration Hub, the company was able to gather feedback on the rethought processes. This enabled iterative improvement of the processes and ensured that the processes met the requirements and expectations of all stakeholders. The ability to leave comments, ask questions, and make suggestions for improvement helped foster a sense of ownership and participation among employees.

In addition, SAP Signavio supported the establishment of a multistage approval workflow for the release of finalized processes. After a process was rethought and improved through feedback, it was released through a multistep approval workflow. This ensured that each process went through a thorough review and approval before it was implemented. The approval workflow contributed to quality assurance and ensured that the finalized processes met business requirements and stakeholder expectations.

12.7.2 Company B: Process-Oriented Transformation Focused on Realizing Initially Defined Added Values

Company B is a large globally operating company in the automotive industry. The company is an innovation leader and sells the latest technologies, products, and services, tailored to the specific needs of its customers. Nevertheless, the company operates in a very competitive market environment. To be successful, the company must offer the best products, guarantee high delivery reliability in logistics, offer a fair price, and thereby ensure maximum customer satisfaction. All these goals must also be aligned with the objective of sustainability: ecological, social, and financial requirements must always be reconciled without compromising on quality and customer satisfaction. Lean and efficient processes combined with cost reduction in shared service processes are thus key success factors for the company. A business case is always required for decisions on major projects at Company B; a continuous comparison with the performance of its competitors is pursued in the same way as internal benchmarking between different regions and organizational units.

Company B operates in four different regions—Europe, North America, South America, and Asia. In the first three regions, there are different, independently developed SAP ERP systems in use. No common development template has been set up; this is why processes are implemented very differently in each region. In Asia, there is a non-SAP ERP system realizing core business processes that was introduced 18 years ago and has been continuously adapted and further developed to meet customer-specific requirements over time. The company is now planning for their SAP S/4HANA transformation with the goal of global process standardization and harmonization, accompanied by a complete process redesign. A new SAP S/4HANA template is planned to be set up, which will be rolled out to the regions worldwide and thus replace the existing systems.

Even though the target of the transformation project is to rethink current business processes, it's equally important to run a smooth project without taking too many risks. Over many years, some investments have been made in the landscape and countless in-house developments have been realized. Therefore, it's also necessary to create transparency about the process flows in the as-is landscape and to take regional specifics into account.

Analyze

Based on the objectives just described, Company B decided to conduct a detailed process analysis. The three SAP ERP systems were all connected to SAP Signavio Process Insights. In addition, the order-to-cash and procure-to-pay processes were set up in SAP Signavio Process Intelligence. For this purpose, data from the three different SAP ERP systems were loaded and supplemented with data from the other non-SAP ERP system.

Three initiatives were set up for working with the extracted data:

- **Calculation of a business case for the upcoming transformation project**
  Company B follows the top-down/bottom-up approach, which consists of three parts:
  - The first step is to calculate the full value-added potential.
  - In the second step, an SAP S/4HANA business case is calculated.
  - The last step is the bottom-up validation and detailing of the data (as already described in Section 12.3).
- **Detailed process analysis**
  The data from SAP Signavio Process Insights is visualized in SAP Signavio Process Intelligence, leveraging the SAP Signavio plug-and-gain approach. This detailed process analysis is used to identify process variants, execute an internal benchmarking analysis comparing systems and company codes, and identify process inefficiencies in different areas.
- **Further detailed analyses**
  In the course of the project, the data available in SAP Signavio Process Intelligence via the SAP Signavio plug-and-gain approach is extended, and additional detailed process mining analysis is performed to understand current process execution in its entirety, including all company-specific peculiarities.

Benchmarking data from comparable companies was used for an assessment of the overall value-added potential to be achieved. Objective analyses were used to compare the balance sheet of Company B with that of competitors and companies with a comparable business model.

To calculate the full value-added potential, Company B first used the SAP Value Lifecycle Manager tool (Section 12.2.2). SAP Value Lifecycle Manager provides easy access to financial and industry-specific data, including analyses from the service provider *S&P Global Market Intelligence*. This data is used to identify the top-down value-added potential. Financial data and process metrics from comparable companies are aggregated for comparison with the company's own results. A Move the Needle analysis then shows the potential added value that can be created by improving a single metric. For each metric, the company is compared with other relevant companies in the automotive industry. The results of the analysis provide an indication of how realistic the assumed improvement actually is.

Significant potential was identified in the following areas (see Figure 12.20):

- **Operating margin**
  Compared to the competitors analyzed, Company B performs the worst; a one percentage point increase in margin would improve operating profit by approximately ~ 203 M EUR.
- **Sales overheads**
  The share of sales overheads compared to company sales is relatively high. An improvement of one percentage point would improve the operating result by approximately ~29.7 M EUR
- **Retention period of stocks in inventory**
  The average length of time stocks remain in inventory is associated with large amounts of tied up capital. A reduction in the length of time stocks remain in inventory would free up ~57 M EUR of capital.

KPI	Poorest Performance	Performance Company B	Best Performance	Improvement	Potential Added Value (in Millions)
Operative margin (in %)	Company B	4.4 ▼ … 13.1	Competitor A	1%	203.0 M EUR
Sales overhead (as a % of corporate sales)	Competitor B	25.7 … 14.3 ▼ … 7.1	Competitor C	1%	29.7 M EUR
Retention time of items in inventory (in days)	Competitor B	84.1 … 45.5 ▼	Company B	1 day	57.0 M EUR

Figure 12.20 Selected KPIs of the Complete Value-Added Analysis at Company B

In total, across all analyzed KPIs, Company B was able to identify the following potentials in the described Move the Needle analysis:

- Improvement in operating result: ~EUR 500 M
- Improvement in unrestricted capital: ~EUR 100 M

The next step was to understand how much of the identified full value potential could be realized through the implementation of SAP S/4HANA. Data points from other SAP S/4HANA projects served as an indication of what potential could also be realized at Company B. To illustrate the added value of SAP S/4HANA, selected SAP S/4HANA capabilities were considered, and the extent to which Company B's process performance differs from the results of other customers in the automotive industry was compared. Benchmarking data was used again, but, this time, the figures didn't come from the balance sheets of selected companies, but from an internal SAP database.

The key figures from the areas of finance, purchasing, sales and supply chain were used to calculate the business case. For each metric, a comparison was made to see if Company B was in the top 25% of companies, in the bottom 25% of companies, or in between compared to the other companies. If no data was available from Company B for the various metrics, the data was estimated or the average value for the industry was used as the basis for calculating the added value. Based on this, it was possible to determine the extent to which there was potential for improvement.

Figure 12.21 shows the results of the SAP S/4HANA value-added analysis from finance. For various KPIs, the performance of Company B was compared with the best and worst competitors in the automotive industry. Based on this, it was assumed that Company B's performance could be optimized to at least the 25% percentile. The resulting added value is as follows:

- Improvement in operating result: ~EUR 100 M
- Improvement in unrestricted capital: ~EUR 150 M

Key Performance Indicators

KPI Name	Unit	Performance Company B	Best 25%	Average Performance	Poorest 25%
Days sales outstanding	Days	43.6	30	46.2	60
Cost (% revenue): Cost accounting and analysis	Percent	0,051	0,01	0,034	0,064
Cost (% revenue): General accounting and financial statements	Percent	0,058	0,009	0,052	0,075
Cost (% revenue): Accounts receivable	Percent	0,012	0,003	0,041	0,068
Average number of invoices per employee in accounts payable	Documents	11.346	25.000	18.468	8.207
Number of days for the year-end close	Days	18.8	7	13.8	20

Figure 12.21 Selected Key Figures from Financial Accounting as a Result of the SAP S/4HANA Business Case

Figure 12.21 also highlights that a potential improvement in the operating result can be achieved. However, the overall numbers are smaller than before because they are now focusing on the potential value that can be realized with SAP S/4HANA (and not the full value potential anymore). It's interesting to see, however, the high working capital that could be realized. The concrete sum calculated here is even higher than what was assumed in the first step of the business case.

In the next step, the identified added values were validated and further detailed on the basis of PPIs from the system. Business cases had already been calculated at Company B in the past. However, the challenge was always to identify specific areas for action and quantify them in a measurable way. With the SAP S/4HANA project, Company B now chose a systematic data-driven approach that provided a complete picture in terms of potential added value. As a result, the company had a clear basis for argumentation as to how these added values could also be realized in practice.

To measure the PPIs just mentioned, Company B used SAP Signavio Process Insights from the beginning. The three existing SAP ERP systems were connected to SAP Signavio Process Insights and the results were used as an important input to complement the top-down business case with real bottom-up data from actual process execution.

Figure 12.22 shows a value-added calculation based on the following two PPIs:

- **Automation rate for financial accounting documents**
  The PPI measures the percentage of financial documents that were created by a system user, processed via batch input, or triggered automatically by an upstream SAP component within one week.
- **Number of financial postings per week**
  Again, a specific week is used as a reference period to derive an analogy for the entire year.

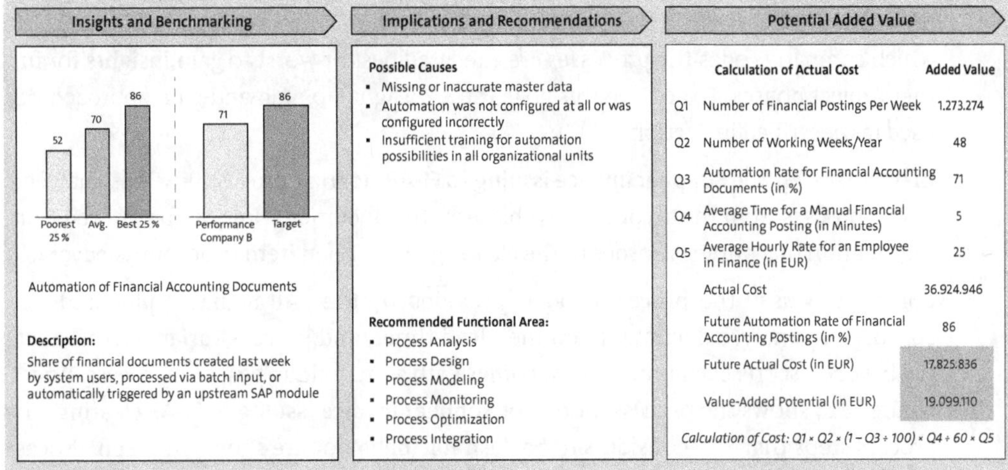

Figure 12.22 Detailed View of a PPI for Automating Financial Accounting Postings

Various assumptions were made for the calculation:

- Number of working weeks per year: 48
- Average time required for a manual posting in financial accounting: 5 minutes
- Average hourly rate for employees in finance: EUR 25

Figure 12.22 shows the current automation rate at Company B (71%) and compares it with the automation rate of other customers in the automotive industry:

- 25% of customers have an automation rate of 52% or worse for creating financial accounting documents.
- 25% of customers manage to automate at least 86% of document creation.
- 50% of customers fall in between, meaning that the automation rate is between 52% and 86%.

Company B is therefore in the midfield compared to other customers in the automotive industry. A realistic assumption now seemed to be that Company B could catch up with the top 25% of customers; that is, the automation rate would have to increase from the current 71% to 86% in the future.

To calculate the potential added value, the actual cost was determined on the basis of the automation rate of 71% and compared with the potential future costs (assuming that 86% of bookings are automated). Using a simple calculation logic, an added value potential of ~EUR 19 M was calculated.

The realization of the identified added values can take place in various ways. In any case, however, it's beneficial to understand the underlying processes and process variants in more detail. We'll focus on this in further sections of the analysis phase.

In preparation for the upcoming SAP S/4HANA transformation, Company B also wanted to understand current process variants in the system in more detail and analyze in which company codes the processes are executed best or worst to gain insights for further project phases. To achieve this goal, the SAP Signavio plug-and-gain approach was used (as described in Section 12.2).

In the first step, the **Supplier invoice issuing to FI-AP clearing** process flow was analyzed. In this process from the accounts payable area, the document flow can be tracked from the creation of a purchase order to the clearing of the open item in accounts payable.

Company B used the process models provided by the SAP Signavio plug-and-gain approach to analyze document volumes, lead times, and process variants to calculate the best-run score. Each process flow comes with a predefined best-run process model; Figure 12.23 shows the process model for **Supplier invoice issuing to FI-AP clearing**. The process steps that SAP considers to be desirable behavior are shown in green. Process steps shown in red represent undesired behavior, and process steps shown in yellow should be evaluated neutrally.

[»]

Product Supplements

Full-size color versions of Figure 12.23, Figure 12.24, and Figure 12.25 are available for download at *https://sap-press.com/5855* under the **Product supplements** section.

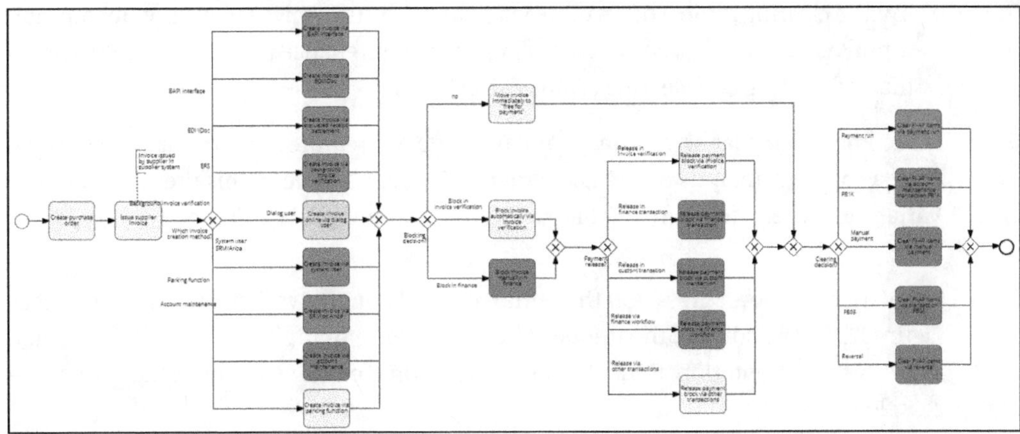

Figure 12.23 Plug-and-Gain Process Model

In a first step, the actual process flow, including typical variants was displayed in a flow diagram (see Figure 12.24). From this diagram, it was possible to deduce the extent to which inefficiencies were holding up the process, at which point manual effort was required, and where process optimization would make sense.

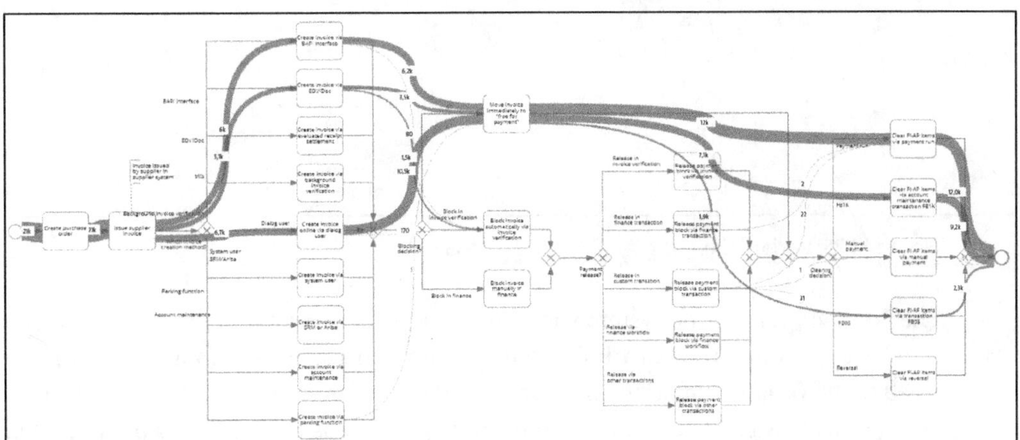

Figure 12.24 Accounts Payable Flowchart

The following observations were made:

- All the cases considered start with the creation of a purchase order.
- After the supplier has issued the invoice, it's posted in Company B's system. This is done partly via Business Application Programming Interfaces (BAPIs), partly via Electronic Data Interchange (EDI)/IDoc, but also very often by manual input of a dialog user (no automation).
- The majority of invoices are approved for payment immediately, although there are some cases where automated invoice verification initially blocks approval.

- Invoice clearing is most often done via a payment run. However, it's not uncommon for invoices to be cleared via manual vendor maintenance (Transaction FB1K) or via manual posting and clearing (Transaction FB05).

It becomes clear that significant manual efforts are required at various points. However, a concrete analysis of a single variant isn't yet possible in this presentation. The Variant Explorer widget in SAP Signavio Process Intelligence could help to go deeper here.

The Variant Explorer gives you the option of displaying all variants transparently (see Figure 12.25). In addition, the frequency of occurrence of individual variants is visualized. This representation allows further conclusions to be drawn about the actual process flow.

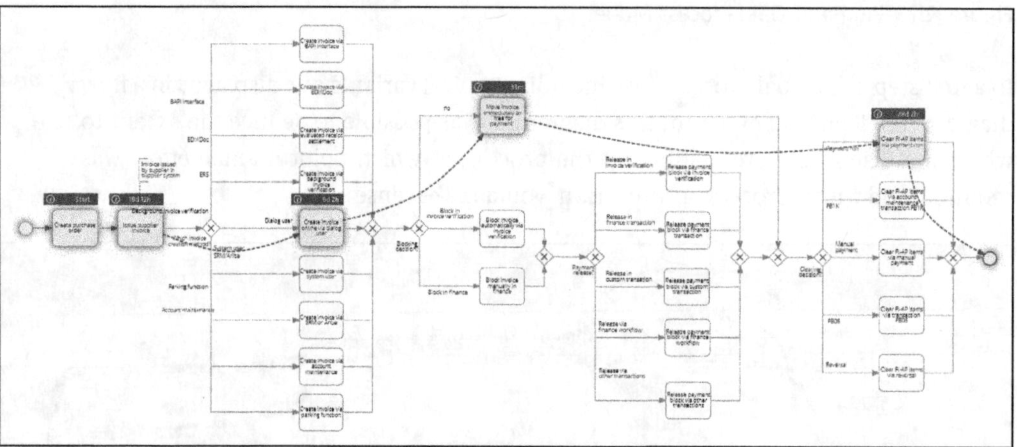

Figure 12.25 Variant Explorer: Analysis of Process Variants

The illustration from the system shows the process of accounts payable (creation of a purchase order up to the clearing of the open item in accounts payable), currently filtered to the variant that occurs most frequently. You can see that, in this case, the purchase order was created manually, but the payment was triggered by an automated payment run. Other variants can be displayed in the system in the same way; Company B has more than 20 different process variants (with different frequencies). A click on the respective process variant enables a detailed comparison.

Now the question arises as to what extent the current realization corresponds to a desired process behavior and how often an undesired behavior occurs. The *Best Run Index* can provide such results (see Figure 12.26):

- The Best Run Index score for the creation of invoices is 100%. This means that invoice creation is already very optimized in the time window under consideration. No manual creation of purchase orders via account maintenance has taken place; the target value of 80% is clearly exceeded.

- For invoice payments, the Best Run Index score is only ~51%. This figure can be explained by the fact that relatively many vendor items were cleared by manual account maintenance with Transaction FB1K. The target value of 80% isn't achieved here, so there is potential for optimization.

Figure 12.26 Calculation of the Best Run Score

The next analysis again focused on understanding the differences between different process execution paths: In which variant and in which organizational unit does the process show no anomalies or the most anomalies? To answer this question, the blockers identified in SAP Signavio Process Insights were counted, broken down by company code, and mapped in SAP Signavio Process Intelligence. If there are no blockers in a company code, the process flow is implemented very well. However, if there are several blockers, this gives an indication of that the process will not be implemented well. Using the context information from SAP Signavio Process Insights thus significantly accelerated the analysis in SAP Signavio Process Intelligence. It was now possible to target the cases that show anomalies.

Figure 12.27 shows that a large number of invoices are processed in company code AB11 without any blockers in the process. This company code can therefore be used as a template for process design in the next steps. Figure 12.28, on the other hand, shows that company code AB03 has many invoices with blockers in the process. This is an example of how the process should not be implemented.

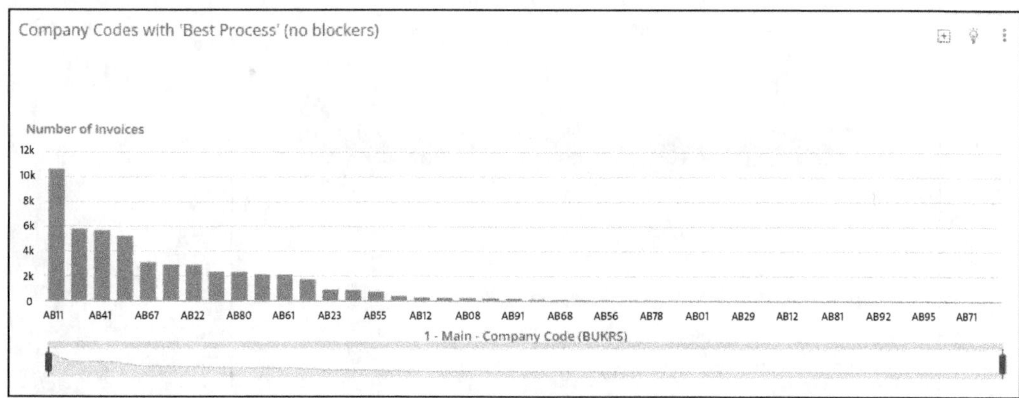

Figure 12.27 Company Codes without Blockers in the Process

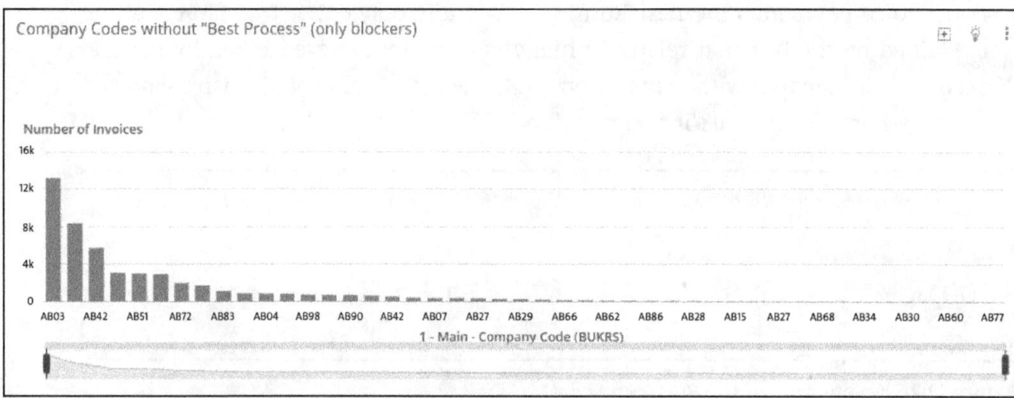

Figure 12.28 Company Codes with Blockers in the Process

In addition to the standard analysis content provided via the SAP Signavio plug-and-gain approach, many other analyses can be executed by leveraging the process mining capabilities of SAP Signavio Process Intelligence. KPIs can be analyzed with regard to various other dimensions: lead times, automations, changes of documents, or delivery reliability can be analyzed fast and tailored a specific situation.

Company B decided to perform more detailed analyses in accounts payable and not only to analyze data from the SAP ERP systems here but also load information from the non-SAP ERP system into SAP Signavio Process Intelligence.

Figure 12.29 shows a section of the dashboard created by Company B in SAP Signavio Process Intelligence, filtered by a region.

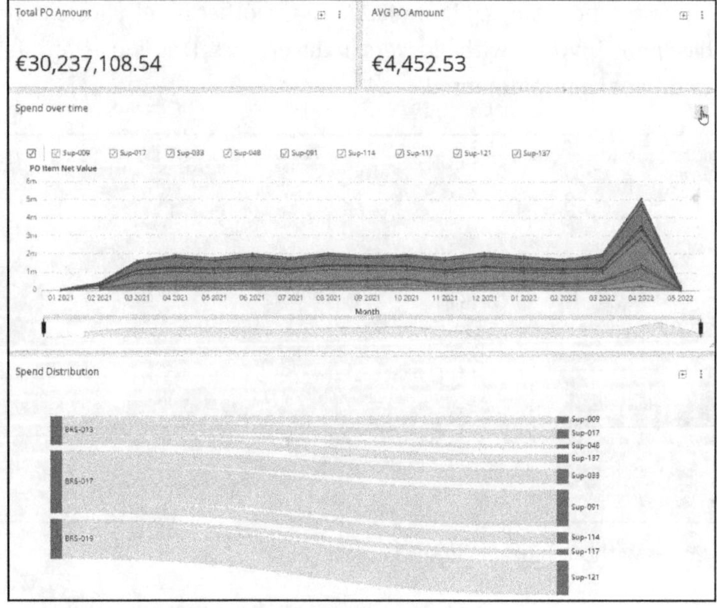

Figure 12.29 Key Figures on Total Expenditure

Here are the following key figures and data:

- First, you can see two key figures on order volumes: the total volume of all orders is ~30 million EUR, and the value of a single order is ~4,500 EUR.
- Furthermore, it visualizes how the volumes of orders placed with different suppliers behave over a certain timeline.
- Lastly, we see another graph that shows which suppliers have the highest order volumes for a selected company code.

Another excerpt from the dashboard created by Company B is shown in Figure 12.30.

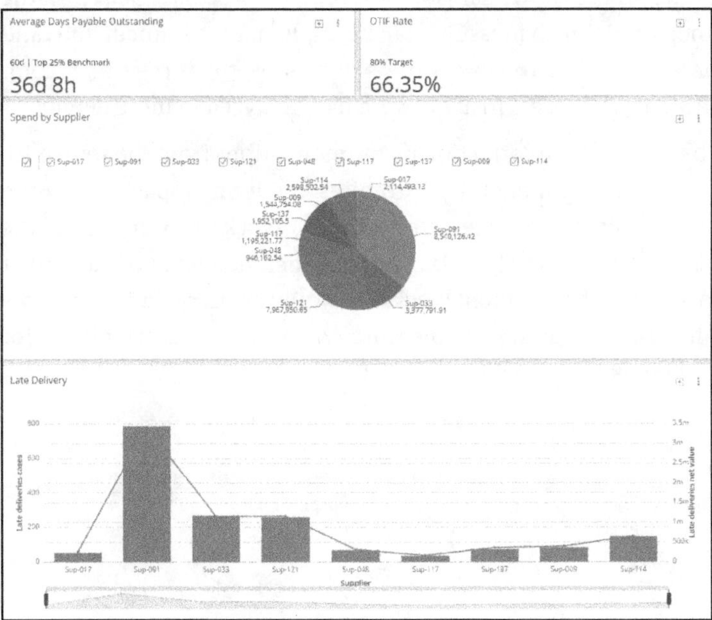

Figure 12.30 Key Figures, Days Outstanding, and Delivery Reliability

The dashboard displays the following:

- The average number of days payable outstanding calculated here as 36 days and 8 hours. Benchmarking data was also added for this key figure: the top 25% of the comparison group have 60 days payable outstanding. This means that Company B has substantial room to improve.
- In the area of delivery reliability, it's evident that Company B is currently not managing to achieve its targets. At 66.35%, delivery reliability is well below the target value of 80%.
- Another graph shows spend per supplier in a pie chart.
- Lastly, a chart was created to analyze which suppliers frequently deliver late and the monetary volumes involved.

The results and dashboards described in this section are only a sample of the analytics set up by Company B. Significant further insights were gained in the areas of order to cash and logistics.

Enhance

During the detailed analysis phase, Company B obtained a very good understanding about current process execution. But some questions came up: How can identified process inefficiencies be improved? Are there on-board tools in SAP ERP or SAP S/4HANA that can be used relatively easily? Are there automations or smart technologies in SAP BTP that can optimize the process flow? Can SAP best practice processes help with the design of the target landscape? Which measures can be implemented immediately, and which are relevant for later phases? To answer these questions, the correction recommendations from SAP Signavio Process Insights were first analyzed in more detail.

Figure 12.31 shows an overview of the correction recommendations from the source-to-pay ordering process. One recommendation promises medium impact with little effort: 5,505 open items (invoices) are subject to a payment block and were due in the past. The recommended action from Figure 12.32 recommends scheduling a regular ABAP report so that invoices with a payment block that would be obsolete from today's perspective are automatically released. This measure can prevent manual efforts for invoice verification and invoice release.

Correction Recommendations (41)

Filter

Finding	Recommendation	No. of Objects Affected	Impact	Effort	Value Driver Affected	Root Cause
20207 open purchase order items wer...	Set delivery completed indicator for purchase order items where goods receipt postings are no longer expected	20.207	■■■	■■□	Reduce Data Management Cost	Old and open transactional d...
12498 purchase order items were foun...	Set final invoice indicator for purchase order items where invoice receipt postings are no longer expected	12.498	■■□	■■□	Reduce Data Management Cost	Old and open transactional d...
10997 open inbound deliveries were fo...	Close inbound deliveries for which goods receipt postings are no longer expected	10.997	■■□	■□□	Reduce Data Management Cost	Old and open transactional d...
8738 purchase order items were found...	Set final invoice indicator for purchase order items where invoice receipt postings are no longer expected	8.738	■■□	■■□	Reduce Data Management Cost	Old and open transactional d...
5505 open items (invoices) blocked for...	Check if payment block can be removed to pay overdue open items	5.505	■■□	■□□	Improve Days Payable Outsta...	Missing end user training
2898 open purchase requisition items ...	Mark purchase requisition items for deletion where conversion into purchase order items is no longer expected	2.898	■■■	■■□	Reduce Data Management Cost	Old and open transactional d...

Figure 12.31 Correction Recommendations for the Source-to-Pay Process

In the area of innovation recommendations, there are also some relevant recommendations. The recommendations regarding the new functions of SAP S/4HANA are of great relevance to Company B. However, interesting recommendations can also be found in the area of SAP Build Process Automation or in the area of machine learning. For example, reference is made to an **Intelligent Approval Workflow**, a functionality in

SAP S/4HANA that analyzes past release schemas for purchase requisitions and makes suggestions for automated releases (see Figure 12.33).

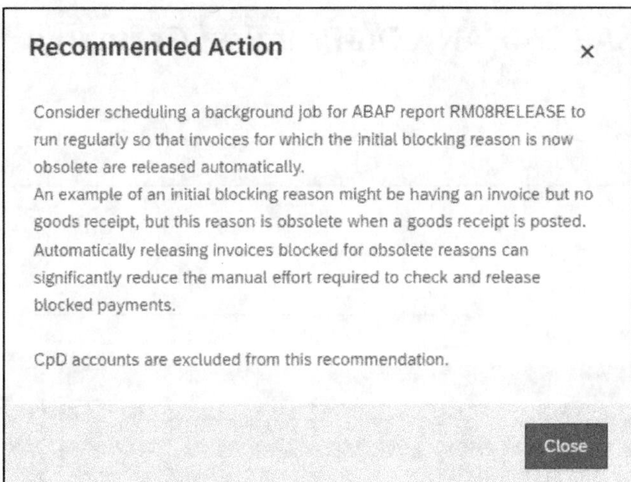

Figure 12.32 Action Recommended by the System

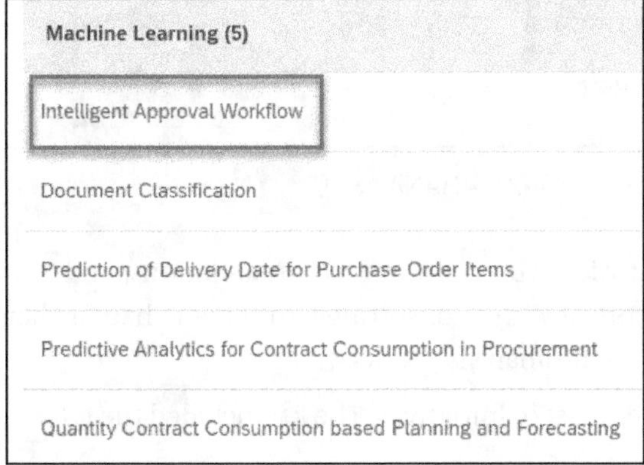

Figure 12.33 Innovation Recommendations

In preparation for a future process design, Company B's project team also obtained an overview of the available SAP S/4HANA best practice processes during this phase. Drilling down from the overview screen in SAP Signavio Process Explorer to the **SAP Product Portfolio** section provides a structured overview of **SAP Best Practices for SAP S/4HANA On-Premise** (see Figure 12.34).

Focusing further on the **Financial Operations** area provides an overview of the available **SAP Best Practice Scope Items** in this area. Scope items are abbreviated by a sequence of three digits/letters (e.g., **Accounts Payable J60**).

For each scope item, there are one or more predesigned BPMN diagrams that Company B can use during process design. Figure 12.35 shows an overview of the available diagrams in the **Accounts Payable** section; clicking on the diagrams on the right-hand side leads to a corresponding detailed view.

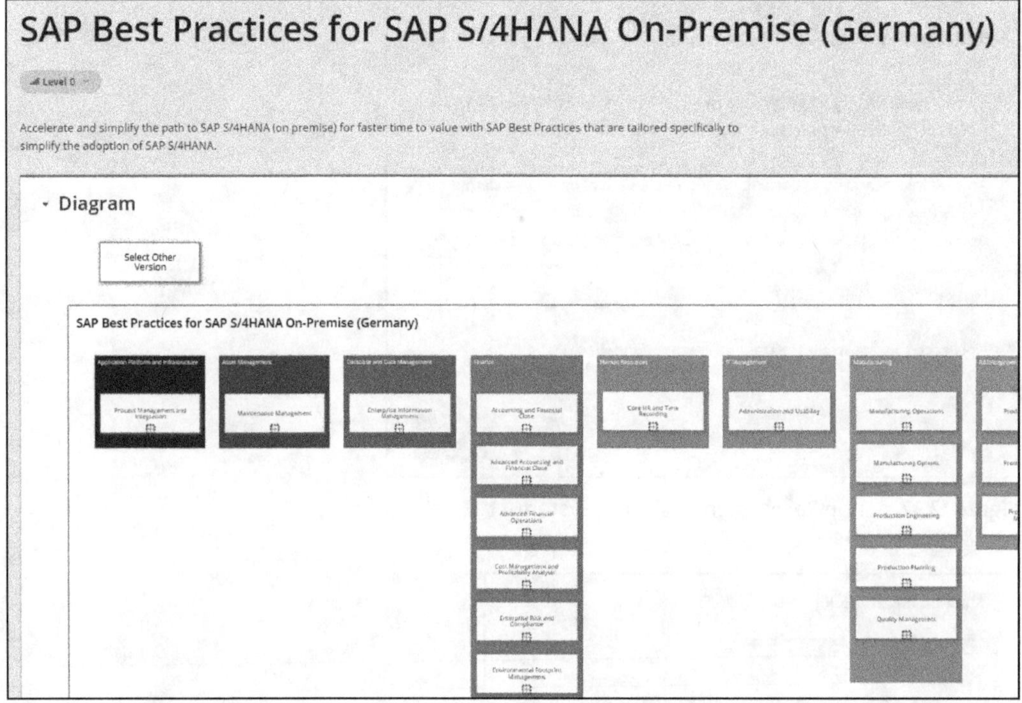

Figure 12.34 SAP Best Practices for On-Premise SAP S/4HANA

In addition to correction recommendations, innovation recommendations, and SAP best practice processes, Company B worked with a consulting partner who had further ideas for process optimization based on the analysis results.

From all the results, Company B created a list of initiatives. The list included the following components:

- Detailed description of the initiative
- Categorization of whether it's an automation, an extension of the existing solution, or a redesign of the process and/or solution
- Categorization in terms of timing of implementation regarding whether the initiative can be implemented immediately in the existing system or as an improvement that should be considered during or even after transformation
- Estimation of the expected effort for the realization

As a result, some initiatives could be implemented immediately, others served as input for the design phase, and still others were given lower priority, that is, not implemented at all.

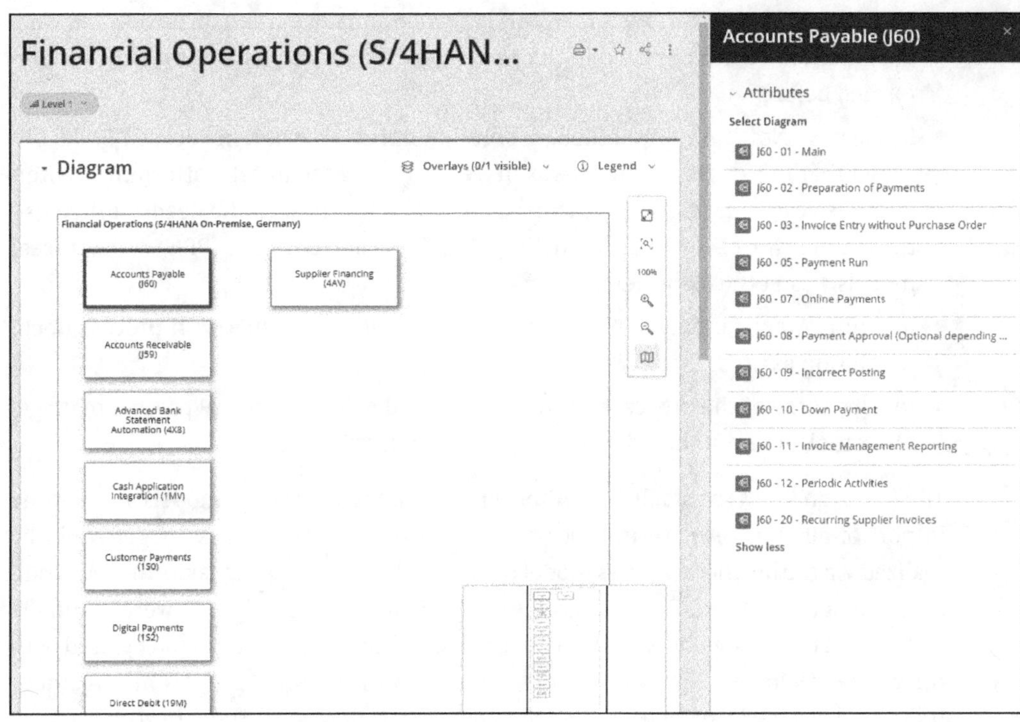

Figure 12.35 Best Practices Financial Operations

Process Design and Solution Design

The next step was to define a target process landscape (process design) and to map it in a corresponding solution design. The topic of process management wasn't completely new at the beginning of Company B's transformation, and some areas had already documented processes in the past; however, this was always managed decentrally and without a uniform approach. A process model derived from the overall operating model with a view of the end-to-end processes on the one hand and the processes in a functional unit on the other (as described in Chapter 10) didn't previously exist. However, the current transformation to SAP S/4HANA requires exactly that: transparency about the future process landscape.

Company B didn't want to lose the knowledge from the previous process models, but it also wanted to use results from the analysis phase and still build new process structures based on best practice processes and avoid deviations from the standard as far as possible. The project team therefore decided on a three-step approach: consolidation of existing models, remodeling based on the results of the process analysis, and design of a new process architecture that is close to the standard. These steps are described in this section.

Let's start with the consolidation of existing models. The topic of process management has so far been approached differently in the various regions:

- Documented process models in BPMN language already existed in Europe and were documented in SAP Signavio in the past. However, process analysis was never performed before.
- In the past, North America relied on solution documentation in Focused Build, but found that this documentation was more suited as a technical platform. The interface was geared more toward the needs of IT and less toward the needs of the business units. An easy-to-use modeling interface that connects PPIs with process models wasn't the focus here.
- In South America, only some PowerPoint and Visio diagrams with process documentation existed.
- In Asia, some of the processes were documented using a non-SAP process management tool, but no standard modeling convention existed.

The initial goal was to establish a uniform platform for all existing models. A one-time import of all North American process models from SAP Solution Manager could be realized via the business process model connector between SAP Signavio and SAP Solution Manager 7.2. PowerPoint diagrams, if still required in the future, were manually transferred to SAP Signavio. For Visio diagrams, the import function could be used with minor rework in SAP Signavio. The diagrams from the external process management tool could also be transferred to SAP Signavio BPMN diagrams. Some rework was also necessary here despite the import function to ensure a uniform and clear representation. Nevertheless, the import saved most of the manual modeling.

The next step was remodeling based on the results of the process analysis. With the variant analysis in SAP Signavio Process Intelligence, it was also possible to automatically create BPMN diagrams that represent the actual process flow. Figure 12.36 shows the Variant Explorer in SAP Signavio Process Intelligence.

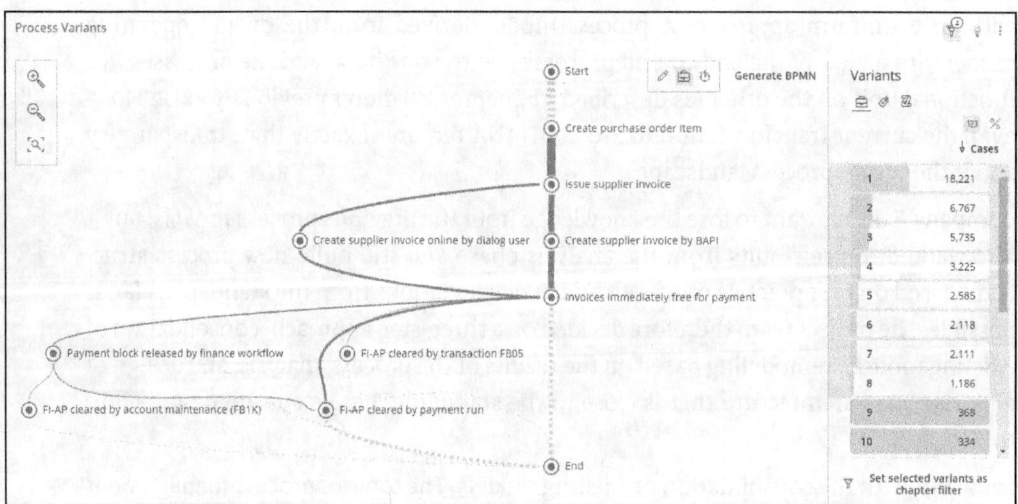

Figure 12.36 Remodeling Based on Already-Existing Process Variants

The eight most common variants were selected in the diagram shown. By clicking **Generate BPMN**, SAP Signavio automatically creates a BPMN process model that depicts the process flow for the eight most common variants. Even though the diagrams created in this way can't be used 1:1, but still require adjustments, this function made process design much easier in further steps, and manual efforts for process modeling could be saved.

The third step was to take all the input from previous steps to design a new process architecture. To build a new process architecture, Company B followed the principles described in Chapter 11, Section 11.2: derived from the company's operating model, a six-level hierarchy was established, divided into a functional view and a process-oriented view. For processes with regional specifics, template processes were first defined, and the corresponding process variants were then derived (see Chapter 11, Section 11.3).

On the lowest level of the process pyramid, different sources were considered as an input to define the actual to-be process flows of Company B:

- **SAP S/4HANA best practice processes**
  The processes provided were essential input for the creation of the processes. Nevertheless, these processes are rather solution-oriented; that is, they show how a (sub)process is mapped in SAP S/4HANA. A functional end-to-end process of Company B in SAP Signavio is therefore based on several SAP S/4HANA best practice processes. In fit-gap workshops, it was ultimately examined to what extent Company B's requirements deviate from the standard best practice models.
- **Existing or imported process models**
  To check during the fit-gap workshops whether all requirements had really been considered, a comparison was made between the newly modeled processes and the previous process models. In areas where Company B's requirements differed significantly from those of other companies, the project team also used previous models as a guide. However, to avoid later in-house developments, the SAP best practice standard process was always used as a guide whenever possible.
- **Process models that resulted from remodeling**
  These models were also used for the completeness check when designing the new target processes. But here, too, the best practice standard process was always leading the target design.

Build and Test

SAP S/4HANA transformation projects for companies the size of Company B usually take several years. This is also what Company B expects for its transformation. The company is currently in the build phase and is implementing the solution based on the previously created process design. Further steps are therefore currently taking place or are planned for the near future.

As already described, Company B has opted for the holistic integrated process transformation approach. This means that the solution requirements are also taken up and realized in a process-oriented manner. To ensure a seamless transition from process management to ALM, Company B uses the Focused Build solution.

Processes defined in SAP Signavio are transferred via the business process model connector application from SAP Signavio to SAP Solution Manager. Changes to the process models in SAP Signavio are also continuously synchronized to SAP Solution Manager. This approach allows for end-to-end alignment between process management, solution design, and solution realization. The synchronization of the process structure from SAP Signavio to SAP Solution Manager lays the foundation for an ALM that is aligned with what is defined on the process layer. All business process requirements are broken down into realizable units. They are implemented and tested in an agile approach, always referencing the originally designed business process.

By using Focused Build, requirements and specifications aren't recorded independently of the process hierarchy but can always be linked to a corresponding process. This makes it possible to track the development progress at any time. Business users can track which of the defined target processes have already been technically implemented and tested and which processes will take some more time.

Deploy and Enable

As soon as the configuration and development of the new solution is complete, the new solution will go live and be rolled out to the organization. SAP Solution Manager 7.2 will again play a key role in this process. With its Change Control Management and Data Consistency Management components, SAP Solution Manager supports not only the technical implementation but also the methodical management of the project.

SAP Enable Now will be used at Company B for the topics of training and enablement. Figure 12.37 shows an exemplary overview of all available learning materials for the direct material procurement process. In addition to content that has been adopted from SAP S/4HANA best practice scope items (recognizable by the reference to J45), other learning materials specific to the process design at Company B have also been added.

Figure 12.38 shows an example of how the interaction between SAP Signavio Process Manager and SAP Enable Now will be realized in the future. There is a process documented in SAP Signavio Process Manager (**Procurement of Direct Materials (New Production)**). When selecting the **Create Purchase Orders** process step, there is now a link to the learning materials for the SAP S/4HANA best practice scope item **Creating Purchase Orders (J45)**. For process steps without a reference to an SAP S/4HANA best practice scope item and other nonstandard processes, the project team would like to create their own written instructions, tutorials, and videos in SAP Enable Now and link them to the corresponding process model.

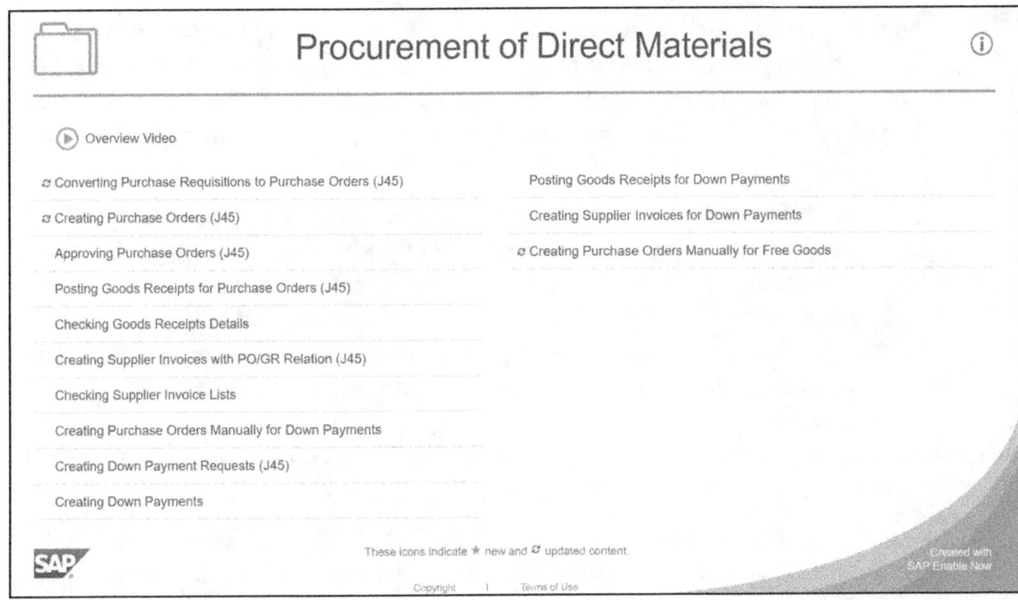

Figure 12.37 Overview of Learning Materials in SAP Enable Now

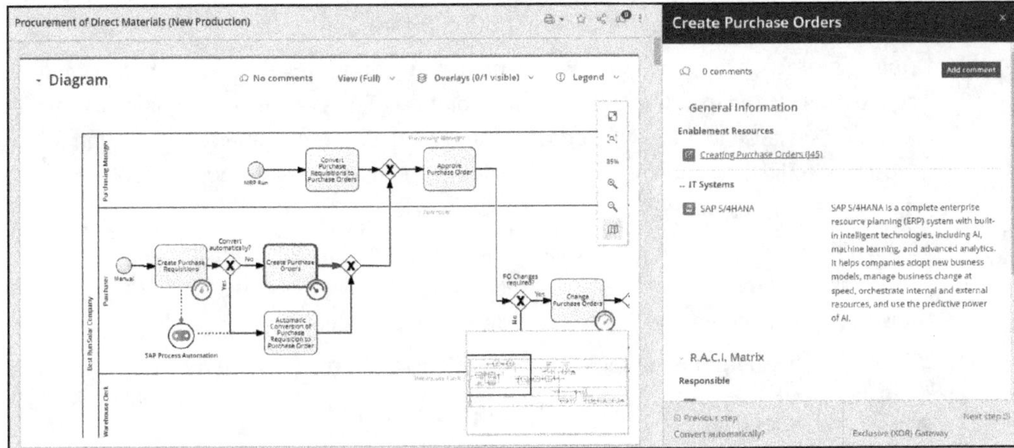

Figure 12.38 Integration of SAP Enable Now with SAP Signavio

How Did SAP Signavio Help Company B?

The use of SAP Signavio has played a major role at Company B from the very beginning. Already when calculating the initial business case with SAP Value Lifecycle Manager, it was possible not only to calculate a Move the Needle analysis and an SAP S/4HANA business case top-down but also to validate it with concrete PPIs from SAP Signavio Process Insights. We've shown the value-added calculation in this example using a selected PPI, but such calculations can be created for most of the PPIs available in SAP

Signavio Process Insights. This approach makes the identified added values more realistic, and levers can be identified to actually realize the added values. The combination of SAP Value Lifecycle Manager and SAP Signavio Process Insights thus enables an end-to-end value-add calculation and validation of an initially top-down calculated business case.

For a detailed process analysis, SAP Signavio Process Insights was initially used, but the data was also further analyzed in SAP Signavio Process Intelligence using the plug-and-gain methodology. Because SAP Signavio Process Insights initially only provides standardized content and doesn't offer the option of considering customer-specific tables, table extensions, and customer-specific fields, these were loaded into SAP Signavio Process Intelligence and extended there accordingly (via the SAP Signavio plug-and-gain approach). In addition, data from the non-SAP ERP system in Asia was added. Thus, process inefficiencies could be identified even better, and these could be evaluated with reference to the specific situation at Company B.

The correction and innovation recommendations from SAP Signavio Process Insights were used to not only identify short-term optimization potential but also identify topics that need to be considered in the process design. The discussed correction and innovation recommendations are only a small selection of the many suggestions from SAP Signavio Process Insights. All suggestions needed to be reviewed and evaluated to determine whether they are relevant for the company.

During process design, Company B used import functions in SAP Signavio Process Manager as well as the interface between SAP Solution Manager and SAP Signavio to have all processes available in SAP Signavio. The remodeling functionality in SAP Signavio Process Intelligence and content from SAP Best Practice Explorer were then used to define the target processes.

For a process-oriented solution design in SAP Solution Manager, the interface between SAP Signavio Process Manager and SAP Solution Manager will be used again. Thus, business departments and IT have a uniform view of the existing processes.

To roll out the solution and train employees, Company B plans to use the integration to SAP Enable Now.

12.7.3 Company C: Solution-Oriented Transformation Focused on Realizing Fast Results and Changes in Selected Areas

Company C comes from the field of discrete manufacturing: The photovoltaic manufacturer from Germany was a global leader in the 2000s, the early years of photovoltaics, and was also very successful economically. However, with the upswing of the Asian solar industry in the beginning of the 2010s, there was ever-increasing competition and price pressure, which continues to this day. Company C, however, still distinguishes itself by a particularly high quality of manufactured solar modules. In addition, a product guarantee of 25 years is offered for the photovoltaic modules, a feature that

Asian competitors generally don't fulfill. As a result, the German premium models continue to be very popular, especially with customers who place increased value on durability and resistance when selecting modules. Nevertheless, there is still a strong price war between manufacturers. Company C has established stable processes, but reducing costs is another very high priority to continue to win against Chinese competitors. Company C uses SAP ERP to handle its core business processes. From the very beginning, it has been important to stay as close to the standard as possible and to create as much transparency as possible about the business processes, but some tidying up of transactional data and master data is long overdue.

Company C conducted various preliminary studies for the upcoming SAP S/4HANA project. The SAP Simplification Items were analyzed in detail, innovations available with the new solution were assessed, and transformation paths were evaluated. The following results have been summarized by the project team:

- SAP's maintenance commitment for the existing SAP ERP system isn't expected to be extended beyond 2027. In this respect, a switch to SAP S/4HANA is inevitable.
- The effort of a complete process redesign, fit-to-standard workshops, and a roll-out to all regions and plants would significantly exceed the expected benefits.
- Nevertheless, it's necessary to make process adjustments in some areas and to align with various business teams.
- Furthermore, a number of areas have already been identified where rapid results can be expected (see the cleanup efforts just described).
- A transition focusing on the technical changes needed is the best path for Company C.

The project team is currently working intensively with SAP Signavio. In the following sections, we describe the experiences of the project team and the conclusions they are drawing from them.

Analyze

Based on the results of previous pre-studies, the activities to be performed before, during, and after the transformation were defined from a technical perspective. However, when evaluating the migration effort and runtimes during the migration, it became obvious that large volumes of data would have to be moved and that it would make sense to clean up this data before the transformation. Furthermore, despite the migration approach, process optimizations are to be realized in selected areas.

Company C wants to keep the effort for data analysis low, but still benefit from potential improvements that can be implemented quickly. SAP Signavio Process Insights enables this approach; therefore, the existing SAP ERP system was connected to SAP Signavio Process Insights.

In the first step, an example from the source-to-pay area was analyzed. Figure 12.39 shows the first of the 22 identified process flows. SAP Signavio Process Insights offers the option of filtering according to various factors. One process with a relatively large number of blockers is the **Purchase order item creation (without account assignment) to invoice receipt creation** process. The process starts with the creation of a purchase order and shows the steps up to the creation of the invoice in the system.

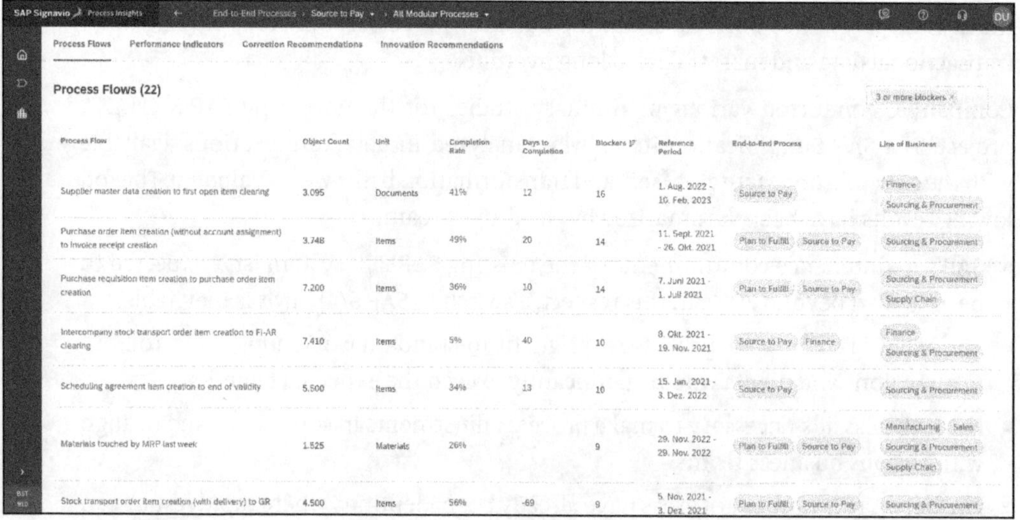

Figure 12.39 Overview of Process Flows in the Source-to-Pay Area

A detailed view of the process shows blockers from different areas (see Figure 12.40):

- Order items were blocked, deleted, or rejected.
- During goods receipt, some order items were only partially delivered or canceled. The reason for that could be inadequate data maintenance activities.
- In process step **Supplier invoices created**, you can see that invoices were created even before the goods arrived.

All these anomalies can be analyzed in more detail by drilling down into the respective context boxes. It's also possible to create benchmarks to filter the process flow according to various criteria, for example, by company code, by material type, by purchasing organization, and so on.

Figure 12.41 shows a comparison of the material types: Which material type has the highest aggregated value? For how many of the purchase order items was a goods receipt posted in the period under consideration (i.e., which materials arrive earlier, and which arrive later)?

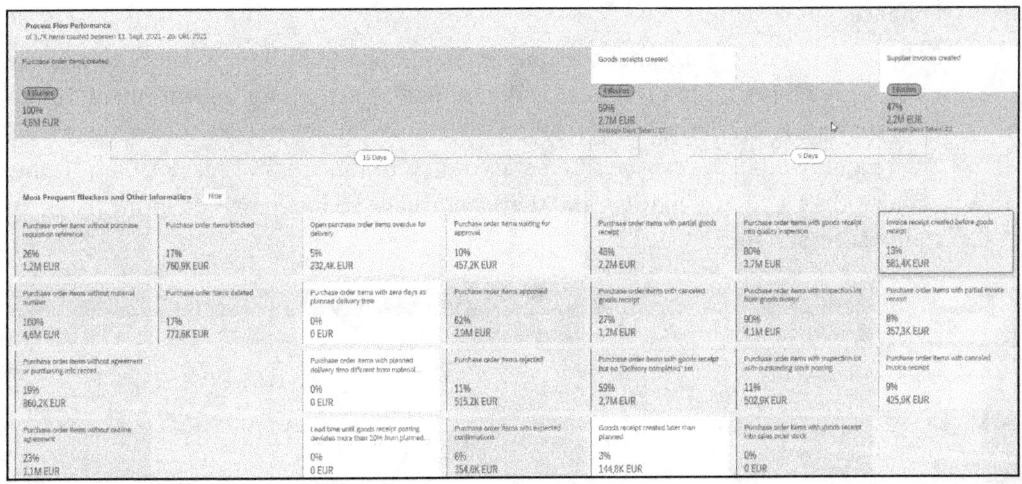

Figure 12.40 Detailed Process Flow in the Purchasing Area

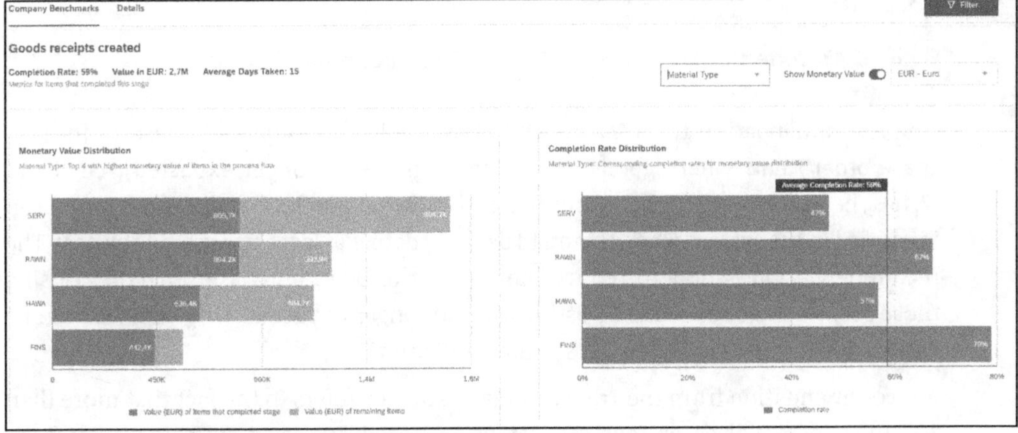

Figure 12.41 Internal Benchmarks

Process experts from purchasing can draw relevant conclusions about possible optimizations from these values. In addition, SAP Signavio Process Insights provides standardized information on how the process can be improved accordingly. We'll go into this in more detail in the next phase (enhance).

Company C decided against a deeper analysis with SAP Signavio Process Intelligence and the SAP Signavio plug-and-gain approach. Such a project would require initial efforts for the setup as well as tie up additional capacities and is therefore not in the scope of the transformation project.

Enhance

SAP Signavio Process Insights not only offers Company C the possibility to analyze existing processes in detail but also lists various suggestions for improvement. Figure 12.42 shows an overview of all correction recommendations in the source-to-pay area. The recommendations can be filtered according to the areas of **Master Data**, **Transactional Data**, and **Automation**, and they can usually be implemented immediately in the existing system.

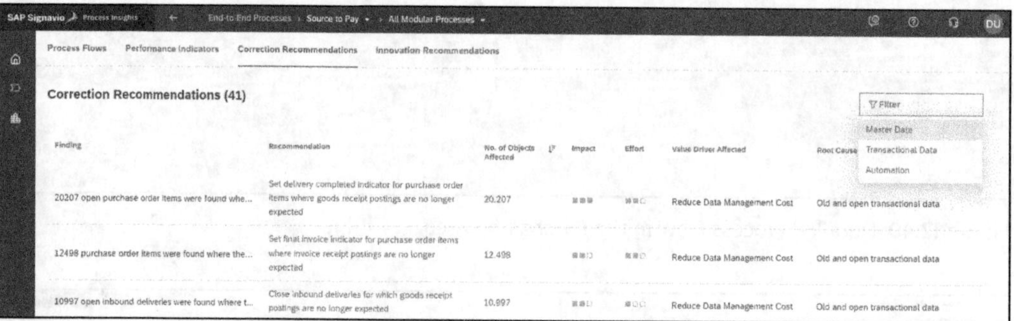

Figure 12.42 Correction Recommendations for the Source-to-Pay Process

One recommendation from **Master Data** is **Set delivery completed indicator for purchase order items where goods receipt postings are no longer expected** (see Figure 12.43). The background to the recommendation is that purchase order items have been identified in the system whose planned delivery date is at least one year in the past. The recommendation is now to reduce the number of open purchase orders by closing those for which goods receipt postings are no longer expected. This behavior can be easily adjusted by a setting in the system configuration.

A recommendation from the **Transactional Data** area refers to the fact that more than 30% of open invoices or invoices not yet released for payment haven't yet been checked. The system now suggests the automated release of invoices if the initial reason for blocking the invoice no longer exists (**Consider scheduling a background job for ABAP report RM08RELEASE to run regularly so that invoices [...] are released automatically**; see Figure 12.44). An example of an invoice block could be that there is an invoice for a material for which no goods receipt has been posted yet. However, this block is obsolete once the goods receipt has occurred. A regularly scheduled ABAP report can help avoid manual effort here: the report checks whether the reasons for blocking still exist and releases invoices accordingly based on the result.

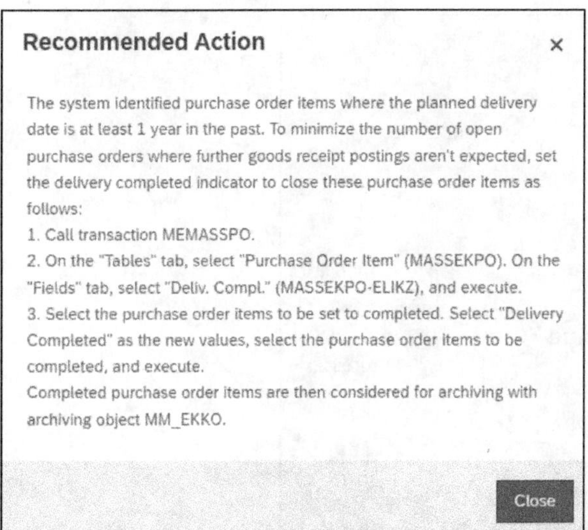

Figure 12.43 Correction Recommendation for the Master Data

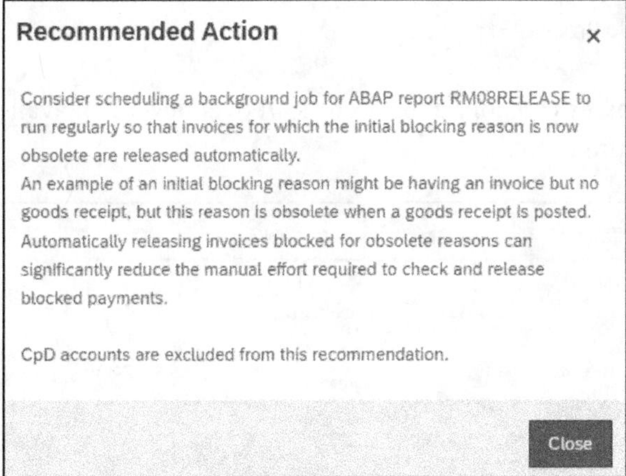

Figure 12.44 Correction Recommendation for Transactional Data

Process Design and Solution Design

Even though Company C has decided to implement a solution-driven transformation with few changes to the processes, improvements in selected processes are still to be realized. To understand opportunities for process standardization and automation, the project team uses the content of SAP Signavio Process Explorer. Figure 12.45 shows an overview of the SAP Signavio Process Explorer content: available process models can be searched, sorted by industries, by solution capabilities, by processes, and by concrete best practices from solutions in the SAP product portfolio.

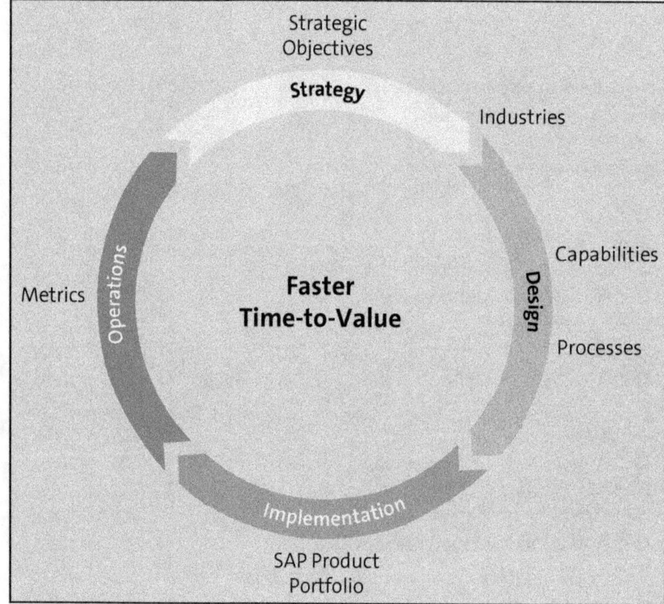

Figure 12.45 Overview of SAP Signavio Process Explorer

The project team selects the **Processes** category and receives an overview of all available end-to-end processes (see Figure 12.46).

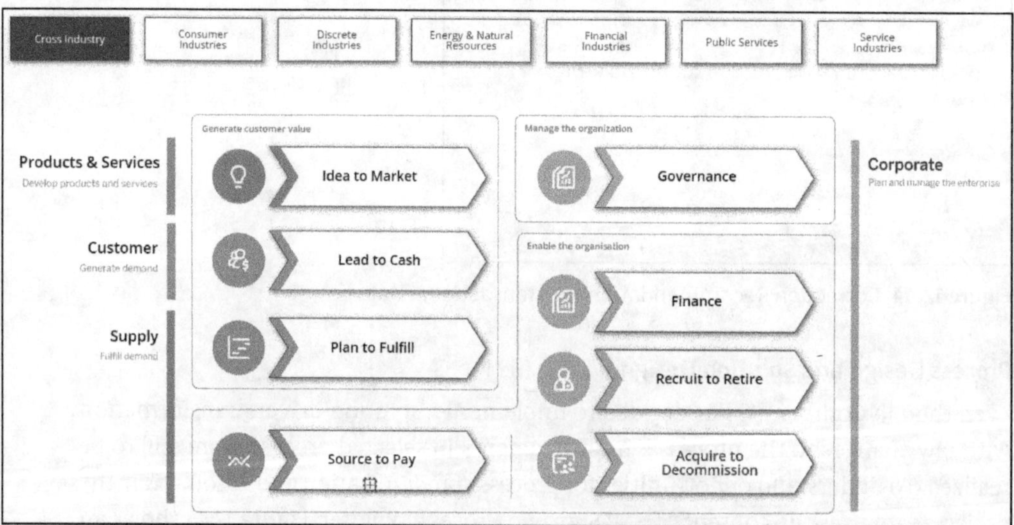

Figure 12.46 End-to-End Processes in SAP Signavio Process Explorer

Process improvements in the source-to-pay area will be identified. The scope of the entire process is now displayed: from supplier selection to planning, actual procurement, handling of disputes, posting, and clearing of the invoice, best practice process models can be found in SAP Signavio Process Explorer (see Figure 12.47).

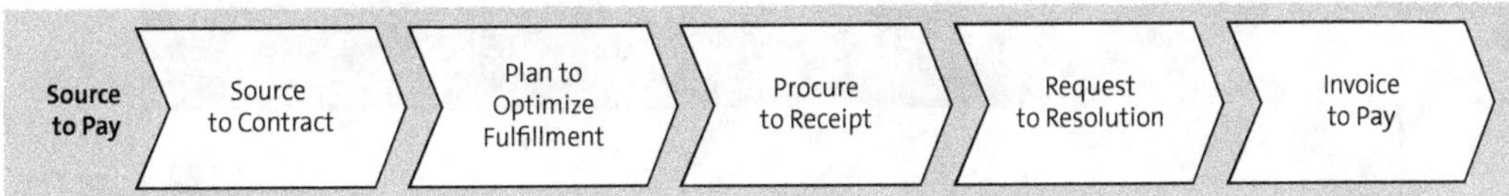

Figure 12.47 Source-to-Pay Process in SAP Signavio Process Explorer

A further drilldown into the **Procure-to-Receive** process (see Figure 12.48) now leads directly to the SAP best practice process **J45 | Procurement of Direct Materials.** The overall goal in the procurement process is to receive the delivery at the right time and to reduce the number of days sales outstanding. For each of the process steps in the best practice process, added value potentials are stored so that users have an indication of the benefits they can derive from implementing the corresponding step.

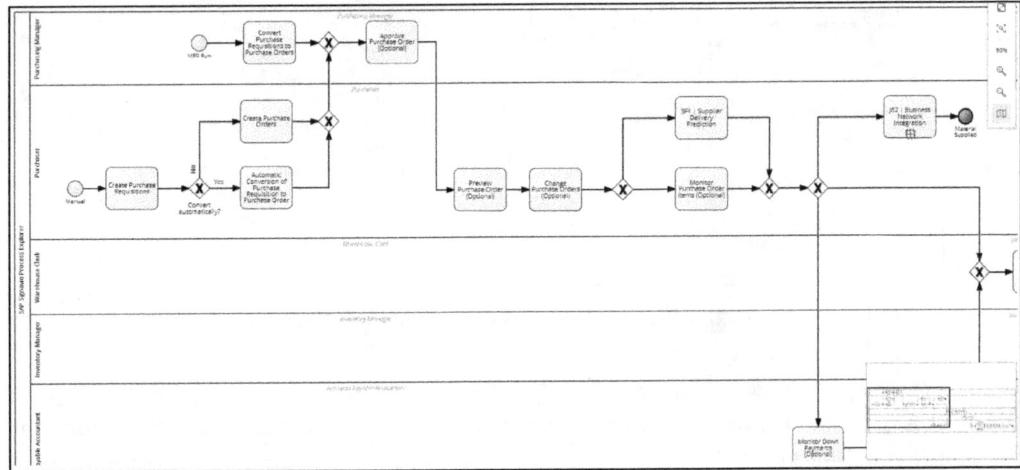

Figure 12.48 Best Practice Process in SAP Process Explorer

The best practice process indicated various process optimization potentials. Following you'll find two examples:

- In the subprocess **3FY | Supplier Delivery Prediction** (see Figure 12.49), the value-added potentials are the reduction of inventory costs (**Reduce inventory carrying cost**), the reduction of revenue losses due to missing stock (**Reduce revenue loss due to stock-outs**), and the improvement of delivery reliability (**Improve on-time delivery performance**). Implementing this subprocess as part of the SAP S/4HANA project could therefore bring significant benefits.
- Subprocess **J82 | Business Network Integration** describes how the customer and supplier can work more closely together by using the SAP Business Network Commerce Automation solution.

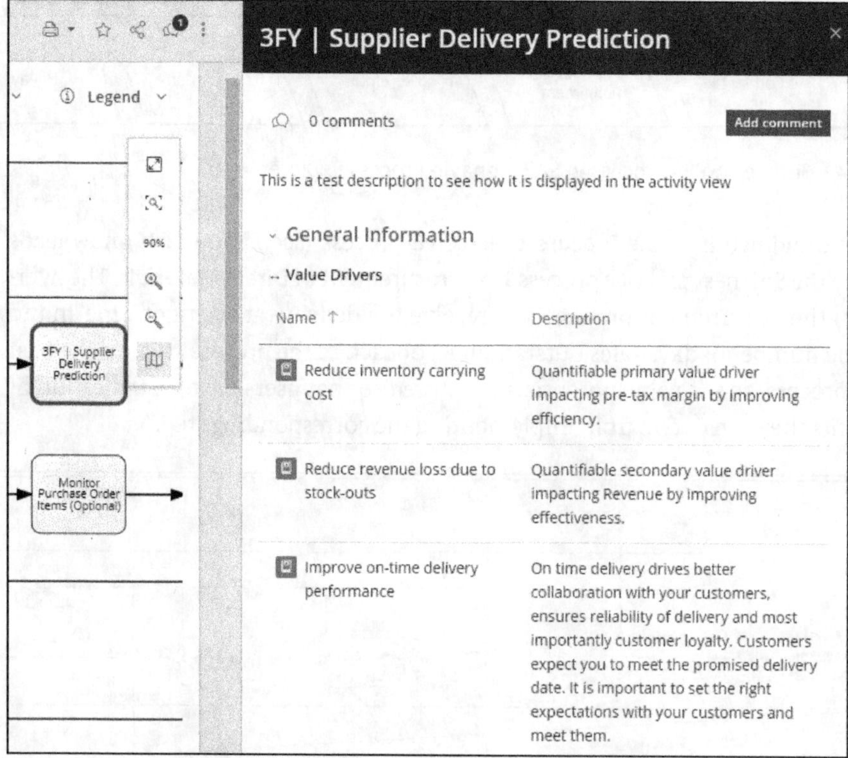

Figure 12.49 Best Practice Process in SAP Process Explorer (Detailed)

Previous activities in the design phase have focused on understanding SAP best practice processes. In the next step, the company is looking at further recommendations provided by SAP Signavio Process Insights, in particular, innovation recommendations.

In the enhance phase, the project team dealt with the correction recommendations, that is, automations that can be implemented directly. In contrast, the innovation recommendations compile proposals for a future solution architecture. The suggestions can relate to functions in SAP S/4HANA, SAP Build Process Automation, or SAP Fiori apps available in other SAP solutions or new technologies such as machine learning scenarios.

In the area of SAP Build Process Automation, the team comes across two recommendations (see Figure 12.50). One of the proposed automation scenarios is that purchase requisitions can be created from a Microsoft Excel file (**Create Purchase Requisitions from Excel**). Selecting this innovation recommendation leads directly to the SAP Intelligent Robotic Process Automation Store, where the bot is described in detail and can be downloaded (see Figure 12.51).

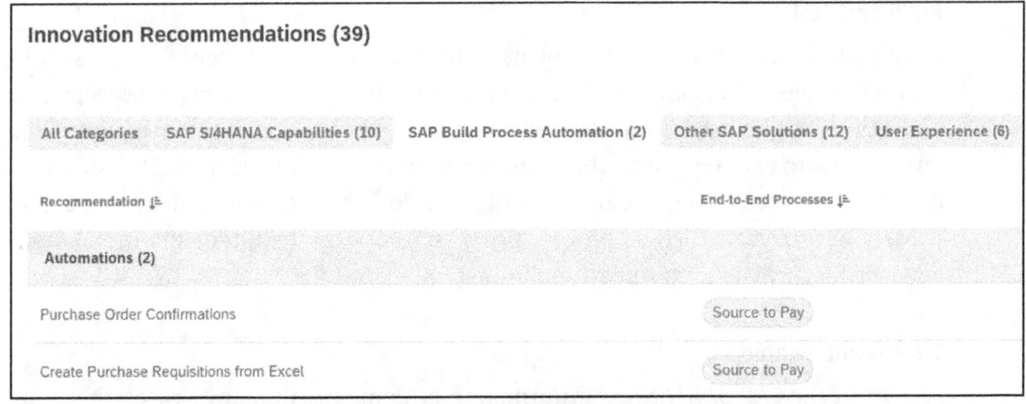

Figure 12.50 Innovation Recommendations for SAP Build Process Automation

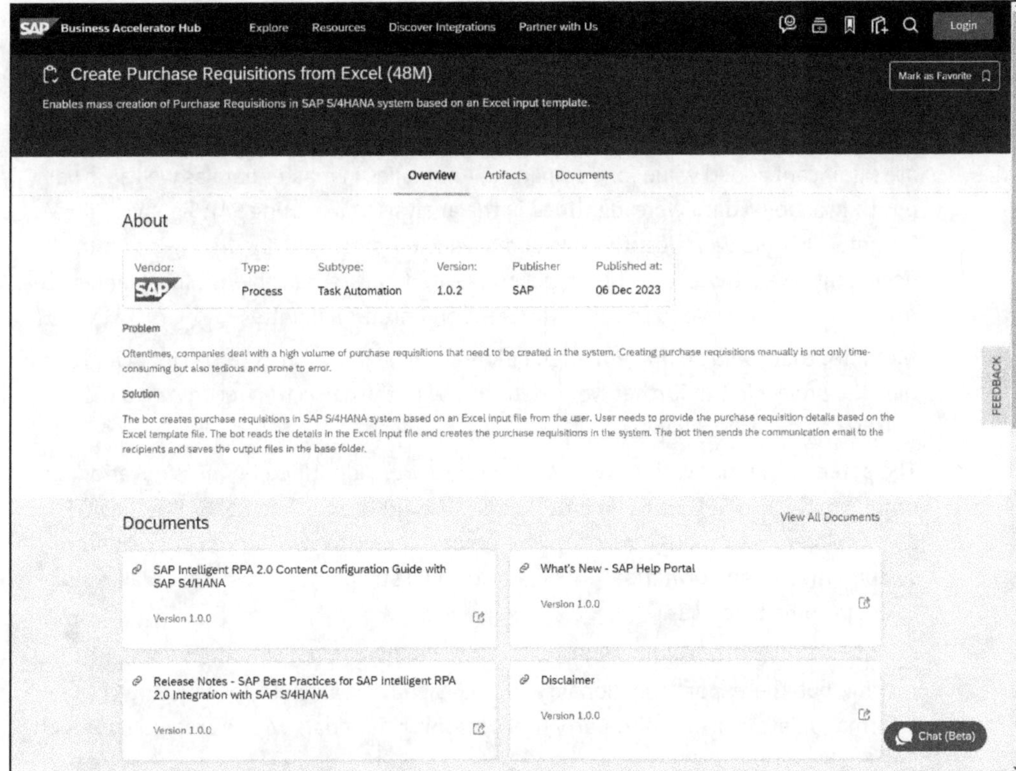

Figure 12.51 SAP Intelligent Robotic Process Automation Store

Thus, the activities described in this section serve to identify optimization opportunities that may be of interest in the context of the project.

Build and Test

As initially described, Company C plans to make significant changes to processes in only a few selected areas. Therefore, the described integration between SAP Signavio and ALM won't be used. Non-SAP tools have been used in the past for requirements analysis and to plan test cases; this approach hasn't been changed due to the solution-oriented transformation. The role of SAP Signavio in that case is to deliver business process information such as process flows and business requirements in solution-agnostic file formats to third-party solutions.

Deploy and Enable

As part of the SAP S/4HANA transformation at Company C, many processes will be taken over from the legacy system to the new system. For many transactions, users won't notice any difference from the previous solution. Nevertheless, there will be changes that will require training. To this end, Company C plans to use SAP Enable Now.

[»]

How Did SAP Signavio Help Company C?

Despite the system-driven transformation approach, SAP Signavio was able to generate significant added value for Company C. Anomalies in master data as well as in various transactional data were identified in the analysis phase using SAP Signavio Process Insights. Blockers were identified in some selected processes. By drilling down to the document level, it was possible to get to the root cause of problems. Internal benchmarks made it possible to compare different organizational units.

Measures that could be implemented before the changeover to SAP S/4HANA and simplify the project in the further were identified with the correction recommendations in SAP Signavio Process Insights.

The potential for innovations within the project were identified by the innovation recommendations in SAP Signavio Process Insights and the contents of SAP Signavio Process Explorer. Even though no complete process redesign was carried out as part of a system-driven transformation, SAP Signavio could still point out where it makes sense to adapt some processes.

Company C plans to use non-SAP solutions for managing business requirements and testing, but the export functionality of SAP Signavio can help to have the relevant information available in third-party tools. Learning materials will be created later with SAP Enable Now.

12.8 Summary

Leveraging SAP Signavio Process Transformation Suite in the context of an SAP S/4HANA transformation project provides significant added value across all phases. Additionally, it's also important to mention that the switch to SAP S/4HANA is the ideal entry point for the introduction of continuous business process transformation management in your company.

In this chapter, we've built on the findings of the previous chapters and explained step-by-step how the individual building blocks can be linked together. With SAP Signavio's end-to-end business process transformation methodology, significant added value can be achieved in the context of SAP S/4HANA transformation projects. With the integration of enterprise architecture, business process transformation management, and ALM, SAP Signavio not only supports the change to SAP S/4HANA but is also the basis for continuous process improvements once the SAP S/4HANA transformation project is closed.

We covered the following implementation phases:

- In the analyze phase, we looked at measuring process performance and discussed how added value can be generated through process and benchmarking data.
- The enhance phase has shown how improvements in the already-existing ERP system can be identified, prioritized, and realized. However, major changes require the design of new business processes and the construction of a future process architecture.
- We described the procedure for prioritizing and segmenting a modular process design and building customer journey maps in the process design and solution design phases.
- For the solution build and test phases, close collaboration between the business units has been outlined as a critical success factor. SAP Signavio linked with SAP Solution Manager can significantly simplify this phase. Focused Build provides the project team with tools for efficient and effective project management.
- Once the developments are completed, the deploy and enable phase is about rolling out the solution and team enablement. We outlined that continuous transparency of achieved improvements can positively contribute to acceptance of the new solution in the organization and that training for employees can be supported via the integration of SAP Signavio and SAP Enable Now.

To present the theoretically explained phases in a comprehensible way, we've also used three practical examples in this chapter to explain how customers from different industries and with different objectives have used or are currently still using business process transformation management to add value as part of their SAP S/4HANA transformation project.

However, the project is only the start for successful business transformation management. The greatest added value is achieved with continuous improvement activities also after go-live. We'll go into this in more detail in the next chapter.

Chapter 13

Extending Business Process Transformation Management Beyond the SAP S/4HANA Project

In this chapter, we address ongoing process optimization after the SAP S/4HANA implementation project. We reflect on lessons learned from the project and focus on future improvements to business process transformation management. In doing so, we consider the role of culture, technology, and end-to-end processes. Finally, we present methods for analysis, optimization, and success measurement. This chapter thus serves as an orientation for ongoing development after the SAP S/4HANA project.

Having dealt in detail with business process transformation management in the context of the SAP S/4HANA transformation in the previous chapter, we'll now take a look at the *continuous process improvement*, which begins after the project has been completed. The goal is to continue to profitably apply the lessons learned, ways of working, and methods you developed and applied during the SAP S/4HANA project in the post-project period. In this chapter, we discuss how business process transformation management can be continuously developed after the successful implementation of SAP S/4HANA. In addition to continuous process improvement, we'll focus on the improvement and further development of business process transformation management. This is to be sustainably and permanently anchored in the company even after the original goal of the SAP S/4HANA transformation has been achieved and serves company-wide process control beyond the boundaries of SAP S/4HANA.

We start this chapter with a look at the transformation that took place and the lessons learned (Section 13.1). In Section 13.2, we focus on the continuous review of a completed project. In Section 13.3, we identify the differences between the use of business process transformation management during the transformation and the phase of continuous process improvement after the project. In the remaining sections, we address specific aspects of business transformation process management that should continue to be applied or further developed.

13.1 Review and Lessons Learned

During the SAP S/4HANA transformation, a lot of experience has been gained as well as valuable insights into the change process in the company. It makes sense to analyze these to sustainably anchor business process transformation management in the company beyond the SAP S/4HANA project and make it usable for future process improvement initiatives. The successful implementation of SAP S/4HANA is an important milestone that enables the company to optimize its business processes and make them more efficient. For example, many processes have been harmonized, standardized, and automated; redundant systems have been eliminated; and improved data flow has been enabled. It's therefore important to appreciate the value contribution of business process transformation management as part of the transformation.

Some of the lessons we've learned from this transformation process relate to *planning and preparation*. Transformation requires careful planning and coordination between the various teams and stakeholders. It has become clear that involving all stakeholders from the outset is essential to ensure that all requirements have been taken into account and all potential challenges to business process changes have been identified early on.

Another important aspect that emerges from the hindsight of the SAP S/4HANA transformation is the importance of *training and employee development* based on processes that are uniformly defined throughout the company for the success of the transformation. The implementation of new systems and processes leads to changes in the workflows of employees. It's therefore important that they receive sufficient training and support.

The transformation also demonstrated the importance of clear *communication and leadership* throughout. Through challenges and unforeseen events, it was important that the leadership team communicated clearly—again based on consistently defined processes across the company—and maintained a positive attitude to keep employees engaged and motivated.

In addition, the importance of a *robust technical infrastructure* was emphasized. Challenges related to the integration of SAP S/4HANA with other SAP and non-SAP systems, data migration and security, and system performance and stability were identified. These technical challenges require careful preparation and planning, as well as the ability to respond flexibly and adaptively to unforeseen issues. Process management has a significant impact on addressing these issues. By systematically analyzing and describing business processes, process management can ensure that all process-related interfaces between SAP S/4HANA and other systems are identified and taken into account during the changeover. It also enables orderly data migration and data security by providing the opportunity to define clear policies and procedures for handling sensitive information. In addition, process management helps improve system performance and stability by continuously monitoring the efficiency of operations. As a result, companies are able to identify problems early on, especially in processes that are important to them, and respond to them quickly.

Another important lesson relates to the *governance and management of the transformation process*. Establishing effective governance structures and management practices has proven essential to guide the transformation process through process management and to ensure that this establishment is in line with the strategic goals of the company. Finally, the need to continuously measure and monitor the progress of the transformation also became apparent. This is best done by being transparent about how many identified process adjustments have already been designed and implemented and how many still need to be worked on. These and many other insights are valuable lessons that inform future process improvement initiatives. They illustrate that transformation is much more than a technical implementation. It's a holistic process that requires planning, communication, training, leadership, and continuous improvement.

A thorough *lessons learned analysis* is the starting point of implementation or continuation of process management after the SAP S/4HANA project. It helps companies strengthen their change management and continuous improvement capabilities. It's of particular importance to recognize that the process transformation structures implemented as part of the SAP S/4HANA project are merely the starting point. They form a solid foundation on which to build. However, it's important not to view them as the ultimate goal. Rather, they should be seen as a first step that provides room for continuous improvement and adjustment. Companies should take the opportunity to build on these structures to continue to optimize their business processes and adapt to the ever-changing requirements and opportunities of the digital age. Business process transformation management lays the foundations for flexibility and agility to realize innovations in SAP S/4HANA in the future, to exploit the full potential of the system, and thus to support the digital transformation of the company in the best possible way. In fact, the further development and deepening of these structures after the project can play a crucial role in the company's continued success. Looking back, the SAP S/4HANA transformation provides an excellent opportunity to leverage the knowledge learned and use it as a springboard for future improvements. It enables companies to continuously improve their processes and systems, increase their efficiency, and ultimately enhance their competitive advantage.

In this sense, the SAP S/4HANA transformation not only serves to implement a new ERP system, but can also become a catalyst for a culture of continuous improvement and learning within the company—a process that strengthens the company and equips it for future challenges and opportunities.

13.2 Focus on Continuous Improvement

Continuous process improvement after the implementation of your SAP S/4HANA project brings new challenges that differ significantly from the project phase. In the SAP S/4HANA project, the focus was on the implementation of SAP S/4HANA and the associated process changes. This is a clearly structured process aimed at achieving the defined project goals. The mandate for change is therefore clearly defined in this phase and legitimized by management. The willingness to change is usually high, as all those involved pursue the common goal of making the change to SAP S/4HANA successful.

The continuous process improvement phase begins after the implementation of SAP S/4HANA and extends over the entire process lifecycle. The goals of this phase aren't necessarily clearly defined. Therefore, a culture of constant reflection, learning, and adaptation is required. It's a continuous process of constantly looking for ways to improve the efficiency and effectiveness of processes and increase the benefits of SAP S/4HANA, although the mandate for change in this phase is often less clearly defined. Readiness for change can vary and often depends on factors such as the perceived need for change, the company's ability to successfully implement change, and the overall culture of the organization.

From an organizational perspective, the continuous process improvement phase also differs from the optimization phase during the SAP S/4HANA project. During the optimization phase, the project organization is usually highly hierarchical and functional, with clear roles, responsibilities, and accountabilities. In the continuous improvement phase, on the other hand, the organizational structure is often flatter and more flexible, but at the same time, it allows and requires more room for creativity and initiative.

In addition to organizational and process differences, corporate culture plays a crucial role in continuous improvement. Establishing a *culture of continuous improvement* is often the foundation for successful business process transformation management beyond a specific SAP S/4HANA project. A culture of continuous improvement encourages innovation and creativity, supports experimentation and risk-taking, and recognizes mistakes as opportunities to learn and improve. It encourages employees to question processes and ways of working and to look for better solutions. In such a culture, ideas for improvement are not only accepted but are actively expected and encouraged. Employee engagement is an essential component of such a culture. When employees feel that their ideas are heard and valued and that they can make a difference, they are more likely to actively participate in the continuous improvement process. Training and development opportunities can help strengthen employees' skills and competencies and better prepare them for the demands of continuous improvement. Leadership also plays a critical role. Leaders must recognize and exemplify the value of continuous improvement, communicate the messages and goals of continuous improvement, and support employees in implementing their ideas and suggestions for improvement.

Change management is also an important aspect here. As changes are no longer carried along in the context of moving to a new system, the importance for success actually increases. Continuous improvement often requires changes in ways of working and processes, and these changes can create resistance. Effective change management can help overcome this resistance and ensure that changes are effectively implemented and accepted.

Another important difference concerns the way in which improvements are identified and implemented. During the optimization phase of the SAP S/4HANA project, the identification of improvements is usually centralized and often strongly focused on the technical and functional aspects of the new system. The project team works closely with the business departments to optimize processes and derive the maximum benefit from the features and capabilities of SAP S/4HANA. External consultants are often involved, bringing in specialized knowledge and experience. In contrast, continuous improvement after implementation is usually more decentralized and participative. Improvement ideas can come from all employees and are collected and evaluated in a bottom-up process. This requires a high level of employee participation and engagement, as well as a culture that encourages innovation and continuous improvement. It's no longer just about understanding and making the most of the new system, but rather about continuously questioning and improving business processes.

Another important difference concerns how change is implemented and authorized. During the implementation phase, there is usually a strong mandate for change. The organization understands that implementing a new system will require extensive changes in processes and ways of working, and there is a strong willingness to accept and implement these changes. After implementation, this may change. This willingness to change may decrease, especially if the organization feels that the biggest challenges have been overcome. The mandate for change may weaken, and it may become more difficult to gain support for continuous improvement initiatives. This requires strong leadership and ongoing communication to emphasize the importance of continuous improvement and rally the organization to the process.

These and other differences lead to novel and different challenges during continuous improvement after SAP S/4HANA implementation. Continuous improvement requires specific strategies and approaches that differ from those used during the implementation phase. At the same time, however, it also presents a tremendous opportunity to maximize the benefits of SAP S/4HANA and strengthen the organization's competitiveness and performance in the long term.

In the following sections, we now look at selected aspects with regard to the further development of business process transformation management, taking into account the retrospective and the differences in the use of business process transformation management in comparison during and after the SAP S/4HANA transformation outlined in this section.

13.3 Process Governance after Conclusion of the SAP S/4HANA Project

In the continuous improvement phase on which we're now focusing, process governance plays a central role. The governance structures and management practices established during the SAP S/4HANA transformation form a solid foundation that can also be of great benefit after the project and should be transferred into a sustainable exercise of governance. However, it's important to regularly review and adjust these structures and practices to ensure that they optimally support the continuous process improvement process.

The governance models introduced during the transformation phase (see Chapter 12, Section 12.4) were a fundamental prerequisite for defining clear responsibilities and decision-making structures. But during this phase, the initiatives were time-limited and aimed at specific results or processes to be considered during the transformation. In the continuous improvement phase, the improvement initiatives or projects to be defined are often less structured and require a more flexible, adaptive approach. This means that governance structures and management practices must evolve and adapt to support these new requirements. This may include making stakeholder management more focused or refining processes for prioritizing and selecting improvement projects.

One example is the role of the *steering committee* established during the SAP S/4HANA transformation. The steering committee is a central body that steered the decisions of the project and monitored progress. It consisted of executives and key stakeholders and met regularly to monitor progress and make decisions. However, in the continuous improvement phase, the steering committee is less effective in its original form. Improvement initiatives following the SAP S/4HANA transformation project are now less structured and require a more flexible approach. This requires rethinking and possibly expanding the original structures and roles. At this point, the question arises as to whether the original steering committee is still the appropriate body to manage and accompany the continuous improvement process. During the transformation phase, the steering committee consisted mainly of executives and key stakeholders. However, in the continuous improvement phase, it might be useful to involve employees from different, broader areas of the company to gain a broader perspective and foster commitment to the improvement process. In addition, new roles within the organization might be needed to specifically help promote and support continuous process improvement. If not done so far as part of the SAP S/4HANA transformation, a business process transformation office (PTO) should be established (see Chapter 11, Section 11.1) whose role is to provide a framework for continuous process improvement and to help other stakeholders identify and prioritize improvement opportunities and monitor the implementation of these improvements.

In the context of continuous improvement, change management must also be revised or expanded. During the transformation phase, the steering committee focused mainly on managing the progress of the project. However, in the continuous improvement phase, the newly established continuous process improvement methods could play a stronger role in managing change, for example, by coordinating training and development programs, fostering a culture of continuous improvement, and communicating changes and their impact on the business. This example shows that the governance structures and management practices established during the SAP S/4HANA transformation are a valuable starting point, but they need to be adapted and extended to meet the requirements of the continuous improvement phase.

Note that the aspects discussed here—governance structures and change management—are only individual parts of success. They are undoubtedly crucial to the success of the continuous improvement process, but they aren't the only factors to be considered.

There are a number of other elements that are important for successfully maintaining an effective continuous improvement process. These include, for example, the ongoing training and development of employees in the context of business process transformation management; establishing a culture of continuous improvement; providing the necessary tools, methods and resources; and ensuring effective communication at all levels of the company.

Every company is unique, and the exact design of the continuous improvement process is influenced by a number of factors, including the size and structure of the company, the industry in which it operates, and the specific challenges and opportunities it faces. It's therefore critical that organizations consider the aforementioned issues, take a flexible approach to the continuous improvement process, and be willing to continually rethink and adapt their approaches and practices to meet changing needs and circumstances.

13.4 Consideration and Optimization of End-to-End Processes

Once the transformation project is complete, the focus shifts from redesigning and optimizing the processes supported by SAP S/4HANA to end-to-end processes that aren't mainly supported by SAP S/4HANA. During the project phase, companies typically focus on the part of their end-to-end processes that are supported by SAP S/4HANA and may neglect subprocesses that impact customers, suppliers, and other stakeholders. However, looking at end-to-end processes is necessary to improve overall performance through a better understanding of the relationships and interactions between all processes.

A key aspect of looking at end-to-end processes is understanding that the flow of value often extends across departmental boundaries. This means that optimization measures that focus only on individual departments or functions may not be sufficient to improve overall performance. Instead, companies should try to take a holistic look at their processes and identify optimization opportunities along the entire value stream. This makes it necessary to consider other IT systems in addition to SAP S/4HANA. For example, a company might find that a process such as ordering raw materials is running efficiently internally but is causing delays in the supply chain because suppliers are unable to keep up. By broadening its focus and looking at the entire process from order to delivery, the company could identify opportunities for improvement, such as working more closely with suppliers or implementing a demand-driven sourcing strategy.

Equally important is the optimization of processes that may not have been considered in detail during the SAP S/4HANA implementation, especially in areas outside the SAP S/4HANA scope, or of those processes that were analyzed but not implemented due to the chosen transformation approach or, for example, time or financial restrictions. During the implementation phase, many companies focus on optimizing the processes that are directly linked to the new system and through whose adaptation the greatest added value is generated. However, there are often processes that are indirectly affected by the implementation or that weren't considered during implementation due to a cost-benefit consideration. An example of this could be a process in the area of human resources management. During the implementation of SAP S/4HANA, the focus could be on optimizing payroll. However, other processes such as HR development or performance management might have been overlooked. After implementation, it's important to look at these processes again and look for opportunities for optimization.

Another area that is often overlooked is customer interaction. While many companies are optimizing their internal processes as part of the SAP S/4HANA implementation, they often overlook the opportunity to design their processes to improve customer service. This could be achieved, for example, by introducing new technologies to improve the customer experience or by revamping customer interaction processes, such as improving the ordering or complaints process.

It's still important to consider process optimization in a broader context. This means that it's no longer necessary to consider the impact of process improvements that affect the SAP S/4HANA system but instead the impact on the strategic goals of the company, on employees, and on other stakeholders. For example, an optimization that results in cost savings could potentially have a negative impact on employee satisfaction or product quality. Therefore, it's important to strike a balance and ensure that process improvements support the company's overall goals.

13.5 Analysis and Optimization of Implemented Processes

In the run-up to the SAP S/4HANA project, the analysis focuses primarily on the company's existing processes. The goal is to identify weaknesses and bottlenecks as well as opportunities for process optimization to gain a comprehensive understanding of the current business processes. The analysis aims to determine how the implementation of SAP S/4HANA can improve processes and increase operational value. The specific requirements of the company are also taken into account to develop tailored solutions.

After the SAP S/4HANA implementation, the focus of the analysis changes. Now the focus is on monitoring the effectiveness of the implemented improvements and ensuring that the processes are actually executed as planned. More comprehensive analysis is built and performed to verify process compliance and identify potential deviations or issues. This includes, for example, monitoring process metrics to assess the efficiency of the processes. If necessary, further specific analyses can be carried out to identify and make targeted optimizations.

Overall, the analysis prior to the SAP S/4HANA project is more exploratory, as it aims to identify potential areas for improvement. After implementation, the analysis focuses more on monitoring and optimizing the newly implemented processes to ensure that the company realizes the expected benefits of SAP S/4HANA. It's a continuous process that allows the company to adapt its processes and get the maximum benefit from SAP S/4HANA.

The analysis and optimization of implemented processes is a continuous process that is supported by the use of specific tools. SAP Signavio offers a variety of options for this purpose, such as linking process performance indicators (PPIs) with process steps or the integration of live insights into the process models (see Figure 13.1).

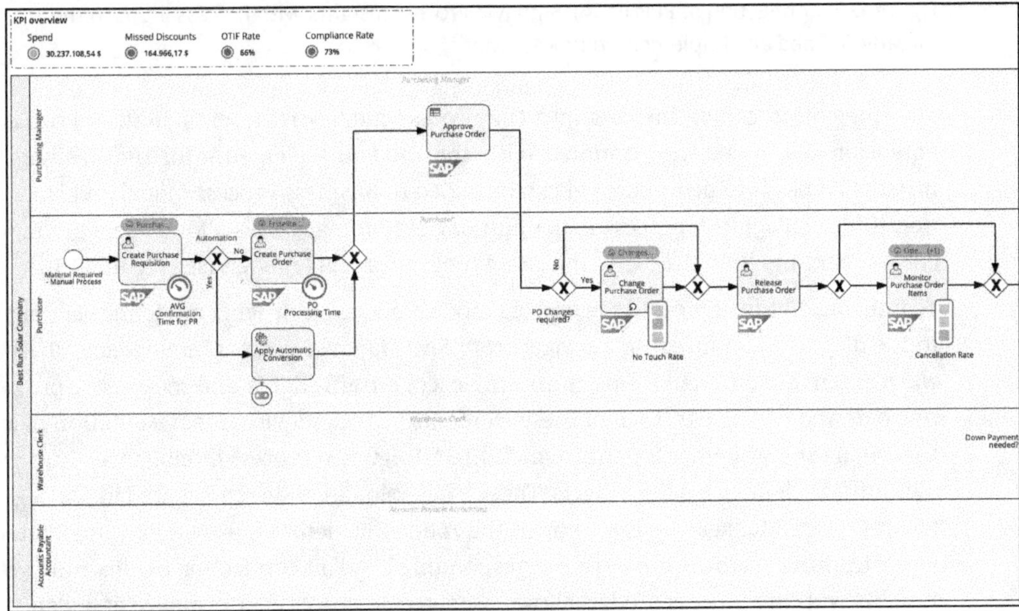

Figure 13.1 Extended Process Model with Graphically Displayed Analysis Results from SAP Signavio Process Insights and SAP Signavio Process Intelligence

A particularly interesting feature is the ability to link PPIs from SAP Signavio Process Insights directly to process steps. This is done via deep links that can be added to the standard SAP Signavio Process Insights PPIs in user-defined attributes as shortcuts (see Figure 13.2). Through these links, users can easily access relevant PPIs and take them into account in their process analysis and optimization.

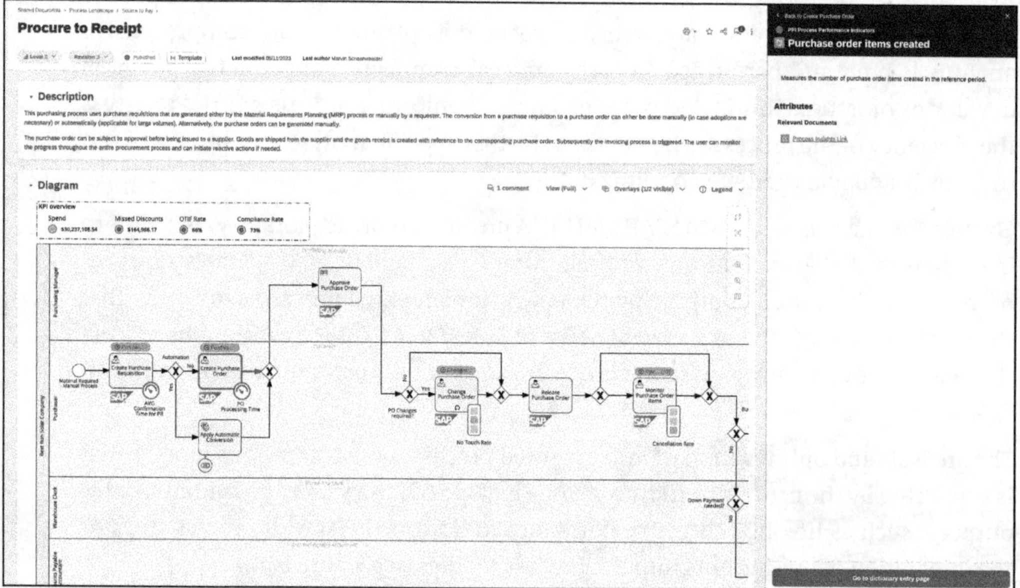

Figure 13.2 Connection of PPIs in SAP Signavio Process Insights with Process Steps in the Already Defined and Implemented Process Model

The integration of live insights into the process models represents another crucial extension of the analysis options. With the modeling elements for live insights, insights and KPIs to be monitored can be added to Business Process Model and Notation (BPMN) diagrams, process maps, and navigation maps. This allows users to view current data and PPIs directly in the context of the process models (see Figure 13.3).

To integrate live insights into the process models, add a live insights modeling element to the diagram, and link it to a widget from SAP Signavio Process Intelligence. These widgets can take different forms, from simple icons, traffic lights, and cockpits, to progress bars and ring charts. Each of these elements provides a visual representation of a particular aspect of process performance. In SAP Signavio Process Intelligence, thresholds must be defined for the widgets. These thresholds are then represented in SAP Signavio Process Manager by the color of the modeling element, providing immediate visual feedback on the status of the corresponding PPI. This connection of PPIs and live insights with the process models allows users to see process performance at a glance and make informed process optimization decisions based on this information. This enables continuous and data-driven process improvement based on actual performance data, not just theoretical models.

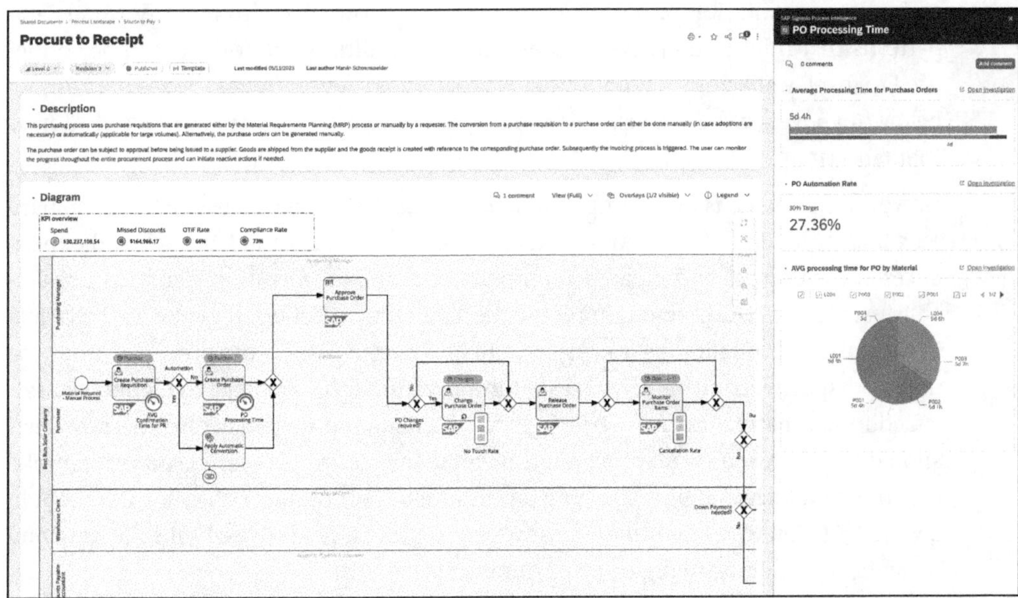

Figure 13.3 SAP Signavio Process Intelligence Analysis Results Associated with Specific Process Steps in the Already Defined and Implemented Process Model

However, using these tools requires careful planning and execution. It's important to choose the right PPIs, to configure the widgets sensibly, and to place the live insights in such a way that they offer real added value for process analysis and optimization. In addition, it must be ensured that all users have the necessary permissions to access the linked content.

The successful application of these tools and the resulting data-driven process improvement provide a solid starting point for further analysis and optimization of the implemented processes. This is not only about the technical aspects but also about a deeper understanding of the process performance and the underlying factors that influence this performance. The methods and approaches used to analyze the newly implemented processes are diverse and depend on the specific requirements and objectives of the company. They include both qualitative and quantitative methods and can be applied to different levels of the process hierarchy, from individual process steps to the entire process flow. A key aspect of process analysis is identifying and selecting the right PPIs. PPIs are quantitative measures that indicate how well a process is achieving its objectives. They can relate to different aspects of the process, such as efficiency, effectiveness, quality, or speed. By selecting the most relevant PPIs, the aspects of the process that are most important for achieving business goals can be more closely monitored and analyzed.

Once identified, these PPIs can be linked and analyzed using the deep links to the standardized PPI in SAP Signavio Process Insights or in the specifically defined SAP Signavio Process Intelligence widget. This connection directly in the process model provides a visual representation of performance data and allows users to identify trends and patterns that might otherwise be overlooked. In addition, the widgets can be configured to display specific thresholds that will trigger an alert if the process is performing below or above a certain level. This can help identify problem areas early and address them in a timely manner.

However, the analysis of the implemented processes should not just be based on hard data and numbers; it's also important to take into account the feedback and experiences of the process participants. Interviews, surveys, or workshops can provide valuable insights into the practical implementation of the processes that may not be visible in the data. This qualitative analysis can help you understand the processes from the users' perspective and identify areas for improvement that can lead to higher user satisfaction and better process performance. The data and insights from these reviews should then be used to adapt and optimize the processes. This can be done by implementing new technologies, changing process steps, or training employees. Care should always be taken to ensure that adjustments to processes are based on solid data and insights, not just guesswork or assumptions.

In summary, the analysis and optimization of implemented processes is a continuous process that requires careful planning, regular reviews, and continuous adjustments. By using technologies such as SAP Signavio and a corresponding methodology that also extends existing analysis functions, as described by way of example, companies can effectively analyze and optimize their processes to increase their performance and achieve their goals.

13.6 Evaluation and Prioritization of Process Improvements

Now that we've shed light on the importance and methods of analyzing processes, it's time to focus on the next crucial step: evaluating and prioritizing process improvements. During an SAP S/4HANA transformation, processes are also analyzed to define optimized and standardized workflows. In the process, one prioritizes major optimization potentials and quick wins during the transformation. After the transformation, one evaluates the effects and continuously optimizes the processes in small steps to achieve further improvements. Although the *evaluation and prioritization of process improvements* here isn't fundamentally different from the evaluation and prioritization of process improvements during the transformation, the approach requires consideration of some aspects after the go-live has taken place. Before we look at the key steps in assessing and prioritizing process improvements, we list them here:

- **Continuous release management of processes and systems**
  An implementation of improvements isn't only done once at go-live, but also during the productive operation of the SAP S/4HANA system and therefore requires release management of the process and system.

- **Process improvements versus process innovations**
  Process improvements compete with new requirements (process innovations), in which a greater benefit for the company is often seen. Therefore, you need to clearly demonstrate the added value of the planned improvements.
- **Effects of process changes**
  The impact of process changes may need to be considered in more detail, as only selective regression testing can be performed due to capacity constraints.
- **Stakeholder involvement**
  The circle of contacts expands as additional end-to-end processes are added in addition to processes that are mapped in SAP S/4HANA.

The conception of process improvements is based on the knowledge gained from the process analysis. However, this often results in more improvement suggestions than can be implemented in terms of time and resources. Therefore, it's critical to use a systematic method to evaluate and prioritize these suggestions to ensure that the actions with the greatest potential benefit to the business are implemented first.

A first step in this process is the precise evaluation of each improvement proposal. Important criteria for this evaluation can be the expected benefit of the improvement, the effort required for implementation, the risks associated with implementation, and the strategic importance of the affected processes. The expected benefit can consist of both quantitative variables such as cost reduction or time savings, but also qualitative aspects such as increased customer satisfaction or improved employee job satisfaction.

The evaluation of the individual suggestions for improvement is followed by prioritization. Here, it can be helpful to use a structured model such as a *prioritization matrix*. In this matrix, for example, the expected benefit and the implementation effort can be compared to create a ranking of the improvement proposals.

Although expected benefits and implementation effort are key criteria for prioritizing process improvements, you should also consider other factors. One of these factors is the urgency of an improvement. Sometimes processes can be so problematic that their improvement or revision can't be postponed, regardless of the potential cost or effort. In such cases, it's important to act quickly to prevent more serious impacts on the business. Another important factor is the alignment of a planned improvement with the overall business strategy. It's essential that process improvements support the strategic goals and vision of the company. Improvements that are more closely aligned with the corporate strategy should generally be given a higher priority than those that are less well aligned. In addition, external factors, such as regulatory requirements or market changes, can also influence prioritization. Process improvements that help ensure regulatory compliance or improve the company's competitiveness may be a high priority despite high implementation costs or low direct benefits. An example of a prioritization matrix with different factors is shown in Figure 13.4.

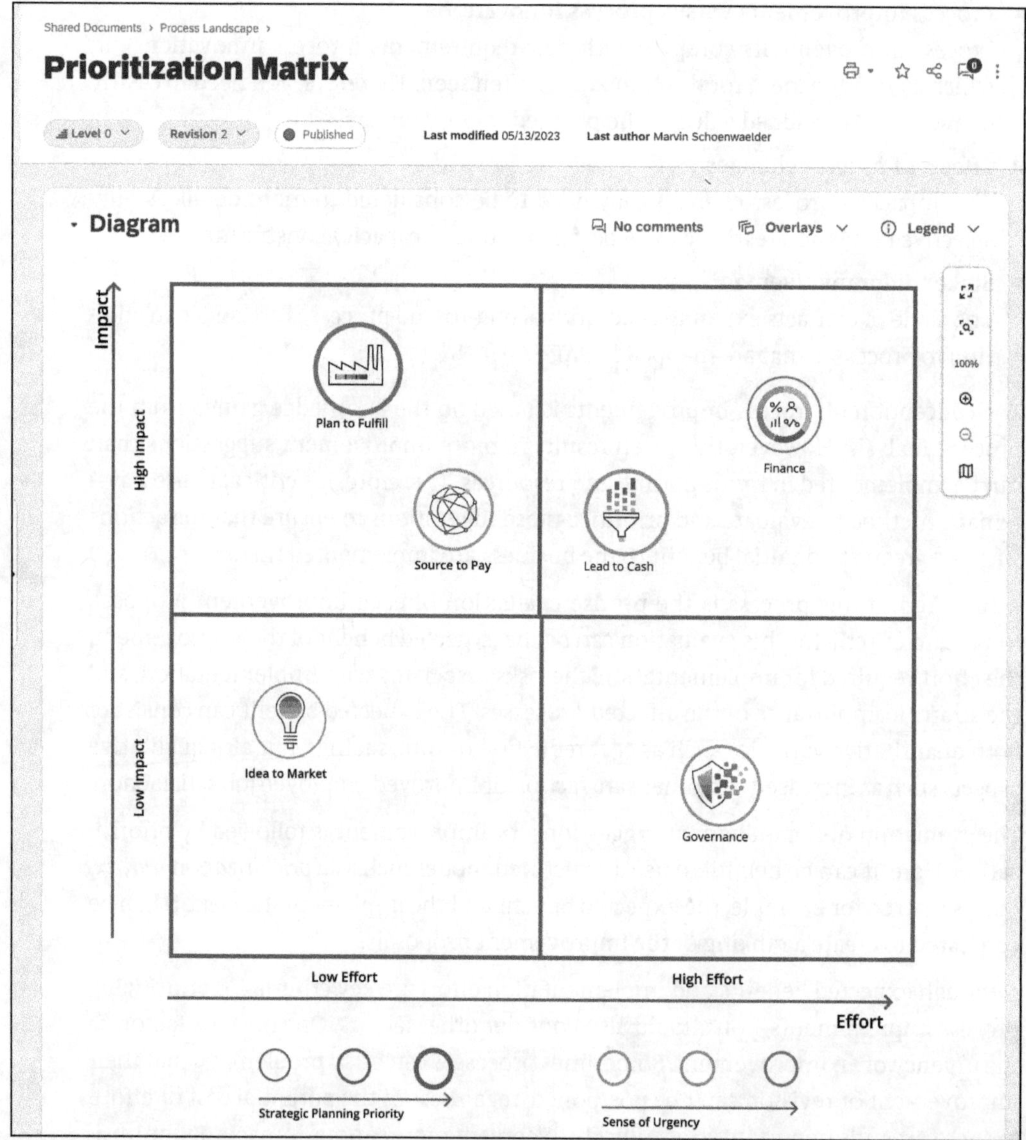

Figure 13.4 Example of a Prioritization Matrix

The evaluation and prioritization of process improvements is a complex process that requires a thorough consideration of various aspects. It's important that this process is carried out in a structured and systematic manner to ensure sound and comprehensible decision-making. Note that the evaluation and prioritization of process improvements isn't a one-time process but is repeated on a regular basis. This makes it possible to respond to changes in the business environment and ensure that the selected improvement measures remain relevant and valuable to the company.

13.7 Measurement and Presentation of Progress

The *progress measurement* is an important basis for continuous process optimization. It's important to continuously monitor how well the initiated process changes or implemented improvements are working and whether they are delivering the desired results. To this end, PPIs are once again of central importance. They make it possible to assess progress in an objective and measurable way.

The presentation of progress is also vital and should be designed in such a way that it's easy to understand and provides a quick overview of the current status of process optimization. Various presentation methods and tools can be used for this purpose, such as dashboards, reports, or visual process diagrams.

Dashboards are highly effective for charting progress and monitoring performance in process optimization. They provide a visual interface that allows users to view and analyze various PPIs, and they provide the ability to update and track data. One of the greatest strengths of dashboards is their ability to show the state of a process before and after an improvement initiative. You can compare performance metrics before the change was implemented to current metrics to see the effect of the improvement effort. This is not only helpful for internal evaluation but can also be very effective in communicating progress and improvements to stakeholders. Another great advantage of dashboards is their interactivity. Users can apply different filters or views to explore specific aspects of the data. For example, they can filter the data by department, time period, or specific PPIs. This gives users the ability to view and analyze the data in the way that is most relevant to them. In addition, by viewing the data over a period of time, users can see how performance has changed over time and if there are any particular trends or patterns that should be noted. This can help identify new opportunities for improvement or identify potential problems early on.

All these functions are integrated into SAP Signavio Process Intelligence (see Chapter 4), which makes it an extremely effective tool for monitoring and displaying progress in process optimization. In addition to the benefits of the dashboard functions, the created analysis widgets can then also be embedded back into the process model and displayed to users directly in the process model.

Finally, we would like to emphasize that the measurement and presentation of progress is not only used to evaluate the success of the improvements made so far but also to identify new opportunities for improvement. This enables users to continuously monitor and analyze the performance of processes to identify and exploit new opportunities for improvement.

13.8 Summary

In this chapter, we looked back at the transformation project, analyzed which areas should be further developed in the continuous process improvement phase, and took an in-depth look at the analysis and optimization of implemented processes. The use of data-driven process analysis tools, especially SAP Signavio Process Insights and SAP Signavio Process Intelligence, was in focus.

We looked at the various ways in which process diagrams can be enhanced by linking them to data-driven analytics, analytics widgets, and PPIs. A key focus was on linking PPIs and live insights directly to process steps to enable comprehensive and continuous process improvement. We emphasized the importance of careful planning and execution when using these tools, from selecting the right PPIs to configuring widgets wisely to placing live insights. In addition, the need for continuous review and adaptation of processes to respond to changes in the business environment, new technologies, or changing user requirements was emphasized. In the discussion on the evaluation and prioritization of process improvements, criteria and methods were presented that can help in the decision-making process. It became clear that, in addition to the expected benefits and implementation effort, other factors such as urgency and strategic relevance should also be considered. Finally, we looked at ways to measure and demonstrate progress in process optimization. SAP Signavio Process Insights and SAP Signavio Process Intelligence were highlighted as effective tools to monitor and show this progress.

In summary, continuous analysis and optimization during the transformation of implemented processes as well as the expansion of the process scope under consideration are essential components of the further development of business process transformation management. By applying the tools and methods presented in this chapter, companies can continuously improve their processes and thus increase their performance and achieve their goals more effectively.

Chapter 14
Getting Started with Business Process Transformation Management

SAP Signavio Process Insights, discovery edition and the business process transformation starter pack for RISE with SAP are good places to start for getting into the topic of process-oriented SAP S/4HANA transformation. In this chapter, we'll introduce and compare the different offerings.

In Chapter 12 and Chapter 13, we described use cases and benefits of business process transformation management supporting the move to SAP S/4HANA aligned with the SAP Signavio transformation methodology. However, you probably haven't yet made the decision to run a process-oriented transformation project, or you're not sure whether you want to use SAP Signavio as part of your project. In such cases, you might need real data-driven results from your system landscape, or you might need a system environment customized according to your needs, which can be used to convince other stakeholders in the company of the added value of this approach.

This chapter provides an overview of possible starting points for business process transformation management. In Section 14.1, we show how you can analyze processes in your SAP ERP system already today and have optimization proposals created automatically, even without an SAP Signavio subscription. Furthermore, in Section 14.2, we present the business process transformation starter pack included in RISE with SAP contracts. The package entitles you to the usage of selected SAP Signavio components, making sure you can gain initial experience with the solution.

14.1 Analysis of SAP ERP Systems

The ability to automatically monitor and analyze process key figures in SAP ERP systems was already launched in 2008 in connection with a new concept in SAP Solution Manager. *Business process integration and automation management* includes the following solutions:

- **Business process monitoring (BPMon)**
  Proactive and process-oriented monitoring of the most important business processes.

- **Interface monitoring (IFMon)**
Checks for errors in interface processing and identification of unprocessed documents.
- **Data consistency monitoring (DCMon)**
Identification and correction of inconsistencies of individual systems in a distributed system landscape.

BPMon in SAP Solution Manager already provided a wide selection of preconfigured key performance indicators (KPIs) from various areas (partly process KPIs, partly more technical KPIs. However, as the term monitoring suggests, the approach was and still is mainly known to IT administrators who focus on the technical monitoring of their SAP system landscape (in particular, the SAP ERP, SAP Financial Supply Chain Management [FSCM], SAP Extended Warehouse Management [SAP EWM], SAP Transportation Management [SAP TM], SAP Customer Relationship Management [SAP CRM], SAP Supplier Relationship Management [SAP SRM], and SAP Advanced Planning and Optimization [SAP APO] solutions). Problems can be identified and solved before critical situations occur. Based on previously defined threshold values, the system issues warnings (*alerts*) that are relevant for specific business processes and business process steps.

[»]

KPI Catalog

The KPI catalog (*https://launchpad.support.sap.com/#/kpicatalog*) is a cloud service provided by SAP through SAP ONE Support Launchpad. It contains definitions, technical documentation, and detailed descriptions of KPIs and related metrics available in SAP applications.

The *SAP Business Process Analytics* tool enhances BPMon in SAP Solution Manager, presenting analysis results in a modernized analytical application. The data is technically collected with SAP BW InfoProviders and can now be extracted and analyzed continuously. In addition, trend analysis can be created based on historical data. SAP Business Process Analytics can still be configured via SAP Solution Manager. However, the solution is now called *business process improvement for SAP solutions*. The solution also provides *business key figures* (BKFs) that can be displayed in regular technical reports such as *SAP EarlyWatch Alert*. The configuration of the BKFs always requires manual effort from the IT teams. Even if the configuration is completed successfully, most business users are typically not even aware of the value of the findings from the process KPIs; the results were rarely brought to the attention of the business department.

Figure 14.1 shows a comparison between the available content in the on-premise business process improvement solution in SAP Solution Manager and the new cloud SAP Signavio Process Insights solution. In SAP Solution Manager, 18 process flows with associated metrics and additional process performance indicators (PPIs) are already available. With SAP Signavio Process Insights, more metrics have been added; in Q1 2024, the cloud solution includes 102 different process flows with even more coming in the future. Furthermore, the results are presented graphically, and recommendations for improvements are displayed (correction and innovation recommendations) based on the findings from the analysis. In addition to the new content, SAP Signavio Process Insights is also much easier to set up and configure.

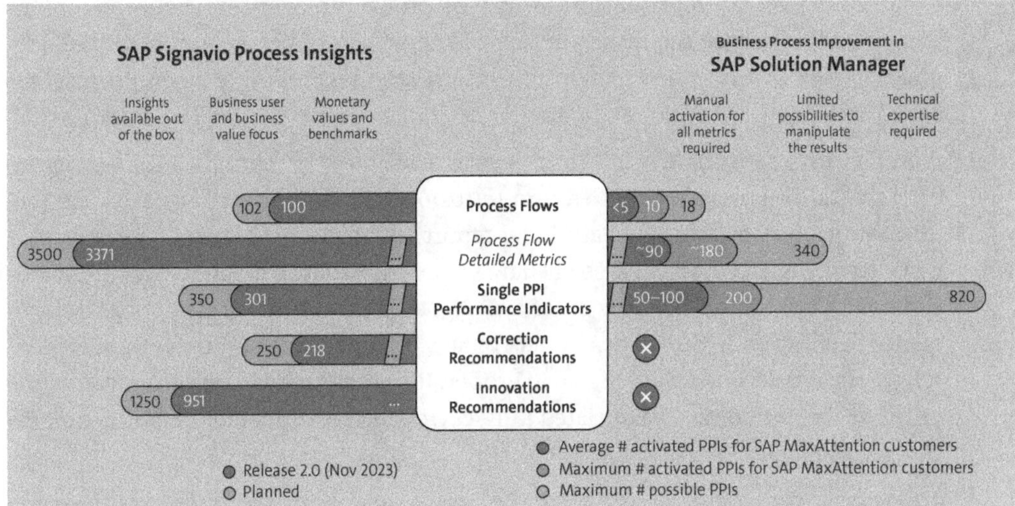

Figure 14.1 Content Comparison: SAP Signavio Process Insights versus SAP Solution Manager

SAP EarlyWatch Alert

SAP EarlyWatch Alert is a regular status report included in standard SAP maintenance contracts that contains key figures on essential administrative areas of an SAP system and provides an assessment of performance and stability. It also lists key figures such as response times and indicates detected configuration problems. When configured appropriately, SAP EarlyWatch Alert runs automatically, allowing users to proactively respond to problems before they become critical. The results can be sent automatically via email on a weekly basis, for example.

Since 2017, SAP has offered various free-of-charge analysis reports, all supporting discussions around business process transformation in the context of an SAP S/4HANA transformation project. With each increment, the results were easier to consume and more focused. The following two reports were available for all SAP customers with an SAP maintenance contract until February 2024:

- The SAP Innovation and Optimization Pathfinder report identified relevant SAP innovations, process improvements, and IT optimization potential for a core SAP system, regardless of whether the ERP system is running on SAP ERP or already using SAP S/4HANA.

- The Process Discovery report was the right choice if a customer was running an SAP ERP system and wanted to get actionable recommendations regarding process improvements and different transformation scenarios that could be relevant in the context of an SAP S/4HANA transformation.

A very high number of customers requested the reports in the past and received results based on their individual system usage. In many of those cases, the evaluation of the results hasn't yet taken place. This is why we'll provide an overview about the content and how to work with the results in this chapter, even though it's not possible to request a new SAP Innovation and Optimization Pathfinder report or a new Process Discovery report.

Starting with a beta phase in 2023 and general availability in February 2024, SAP Signavio Process Insights, discovery edition (*https://www.s4hana.com/*) replaced the SAP Innovation and Optimization Pathfinder report and the Process Discovery report. The new offering is also available free of charge to all customers with an SAP maintenance contract. It's a limited free version of SAP Signavio Process Insights that can be connected to either an SAP S/4HANA system or an SAP ERP system. It provides similar content to the full version of SAP Signavio Process Insights, but the scope is limited to one process flow, including all associated KPIs, correction recommendations, and innovation recommendations.

The results of all three reports show the added value potential of process improvements in an SAP ERP system. However, to view and prioritize the results in more detail, it's necessary to use additional components from the SAP Signavio solution portfolio (see Figure 14.2).

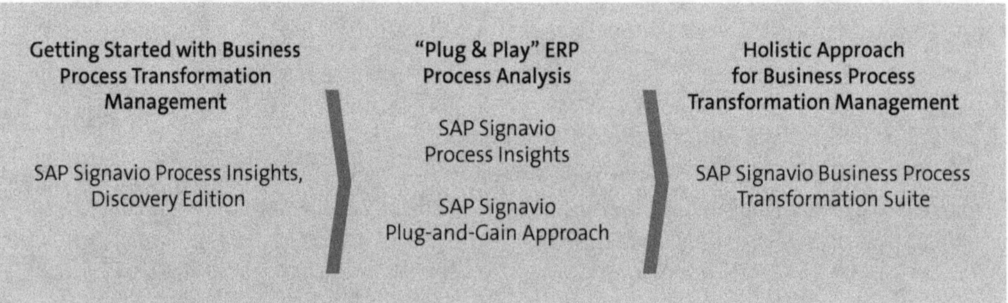

Figure 14.2 Process Discovery as the First Step

The reports not only provide system-specific results and recommendations but can also be used for benchmarking. The data collection mechanisms have been in place for many years, and the result is now used not only for the three free reports but also for populating benchmarking data in the paid full version of SAP Signavio Process Insights. The SAP database now contains data points from more than 15,000 analyses. A choice can be made between 29 industries as a basis for comparison. In the result, all process key figures are then compared with the key figures of other customers in the selected industry. In addition to your own key figures, you can see in the report whether your results place you in the best or worst quartile compared to other SAP customers. Accordingly, the results are displayed in red (left-hand side), yellow (middle), or green (right-hand side, see Figure 14.3).

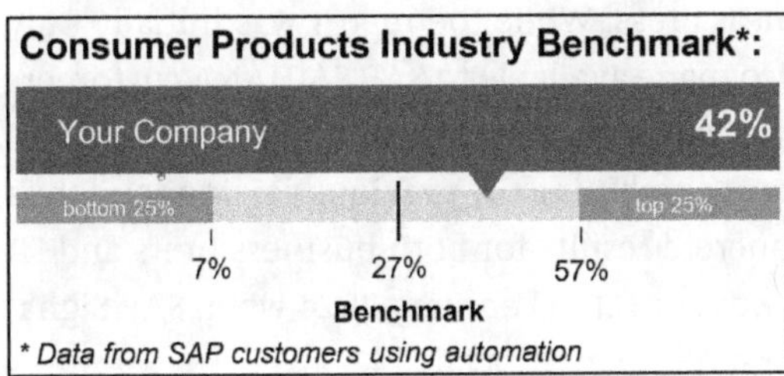

Figure 14.3 KPI Benchmarking

For SAP Signavio Process Insights, discovery edition, all results are made available in two formats: an online version allowing users to filter data and drill down to the line-item level and a management summary in PDF format. The two formats were also available for the SAP Innovation and Optimization Pathfinder report and the Process Discovery report, however, since April 2024, the online version (also known as "Spotlight by SAP") isn't available anymore. This is why the following paragraphs will explain how to work with the PDF version of the results.

The data collection mechanism was and still is based on the same technology that was already used for the services running on SAP Solution Manager: BPMon and SAP Business Process Analyzer. However, the technology can also be used without configurations in SAP Solution Manager. Instead, the *SAP Readiness Check for SAP S/4HANA framework*, which was introduced in Chapter 10 in the context of preparing for an SAP S/4HANA transformation, includes a functionality for business process improvement analysis, which makes the use of SAP Solution Manager in this context obsolete.

Even though the used functionality business process improvement analysis is part of the SAP Readiness Check for SAP S/4HANA framework, the results serve different use cases: While the SAP Innovation and Optimization Pathfinder report, the Process Discovery report, and SAP Signavio Process Insights, discovery edition report focus primarily on the identification of process improvement potential (target group is business departments or interface functions between business departments and IT), SAP Readiness Check for SAP S/4HANA identifies the most important technical aspects in the context of a system conversion from SAP ERP 6.0 to SAP S/4HANA (target group is IT). Implications due to the use of the SAP HANA database technology as well as to required custom code remediation efforts due to a new and adapted source code in SAP S/4HANA are outlined in SAP Readiness Check for SAP S/4HANA.

In the following sections, we'll now take a closer look at the three reports for business process improvement.

14.1.1 SAP Innovation and Optimization Pathfinder

The SAP Innovation and Optimization Pathfinder report was first made available to SAP customers and partners in 2017 to identify relevant innovations, improve business processes, and highlight IT optimization potential. In 2018, the report was expanded to include a line-of-business edition that added further process performance insights and optimization potential for specific business areas. While the report was initially only available to SAP ERP customers, it was also made available to SAP S/4HANA customers in 2019. With the introduction of SAP Innovation and Optimization Pathfinder 2.0 in 2023, SAP started to provide a unified report combining the previous versions and thus representing a holistic approach with concrete results for both business units and IT. The availability to generate new reports was sunset in February 2024 when SAP Signavio Process Insights, discovery edition became generally available.

Figure 14.4 shows an overview page from SAP Innovation and Optimization Pathfinder. The following topics are highlighted:

- **Recommendations by line of business**
  Based on the usage of your ERP system, system-specific recommendations are shown, differentiated by six business units: finance, sourcing and procurement, sales, supply chain, manufacturing, and asset management.

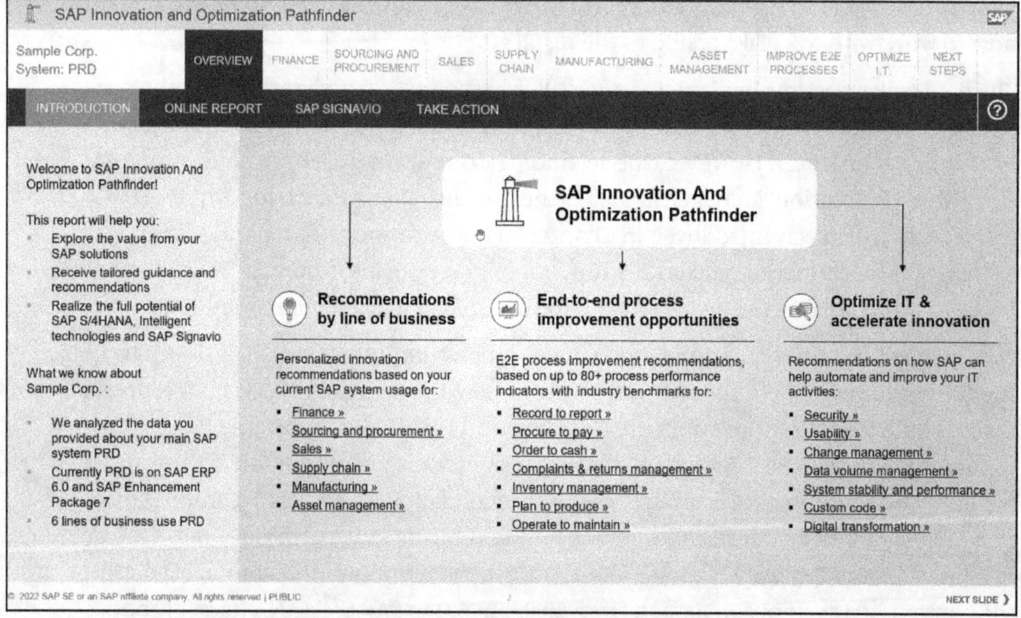

Figure 14.4 SAP Innovation and Optimization Pathfinder (Source: SAP)

- **End-to-end process improvement opportunities**
  Based on 83 PPIs and benchmarking data, industry-specific recommendations are compiled for process improvements in the following areas: record-to-report,

procure-to-pay, order-to-cash, complaints and returns management, inventory management, plan-to-produce, and operate-to-maintain.

- **Optimize IT and accelerate innovations**
  This area highlights recommendations for automating and improving the IT landscape. These cover the areas of security, usability, change management, data management, system stability and performance, customer-specific developments, and preparation for digital transformation.

Recommendations by Line of Business

Exemplary results for the financial area of a company from the consumer goods industry are shown in Figure 14.5. In the section on innovations (**Recommended Finance Innovations**), a distinction is made between four areas:

- **Simplified user experience**
  The availability of SAP Fiori as a new user interface (UI) for business applications is the foundation of SAP S/4HANA. Now, there are some SAP Fiori apps that can already be enabled in SAP ERP at any time. Other SAP Fiori apps require an SAP HANA database as a minimum requirement. With SAP S/4HANA, SAP provides SAP Fiori apps in almost all areas, so working with the classic SAP GUI is no longer necessary. Based on various usage parameters in your ERP system, here is a reference to the SAP Fiori apps that could be relevant for you.
- **Functional enhancements**
  Also based on usage data from your ERP system, functional enhancements that could generate added value, as shown here. A concrete list of the identified transactions as well as an evaluation of how frequently the corresponding functionality is used in your industry can be found in the detailed view. Many of the suggested enhancements can usually already be implemented with existing software licensing. However, the installation of an enhancement package and/or the activation of a business function is then required here.
- **SAP S/4HANA enhancements**
  The **Next-Generation Digital Business** section provides an overview of relevant SAP S/4HANA innovations that may be relevant to you. Again, relevance is assessed based on usage parameters from your system as well as penetration in your chosen industry.
- **Enhancements through SAP cloud solutions**
  In addition to SAP Fiori apps, functional enhancements in the existing system, and SAP S/4HANA innovations, SAP Innovation and Optimization Pathfinder also highlights opportunities for process improvement through SAP cloud solutions. The report lists potentially interesting applications and refers to further online information. Note that recommendations are derived from actual ERP usage on one hand, but general best practice suggestions are also added on the other.

In addition to the section on innovations, the overview also includes a reference per business unit to indicate process improvement potential (shown in Figure 14.5 in the **Improve Finance Processes** section). This topic area is examined in more detail next.

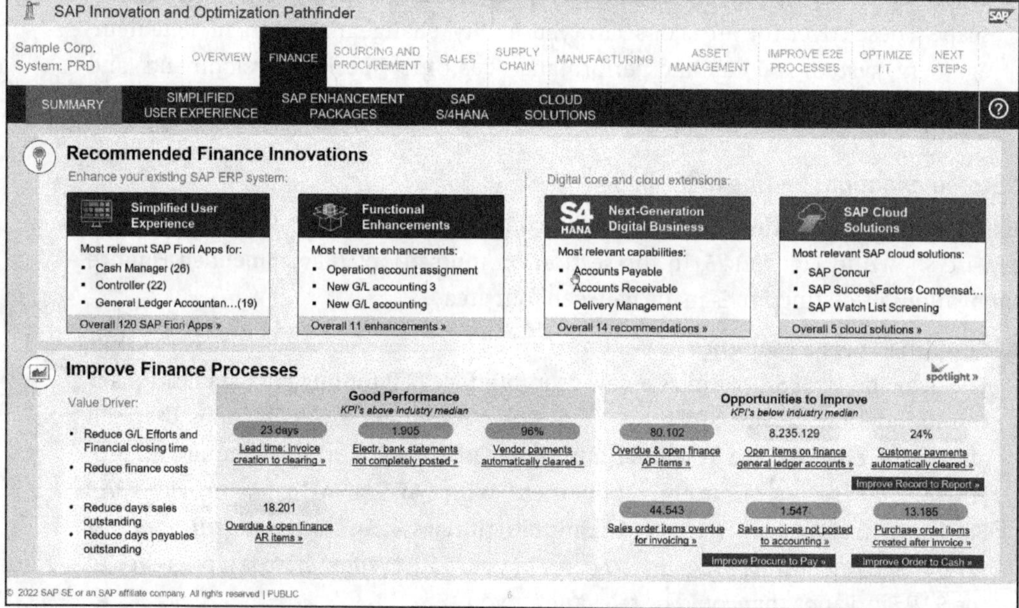

Figure 14.5 Recommendations in the Area of Finance

End-to-End Process Improvement Opportunities

In this section of the report, selected KPIs are used to identify potential for improvement regarding the following categories (see Figure 14.6):

- Reduction of lead times
- Improvement of automation rates
- Error reduction
- Minimization of backlogs

The report identifies specific metrics, compares how you stand relative to other companies in your industry, describes possible reasons for poor process performance, and identifies potential impacts on business operations. The values identified are averages measured in the system. Some metrics are broken down by the five largest company codes, the five largest plants, or the age distribution of documents. Each of the key figures is highlighted in color. A red highlight indicates that the result is in the worst 25% compared to other customers in the selected industry. If the key figures are highlighted in green, the result is in the best 25%. All values in between are displayed in yellow.

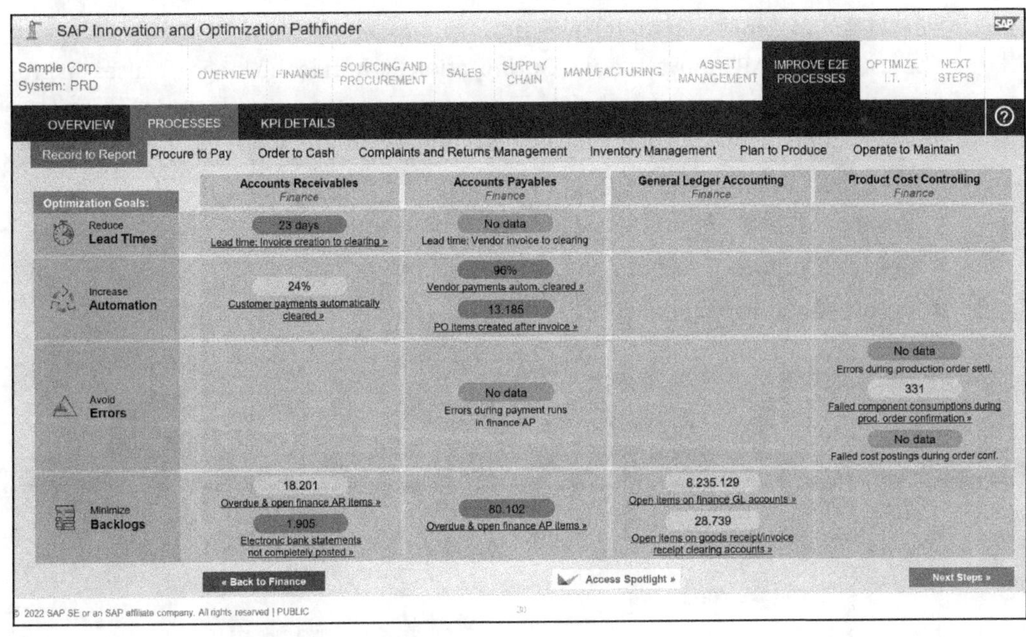

Figure 14.6 Improving End-to-End Processes

> **Product Supplements**
>
> Full-size, color versions of Figure 14.5 and Figure 14.6 are available for download at *https://sap-press.com/5855* under the **Product supplements** section.

It's important to interpret the benchmarking results correctly. Even though the distinction between 29 different industries is already very granular, absolute figures aren't normalized, resulting in the risk of misinterpreting the color coding. For example, a large company is expected to have higher document volumes compared to a smaller company. So, based on this fact alone, large companies are expected to have more red PPIs compared to small companies. The reverse is also true—smaller companies are more likely to have many green PPIs. However, as just mentioned, this only applies to absolute PPIs. For relative PPIs (e.g., percentage of degree of automation), the number of documents involved doesn't play a role.

The PPIs typically provide initial indications regarding the underlying potential for process improvement in your system. However, individual filtering according to multiple criteria, drill down to the document level, or evaluation according to different criteria isn't possible. For more in-depth analyses, an SAP Signavio Process Insights analysis is required.

Optimize IT and Accelerate Innovations

The last area in SAP Innovation and Optimization Pathfinder no longer focuses on process optimizations, but on technical IT optimization opportunities (see Figure 14.7). With the implementation of the measures outlined, various added values can be leveraged:

- Ensure business continuity and increase innovation agility
- Reduce operational cost
- Ensure compliance and security

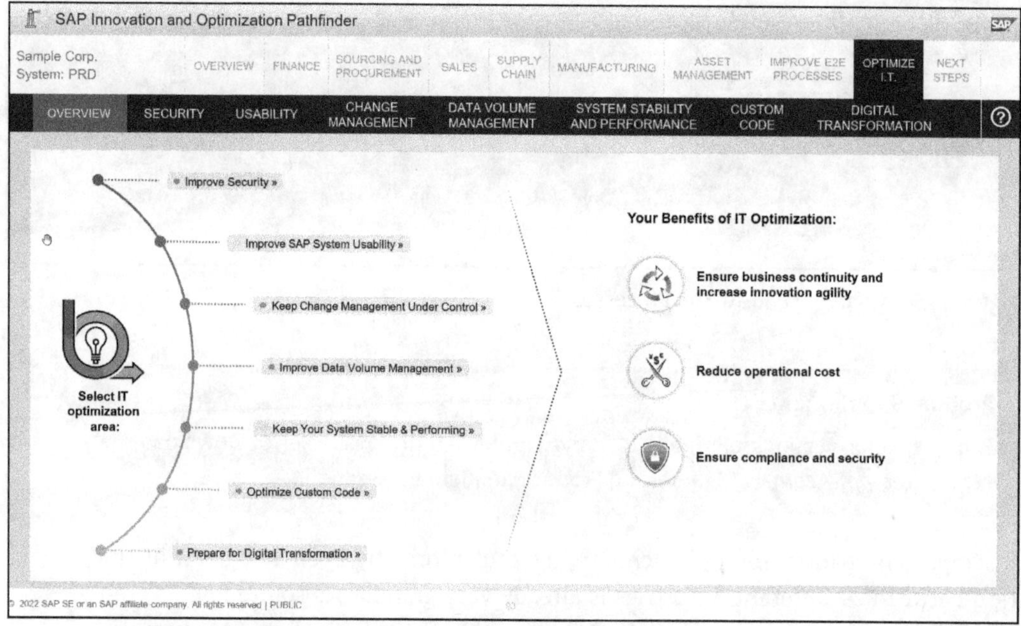

Figure 14.7 Technical Optimization Options

14.1.2 Process Discovery for SAP S/4HANA Transformation

The Process Discovery report for SAP S/4HANA was available from 2021 until February 2024. It was a further development of the SAP Business Scenario Recommendations report that has been available since 2016 (or the Next Generation SAP Business Scenario Recommendation report since 2019). The results of the report support the transformation planning and implementation of your SAP S/4HANA project. In particular, it identifies added values that can be realized through structured business process management as part of the SAP S/4HANA project. Figure 14.8 provides an overview of the dimensions covered in the Process Discovery report for SAP S/4HANA.

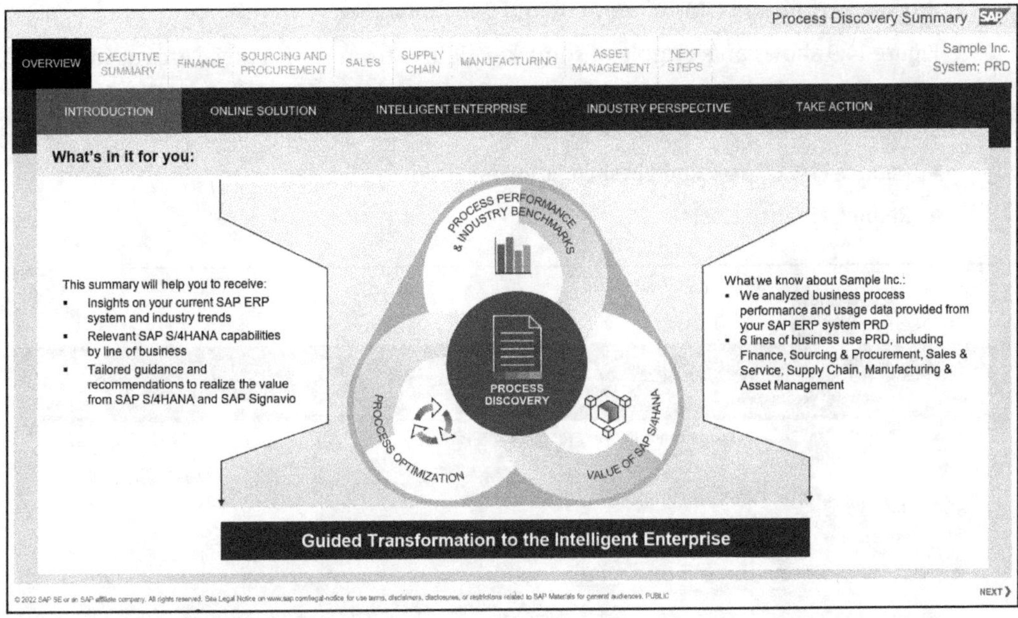

Figure 14.8 Report: Process Discovery

More details about the Process Discovery are given here:

- **Process performance measurement and benchmarking against other companies in your industry**
  Similar to SAP Innovation and Optimization Pathfinder, various PPIs are measured in your system and compared with other companies in your selected reference industry.
- **Identification of process optimization benefits**
  For each PPI, recommendations are provided for analyzing the root cause. Possible reasons for poor process performance are evaluated as well as possible effects on business operations without process improvement measures are outlined. This evaluation is also comparable to the results of the SAP Innovation and Optimization Pathfinder report.
- **Evaluation of the added value of SAP S/4HANA**
  Based on the process analysis, functions in SAP S/4HANA that can lead to process improvement are displayed. Usage data from your system is used to evaluate the relevance, for example, usage intensity and executed transactions that are improved with SAP S/4HANA. Furthermore, the industry benchmark provides indications as to whether a function is used more frequently or less frequently in the industry relevant to you.

We'll go into these three functions in more detail next.

Process Performance Measurement and Benchmarking

Figure 14.9 shows an example of some key figures from the financial area as identified in the Process Discovery report. In the **REDUCE FINANCE COSTS & CLOSING TIME** category, several value drivers were identified:

- **Reduce G/L Efforts And Financial Closing Time**
- **Reduce Finance Costs**

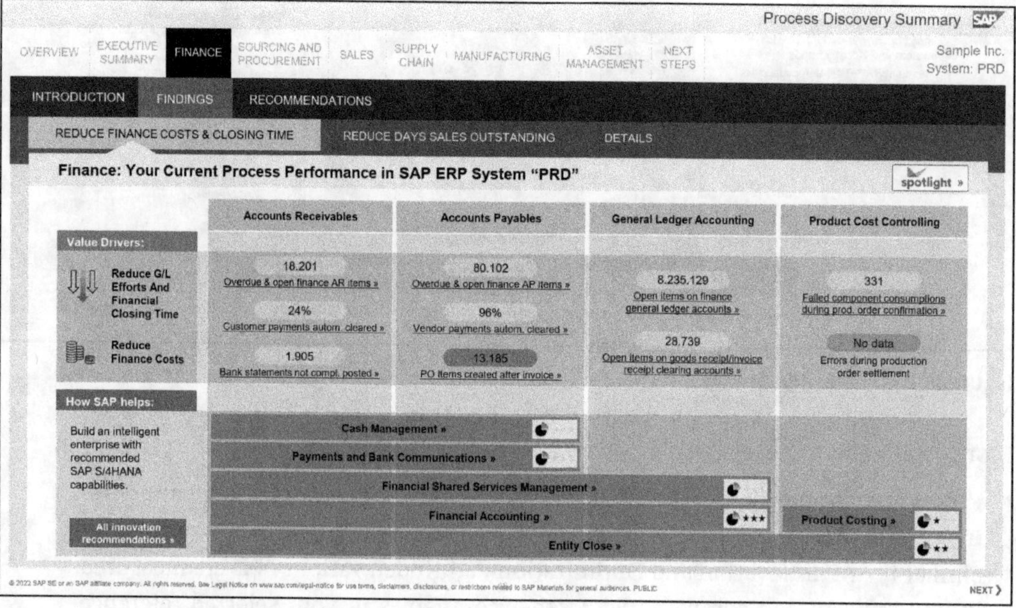

Figure 14.9 Detailed View of the Processes in Financial Accounting

The key figures are broken down by various organizational units:

- **Accounts Receivables**
- **Accounts Payables**
- **General Ledger Accounting**
- **Product Cost Controlling**

Just as with SAP Innovation and Optimization Pathfinder, the coloring of the key figures is derived from the industry benchmarking, which is based on the same assumptions as before.

In addition to the key figures, the lower part of the slide shows an overview that functions in SAP S/4HANA to support you in optimizing the process. The indications broken down by organizational unit are presented in more detail later when we cover the evaluation of the added value of SAP S/4HANA.

Identification of Process Optimization Benefits

For each key figure, the report provides a detailed presentation on one page (see Figure 14.10). There you'll find a concrete description of the KPI, a more detailed overview of the industry benchmarking, and further details such as the age of the analyzed documents or the findings, broken down into the five largest company codes, sales organizations, plants, or material types (depending on the KPI analyzed).

While the key figures in the overview are shown aggregated across all organizational structures, the detailed view is usually very interesting for identifying the cause of inefficient processes. However, just as in the SAP Innovation Pathfinder report, there is no individual filtering option here. Drilling down to the document level is also not possible. Such functions only become possible with the SAP Signavio Process Insights solution, which is subject to a subscription cost.

In addition to the description and further details on what was measured, the detailed view also lists indications of possible causes of conspicuous key figures as well as possible effects on business operations.

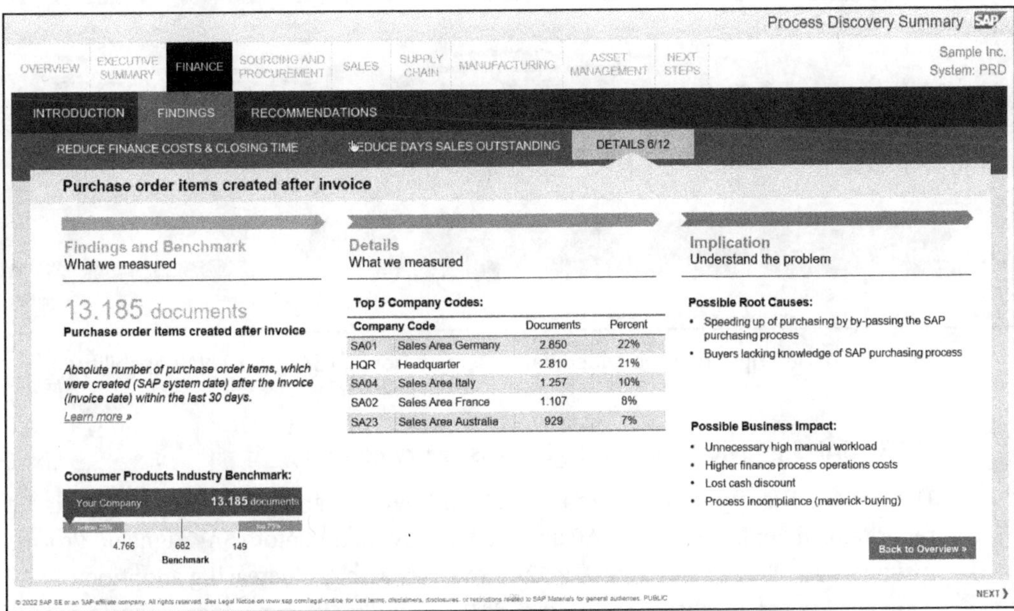

Figure 14.10 Detailed Representation of a Key Figure

Evaluation of the Added Value of SAP S/4HANA

For each suggested recommendation on how to improve a process in SAP S/4HANA, the report provides a detailed overview (see Figure 14.11). This contains the following aspects:

- Description of the functionality in SAP S/4HANA
- Added value of the recommended functionality
- Innovations in SAP S/4HANA with regards to the recommended functionality

Furthermore, the frequency of use in the relevant industry is given, and links are provided to further information on SAP's website. Overall, most of the information mentioned here is available for free on SAP's websites; however, the report derives relevant information based on the process metrics and the industry benchmarks.

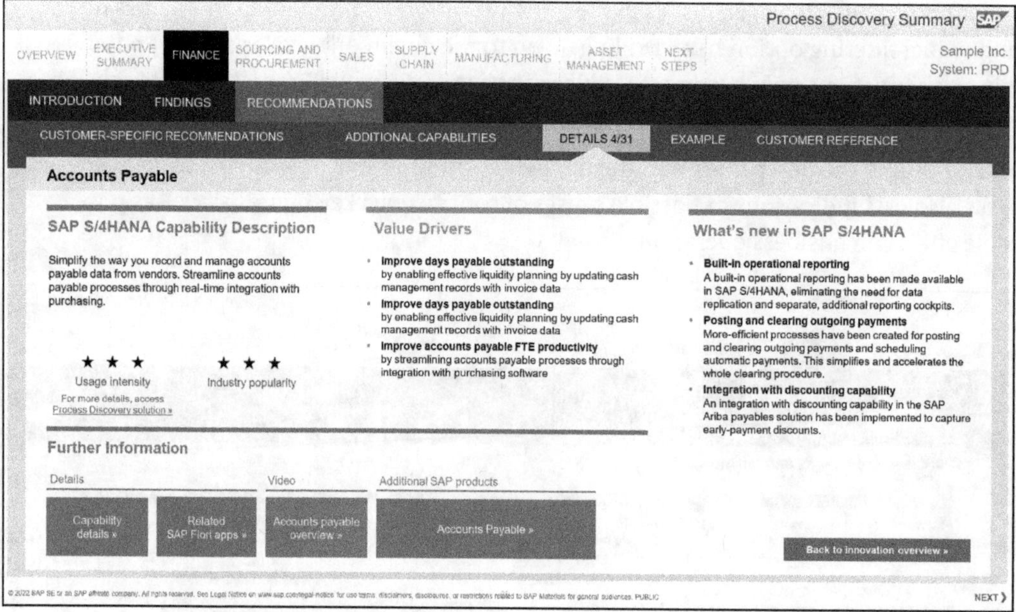

Figure 14.11 Description of the Added Value Derived from an SAP S/4HANA Capability

14.1.3 SAP Signavio Process Insights, Discovery Edition

The Process Discovery report and the SAP Innovation and Optimization Pathfinder report already existed before SAP acquired Signavio and before SAP Signavio Process Insights was released (see Figure 14.12). However, with an increasing adoption of SAP Signavio Process Insights in the market, SAP now makes it even easier for customers to get started with the SAP Signavio solution portfolio by offering SAP Signavio Process Insights, discovery edition, replacing both the Process Discovery report and the SAP Innovation and Optimization Pathfinder report. It became available in beta mode in 2023, and general availability started in February 2024.

SAP Signavio Process Insights, discovery edition is composed of two complimentary offerings:

- **Access to the analysis online**
  Dive into the information so you can understand where the issues are coming from and how to solve them. Named users are required for access.
- **PDF executive summary**
  A five-minute summary is used as a support for presentations and conversations with decision-makers. No access is required.

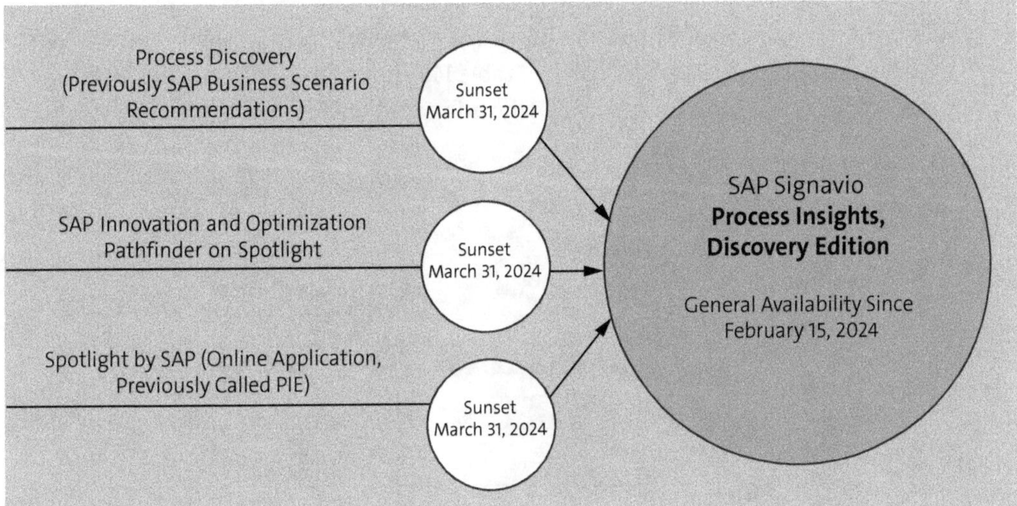

Figure 14.12 Roadmap for SAP Signavio Process Insights, Discovery Edition

SAP Signavio Process Insights, discovery edition is a limited and free-of-charge version of the full product SAP Signavio Process Insights. While the paid solution enables continuous monitoring of business processes and allows users to analyze a wide range of process flows and PPIs across different lines of business, SAP Signavio Process Insights, discovery edition is a one-time analysis of one process flow with associated metrics and a smaller set of PPIs. You can select one out of the three available different process flows:

- Supplier invoice issuing to accounts payable clearing
- Sales billing document creation to accounts receivable clearing
- Purchase order item creation (with account assignment) to invoice receipt creation

The possibilities of analyzing the one selected process flow are the same as in SAP Signavio Process Insights. (We won't discuss the features of SAP Signavio Process Insights because they were already described in Chapter 3 as part of the overview of the business process transformation portfolio.). Hence, SAP Signavio Process Insights, discovery edition provides significantly more content and additional possibilities to analyze the data compared to the Process Discovery report or the SAP Innovation and Optimization Pathfinder report:

- The UI is the same as SAP Signavio Process Insights.
- The usual blockers for the process are displayed.
- Correction and innovation recommendations are available.
- Object-specific filtering is possible.
- Drilling down to the document level is possible.
- Value analysis is available.

Figure 14.13 compares the functionality of SAP Signavio Process Insights and SAP Signavio Process Insights, discovery edition. Table 14.1 shows an overview of the functions of the various initial tools for process analysis in an SAP ERP or SAP S/4HANA system.

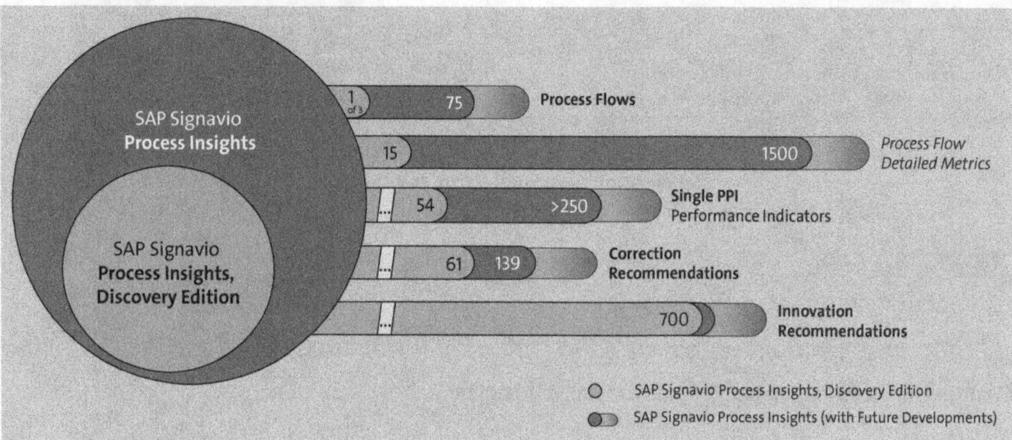

Figure 14.13 Scope of SAP Signavio Process Insights, Discovery Edition

[»]

Technical Requirements for SAP Signavio Process Insights, Discovery Edition

For the execution of SAP Signavio Process Insights, discovery edition, various technical prerequisites must be met in your productive ERP system:

- ERP release: SAP ERP 6.0 (EHP 0 to EHP 8) or SAP S/4HANA
- Function blocks and reports for data collection: ST-A/PI version 01T or higher
- Installation of the latest version of SAP Notes 2745851 and 2758146

A system copy of the productive system can't be used because not all required data points would be available. If you use multiple productive clients in your system, the data collectors must be run separately in each client.

After uploading the relevant documents into SAP's service portal, SAP will set up the tool, provide users for access, and distribute the respective PDF executive summary. The entire procedure is described at *http://www.s4hana.com/*.

		SAP Innovation and Optimization Pathfinder	Process Discovery	SAP Signavio Process Insights, Discovery Edition
Process Analysis	Process Key Figures	83 PPIs	63 PPIs	54 PPIs
	Root Causes and Effects	Yes	Yes	Yes
Proposals	SAP Fiori Apps	Yes	Yes	Yes
	Enhancements in Existing Release	Yes	No	Yes
	Enhancements through SAP S/4HANA	Yes	Yes	Yes
	IT Optimization Opportunities	Yes	No	No
	Qualitative Added Value through SAP S/4HANA	Yes	Yes	Yes
	Quantitative Added Value through SAP S/4HANA	No	No	Roadmap
Look and Feel	Similar to SAP Process Insights	No	No	Yes
	Availability of a PDF Download	Yes	Yes	Yes

Table 14.1 Comparison of Entry Options

14.2 Business Process Transformation Starter Pack

In addition to the cloud ERP system in the form of SAP S/4HANA Cloud Private Edition or SAP S/4HANA Cloud Public Edition, RISE with SAP also includes the business process transformation starter pack. The package includes the usage rights for the SAP Signavio components listed in Table 14.2.

Component	Scope	Note
SAP Signavio Process Insights	50 GB data volume, one-time loading of data from the source system	It's possible to connect either an existing ERP system or a converted or newly implemented SAP S/4HANA system. It's not possible to analyze an SAP S/4HANA Cloud Public Edition system. However, data can only be loaded once, which doesn't allow continuous process analysis.
SAP Signavio Process Manager	3 Process managers	The function can be used to its full extent.
SAP Signavio Process Collaboration Hub	10 Hub users	The function can be used to its full extent.

Table 14.2 Scope of the Business Process Transformation Starter Pack

As you can see, the starter pack doesn't include all SAP Signavio components, as it's only an entry-level package. It allows customers who haven't yet licensed the respective SAP Signavio components to use them. For example, customers who have performed an initial analysis of their as-is processes with Process Discovery in the run-up to the switch to SAP S/4HANA can use SAP Signavio Process Insights to analyze the key figures in the context of process flows and use the extensive filter functions. With the use of SAP Signavio Process Manager, it's possible to transfer the SAP standard content available in SAP Signavio Process Explorer into your own SAP Signavio workspace, refine the standard process models and share them with other project participants.

The contract term of the starter pack is identical to that of the RISE with SAP contract. You can extend the existing usage rights. To do so, in your SAP BTP subaccount, you connect the SAP Signavio Process Insights application using the RISE with SAP service plan (see Figure 14.14). The service plan supports one-time data loading. This can be the analysis of an existing SAP ERP system or the analysis of the SAP S/4HANA system after go-live.

If you want to continuously analyze data with SAP Signavio Process Insights after the one-time load from your source system, this is easily possible. After purchasing an SAP Signavio Process Insights license, you only need to exchange the service plan in SAP BTP (see Figure 14.15).

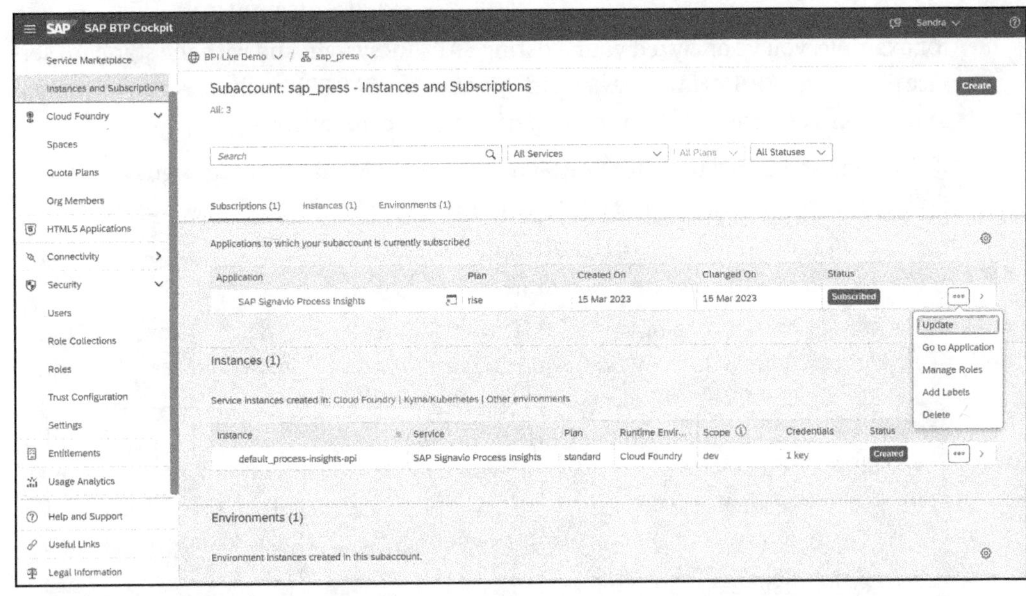

Figure 14.14 Connection to SAP Signavio Process Insights

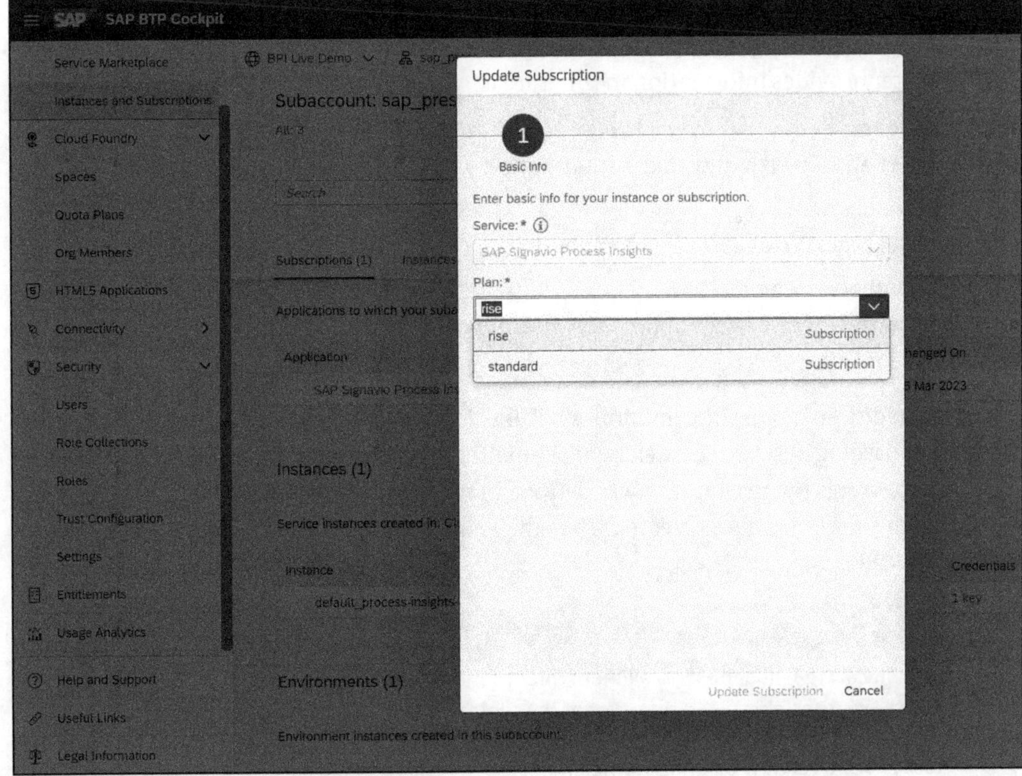

Figure 14.15 Changing the Service Plan

Alternatively, you can leave an existing instance with the RISE with SAP service plan if, for example, you've analyzed your existing SAP ERP system and with the standard service plan your SAP S/4HANA system. In this case, the SAP ERP analysis isn't overwritten and remains as a baseline for process performance comparisons.

If you need additional users for SAP Signavio Process Manager or SAP Signavio Process Collaboration Hub, they can be licensed and added to the SAP Signavio workspace.

[»]

Business Process Transformation Starter Pack

The starter pack has been part of RISE with SAP since the end of October 2021. It enables customers to use SAP Signavio Process Insights, SAP Signavio Process Manager, and SAP Signavio Process Collaboration Hub. The exact scope of use is described in the Supplement of RISE with SAP S/4HANA Cloud and the Supplement of RISE with SAP S/4HANA Cloud Private Edition. You should verify the supplement applicable to your contract to see whether you're eligible. The current supplements can be viewed in SAP Trust Center at *www.sap.com/about/trust-center/agreements/cloud/cloud-services.html?sort=latest_desc.*

To make getting started with SAP Signavio products as easy as possible, SAP Signavio Onboarding Resource Center is available in addition to the extensive documentation on SAP Signavio products in SAP Support Portal. SAP Signavio Onboarding Resource Center provides information in three areas:

- **Get started**
  Get an overview and all the information you need to be able to use the starter pack.
- **Enablement**
  Get access to e-learnings, documentation, product news, and training.
- **Support & Community**
  Learn where to get support for questions and how SAP Signavio Community can help you.

In addition to the information available in SAP Support Portal and SAP Signavio Onboarding Resource Center, regular emails are sent to your main IT contact. The main IT contact is the person who was defined in the contract by your SAP account manager when the contract was created. The emails in Table 14.3, among others, are sent automatically.

Email Name	Shipping Time	Important Contents
Technical Pre-Onboarding Process Insights	Maximum three months before the start of the contract	Description of the necessary technical requirements

Table 14.3 Overview of Onboarding Emails

Email Name	Shipping Time	Important Contents
Process Insights Onboarding	At contract start	SAP Signavio Process Insights Onboarding Video, information on the SAP Signavio Onboarding Resource Center and online training
Prepare Your Users	14 days after start of contract	Configuration of SAP Signavio workspace, information about online trainings
Setting up Process Insights	14 days after start of contract	Step-by-step guide to setting up SAP Signavio Process Insights
BTP Onboarding	At the start of the contract, if no SAP BTP account is available	Information on setting up an SAP BTP global account

Table 14.3 Overview of Onboarding Emails (Cont.)

[«]

Information Sources

Access the online documentation for each of the solutions included in the starter pack here:

- SAP Signavio Process Insights: *https://help.sap.com/docs/BPI*
- SAP Signavio Process Manager: *https://documentation.signavio.com/suite/de/Content/process-manager.htm*
- SAP Signavio Process Collaboration Hub: *https://documentation.signavio.com/suite/de/Content/collaboration-hub.htm*
- SAP Signavio Onboarding Resource Center: *https://support.sap.com/en/product/onboarding-resource-center/sap-signavio.html*

For SAP Signavio, there is a dedicated community with, among other things, blogs and the possibility to ask questions and communicate with SAP experts and customers, which you can reach at *https://community.sap.com/topics/signavio*.

14.3 Summary

Getting started with business process transformation management with SAP is made easy for companies thanks to the various offerings. The option of analyzing productive systems not only makes it possible to get to know the solutions and basic concepts but also to identify potential for improvement. The cloud-based solutions with predefined analysis content also make it possible to get started very quickly and with low resource requirements.

Prior to the project, SAP Signavio Process Insights, discovery edition, as the successor of the SAP Innovation and Optimization Pathfinder report, and the Process Discovery report supports you with a detailed metrics-based analysis of one process flow in your ERP system.

The business process transformation started pack included in SAP RISE with SAP contracts provides even further possibilities. It includes a one-time-load of all process flows and KPIs in SAP Signavio Process Insights to identify areas of improvement as well as entitlements to get started with SAP Signavio Process Manager and SAP Signavio Process Collaboration Hub to define your target processes. Hence, you can identify initial value-added potential and also immediately describe it in the form of target processes. For permanent measurement of process performance, you can switch service plans after purchasing an SAP Signavio Process Insights license, as described in the previous section. If, for example, the number of processes makes it necessary to give additional employees access to the SAP Signavio workspace as project managers or as process participants, this is possible after purchasing the corresponding number of users.

Regardless of which option you use to get started, make business process transformation management and SAP Signavio an integral part not only of your move to SAP S/4HANA but also of your use of it thereafter. Only in this way will you be able to analyze your company and its processes, continuously identify potential for improvement, and also realize the full benefits of SAP S/4HANA for your company.

The Authors

Johannes Strasser is a product expert for RISE with SAP and SAP Signavio at Westernacher Consulting. His consulting focus includes business process transformation and optimization with SAP Signavio. With his profound knowledge of the SAP Signavio Process Transformation Suite, he supports numerous companies throughout Europe in the systematic implementation of SAP Signavio. In this context, he helps his customers to not only sustainably increase their process quality but also the value creation of all business areas.

Michael Sokollek has been with the SAP Group since 1996. Before joining SAP Signavio, he worked in product development, in application consulting, and as an architect. Together with customers, he developed customized SAP S/4HANA transformation roadmaps. Since January 2021, Michael Sokollek has been a senior director in the SAP Signavio Customer Transformation Office. In this role, he supports customers in using SAP's business process transformation portfolio for continuous process improvement in preparation, during, and after an SAP S/4HANA transformation.

Manuel Sänger is part of the SAP Signavio Customer Transformation Office and supports strategic customers in identifying and realizing added value through a process-driven SAP S/4HANA transformation. Before joining SAP Signavio, he worked in SAP transformation consulting, developed SAP S/4HANA roadmaps together with customers, and managed fit-gap workshops within large SAP S/4HANA transformation projects.

Maike Spierling is an expert in the areas of business process transformation and process excellence. She is part of SAP Signavio and supports strategic customers in identifying and realizing added value through a process-driven SAP S/4HANA transformation. Before joining SAP Signavio, she worked in management consulting and supported clients as a process consultant in the context of SAP S/4HANA transformations, among others. In this role, she accompanied customers in the implementation of a process organization, in the use of process analysis methods, and in the continuous optimization of process landscapes.

Marvin Schönwälder is part of the SAP Signavio Customer Transformation Office and supports strategic customers in identifying and realizing added value through a process-driven SAP S/4HANA transformation. Before joining SAP Signavio, he worked in consulting and supported customers in the context of SAP S/4HANA transformations as a process consultant. In this role, he accompanied customers in the implementation of a process organization, in the use of process analysis methods, and in the continuous optimization of process landscapes.

Index

A

B

J

K

L

M

N

O

P

Q

R

S

T

U

V

W

Interested in reading more?

Please visit our website for all new book and e-book releases from SAP PRESS.

www.sap-press.com